AF581624

Clinical, Pathological, and Molecular Characteristics in Colorectal Cancer

Clinical, Pathological, and Molecular Characteristics in Colorectal Cancer

Editor

Stephane Dedieu

MDPI • Basel • Beijing • Wuhan • Barcelona • Belgrade • Manchester • Tokyo • Cluj • Tianjin

Editor
Stephane Dedieu
CNRS UMR 7369 MEDyC
Université de Reims
Champagne-Ardenne
Reims
France

Editorial Office
MDPI
St. Alban-Anlage 66
4052 Basel, Switzerland

This is a reprint of articles from the Special Issue published online in the open access journal *Cancers* (ISSN 2072-6694) (available at: www.mdpi.com/journal/cancers/special_issues/colorectal_cancer_characteristic).

For citation purposes, cite each article independently as indicated on the article page online and as indicated below:

LastName, A.A.; LastName, B.B.; LastName, C.C. Article Title. *Journal Name* **Year**, *Volume Number*, Page Range.

ISBN 978-3-0365-6837-9 (Hbk)
ISBN 978-3-0365-6836-2 (PDF)

Cover image courtesy of Stephane Dedieu

Contents

About the Editor

Stephane Dedieu

Professor Stéphane Dedieu holds a Ph.D. from the University of Bordeaux (France) where he was trained in biochemistry and molecular biology during the early 2000s. He currently co-leads a 40 people research team within a multidisciplinary CNRS labelled unit (UMR CNRS 7369 MEDyC) affiliated to the University of Reims. He has expertise in cancer biology and therapeutic targeting of the extracellular matrix and cellular receptors.

He co-founded Apmonia Therapeutics, a biotechnology company dedicated to the development of anti-cancer strategies and is chair of the scientific and clinical advisory board of the company.

MDPI

Editorial

Clinical, Pathological, and Molecular Characteristics in Colorectal Cancer

Stéphane Dedieu [1,2,*] and Olivier Bouché [3,4]

1 CNRS UMR 7369 MEDyC, Matrice Extracellulaire et Dynamique Cellulaire, 51100 Reims, France
2 Campus Moulin de la Housse, UFR Sciences Exactes et Naturelles, Université de Reims Champagne-Ardenne (URCA), 51100 Reims, France
3 BioSpecT, UFR Médecine et Pharmacie, Université Reims Champagne-Ardenne (URCA), 51100 Reims, France
4 Digestive Oncology and Hepato-Gastroenterology Department, CHU Reims, 51100 Reims, France
* Correspondence: stephane.dedieu@univ-reims.fr

Colorectal cancer (CRC) is the third most diagnosed cancer worldwide, and the second leading cause of death in patients with cancer. Lifestyle, diet factors, obesity, family history, or even pre-existing inflammatory diseases lead to increased risks of developing this heterogeneous malignant disease.

Carcinogenesis steps have commonly involved the growth and expansion of adenomatous polyps from normal colorectal epithelium to adenoma through a multistep process of several years. Many signaling pathways are altered during this transformation course that involves, certainly, hereditary factors, but mainly somatic sporadic mutations affecting both tumor suppressor genes and oncogenes (*TP53*, *APC*, and *KRAS* from the most common recurrent somatic mutations). Chromosomal instability, CpG island methylation phenotype, and microsatellite instability (MSI) frequency are the three main routes leading to tumor transformation and progression.

The most common tumor–node–metastasis (TNM) staging system driving CRC patient treatment is not completely satisfactory because patients with similar histopathology may have various therapeutic responses and relapse frequencies due to differential genetic and epigenetic profiles. New biomarkers such as Immunoscore ®, gene signatures, and postoperative circulating tumor DNA or extracellular vesicles are promising tools used to identify patients with a high risk of recurrence after primary tumor resection.

Patient prognosis has improved over the past few decades in developed countries, due to an improved health path and better awareness of the population to diagnosis, earlier and regular screening, and access to more extensive surgery and more effective targeted therapies. However, the 5-year survival rate of patients with stage IV remains under 10%. Drug development efforts are therefore mainly focused on patients with stage IV metastatic CRC (mCRC). Although surgery is the primary curative treatment of early-stage patients and resectable metastasis, current treatments for unresectable mCRC involve cytotoxic chemotherapies and targeted therapies, either alone or as a combination treatment. Approved targeted therapy includes angiogenesis inhibitors (bevacizumab, aflibercept, and regorafenib), anti-EGFR monoclonal antibodies (cetuximab and panitumumab) in RAS wild-type tumors, and tyrosine kinase BRAF/MEK inhibitors (binimetinib and encorafenib) in BRAF-mutated tumors.

The emergence and success of immunotherapies in other indications could change the game. Despite a wide variety of immunotherapy approaches in early-phase clinical trials, clinical benefits are currently limited to mCRC patients with a microsatellite instability-high (MSI-H) or mismatch repair-deficient (dMMR) profile, representing approximately 15% of patients. However, recent data also suggest an upfront role for immunotherapy in resectable early-stage MSI-H/dMMR CRC.

A better understanding of the specific clinical and/or molecular features in CRC should therefore improve patient stratification and follow-up, together with treatment algorithms,

Citation: Dedieu, S.; Bouché, O. Clinical, Pathological, and Molecular Characteristics in Colorectal Cancer. *Cancers* **2022**, *14*, 5958. https://doi.org/10.3390/cancers14235958

Received: 23 November 2022
Accepted: 30 November 2022
Published: 2 December 2022

especially when there is a significant unmet medical need. This will require the identification of new biomarkers and therapeutic targets to overcome current barriers and limitations.

Through original articles, this Special Issue provides novel findings on biomarkers of prognostic relevance. Interestingly, PD-L1 expression succeeds in discriminating patients with differential prognosis in the consensus molecular subtype (CMS)2/3 [1], considering overall and disease-free survival [2]. Regarding metastasis-driving protein biomarkers, PD-L1 as well as $\alpha2\beta1$ integrin, CD44v6, IGF-1R, and EGF-R exhibit distinct expression patterns depending on the metastatic organ site [3]. Selective pharmacological targeting based on these molecular signatures could thus faciliate the differential treatment of distant metastases according to their specific metastatic locations.

The expression of discoidin domain receptors (DDRs), i.e., collagen receptors with tyrosine kinase activity, was furthermore investigated in a large cohort of CRC patients [4]. DDRs were found highly expressed in colon adenocarcinoma and associated with a molecular profile that could be integrated within the CMS4 group. While its role as a prognosis marker remains uncertain, DDR expression was found to be associated with shorter event-free survival in CRC patients.

By developing a comprehensive liquid biopsy profile in mCRC patients, Sachin Narayan et al. [5] have also highlighted and characterized heterogeneous populations of oncosomes and CTCs. Although studies including larger numbers of patients are needed for clinical validation, this work supports the predictive benefit of liquid biopsy in the follow-up of mCRC. In a complementary view, the study by Izabela Papiewska-Pająk et al. [6] emphasized the importance of the miRNA content of extracellular vesicle released by CRC for supporting tumor progression, which may be useful as a biomarker indicating the stage of CRC.

In addition, an original study analyzed the contribution of the genetic component to CRC risk in the Basque population with a specific genetic history [7], while another assessed the importance of routine immunohistochemistry screening for MMR status in CRC patients in the identification of Lynch syndrome patients [8].

Considering the prognosis heterogeneity of CRC patients with stage II or III, David Viñal and colleagues [9] proposed a simple score using three clinico-pathological parameters available in routine clinical practice (T4, N2, and high tumor budding) to stratify the recurrence risk and patient prognosis.

In a precision medicine approach, the potential of HER-2 targeting in mCRC was reviewed [10], and a comprehensive update was provided on the CXCL12/CXCR4/CXCR7 axis, including pharmacological perspectives [11]. Finally, the review by Rami Rhaiem and colleagues [12] discussed data on the role of RAS mutational status in tailoring the surgical and/or thermal ablation approach of colorectal liver metastases.

Thanks to advances in molecular biology identifying theranostic biomarkers from tissues but also liquid biopsies, the future lies in increasingly personalized management for the therapeutic choice and monitoring of patients with mCRC.

Conflicts of Interest: The authors declare no conflict of interest.

References

1. Ten Hoorn, S.; de Back, T.R.; Sommeijer, D.W.; Vermeulen, L. Clinical Value of Consensus Molecular Subtypes in Colorectal Cancer: A Systematic Review and Meta-Analysis. *J. Natl. Cancer Inst.* **2022**, *114*, 503–516. [CrossRef] [PubMed]
2. Azcue, P.; Encío, I.; Guerrero Setas, D.; Suarez Alecha, J.; Galbete, A.; Mercado, M.; Vera, R.; Gomez-Dorronsoro, M.L. PD-L1 as a Prognostic Factor in Early-Stage Colon Carcinoma within the Immunohistochemical Molecular Subtype Classification. *Cancers* **2021**, *13*, 1943. [CrossRef] [PubMed]
3. Wrana, F.; Dötzer, K.; Prüfer, M.; Werner, J.; Mayer, B. High Dual Expression of the Biomarkers CD44v6/$\alpha2\beta1$ and CD44v6/PD-L1 Indicate Early Recurrence after Colorectal Hepatic Metastasectomy. *Cancers* **2022**, *14*, 1939. [CrossRef] [PubMed]
4. Ben Arfi, K.; Schneider, C.; Bennasroune, A.; Bouland, N.; Wolak-Thierry, A.; Collin, G.; Le, C.C.; Toussaint, K.; Hachet, C.; Lehrter, V.; et al. Discoidin Domain Receptor 1 Expression in Colon Cancer: Roles and Prognosis Impact. *Cancers* **2022**, *14*, 928. [CrossRef] [PubMed]

5. Narayan, S.; Courcoubetis, G.; Mason, J.; Naghdloo, A.; Kolenčík, D.; Patterson, S.D.; Kuhn, P.; Shishido, S.N. Defining A Liquid Biopsy Profile of Circulating Tumor Cells and Oncosomes in Metastatic Colorectal Cancer for Clinical Utility. *Cancers* **2022**, *14*, 4891. [CrossRef] [PubMed]
6. Papiewska-Pająk, I.; Przygodzka, P.; Krzyżanowski, D.; Soboska, K.; Szulc-Kiełbik, I.; Stasikowska-Kanicka, O.; Boncela, J.; Wągrowska-Danilewicz, M.; Kowalska, M.A. Snail Overexpression Alters the microRNA Content of Extracellular Vesicles Released from HT29 Colorectal Cancer Cells and Activates Pro-Inflammatory State In Vivo. *Cancers* **2021**, *13*, 172. [CrossRef] [PubMed]
7. Garcia-Etxebarria, K.; Etxart, A.; Barrero, M.; Nafria, B.; Segues Merino, N.M.; Romero-Garmendia, I.; Franke, A.; D'Amato, M.; Bujanda, L. Performance of the Use of Genetic Information to Assess the Risk of Colorectal Cancer in the Basque Population. *Cancers* **2022**, *14*, 4193. [CrossRef] [PubMed]
8. Lemos Garcia, J.; Rosa, I.; Saraiva, S.; Marques, I.; Fonseca, R.; Lage, P.; Francisco, I.; Silva, P.; Filipe, B.; Albuquerque, C.; et al. Routine Immunohistochemical Analysis of Mismatch Repair Proteins in Colorectal Cancer-A Prospective Analysis. *Cancers* **2022**, *14*, 3730. [CrossRef] [PubMed]
9. Viñal, D.; Martinez-Recio, S.; Martinez-Perez, D.; Ruiz-Gutierrez, I.; Jimenez-Bou, D.; Peña-Lopez, J.; Alameda-Guijarro, M.; Martin-Montalvo, G.; Rueda-Lara, A.; Gutierrez-Sainz, L.; et al. Clinical score to predict recurrence in patients with stage II and III colon cancer. *Cancers* **2022**, *14*, 5891. [CrossRef]
10. Torres-Jiménez, J.; Esteban-Villarrubia, J.; Ferreiro-Monteagudo, R. Precision Medicine in Metastatic Colorectal Cancer: Targeting ERBB2 (HER-2) Oncogene. *Cancers* **2022**, *14*, 3718. [CrossRef] [PubMed]
11. Goïta, A.A.; Guenot, D. Colorectal Cancer: The Contribution of CXCL12 and Its Receptors CXCR4 and CXCR7. *Cancers* **2022**, *14*, 1810. [CrossRef] [PubMed]
12. Rhaiem, R.; Rached, L.; Tashkandi, A.; Bouché, O.; Kianmanesh, R. Implications of RAS Mutations on Oncological Outcomes of Surgical Resection and Thermal Ablation Techniques in the Treatment of Colorectal Liver Metastases. *Cancers* **2022**, *14*, 816. [CrossRef] [PubMed]

Article

PD-L1 as a Prognostic Factor in Early-Stage Colon Carcinoma within the Immunohistochemical Molecular Subtype Classification

Pablo Azcue [1,*], Ignacio Encío [1,2], David Guerrero Setas [3,4,5,6], Javier Suarez Alecha [7], Arkaitz Galbete [2,4,8], María Mercado [3], Ruth Vera [2,6] and Maria Luisa Gomez-Dorronsoro [2,3,*]

1 Department of Health Science, Public University of Navarra (UPNA), 31008 Pamplona, Spain; ignacio.encio@unavarra.es
2 Institute for Health Research Navarra (IdISNA), 31008 Pamplona, Spain; arkaitz.galbete@navarra.es (A.G.); ruth.vera.garcia@navarra.es (R.V.)
3 Department of Molecular Pathology, Hospital Complex of Navarra (CHN), 31008 Pamplona, Spain; dguerres@navarra.es (D.G.S.); mr.mercado.gutierrez@navarra.es (M.M.)
4 Campus Arrosadia, Public University of Navarra (UPNA), 31006 Pamplona, Spain
5 Molecular Pathology of Cancer Group–Navarrabiomed, 31008 Pamplona, Spain
6 Department of Medical Oncology, Hospital Complex of Navarra (CHN), 31008 Pamplona, Spain
7 Department of Surgery, Hospital Complex of Navarra (CHN), 31008 Pamplona, Spain; fj.suarez.alecha@navarra.es
8 Navarrabiomed-Hospital Complex of Navarra (CHN), Redissec, 31008 Pamplona, Spain
* Correspondence: azcue.136628@e.unavarra.es (P.A.); ml.gomez.dorronsoro@navarra.es (M.L.G.-D.)

Citation: Azcue, P.; Encío, I.; Guerrero Setas, D.; Suarez Alecha, J.; Galbete, A.; Mercado, M.; Vera, R.; Gomez-Dorronsoro, M.L. PD-L1 as a Prognostic Factor in Early-Stage Colon Carcinoma within the Immunohistochemical Molecular Subtype Classification. *Cancers* **2021**, *13*, 1943. https://doi.org/10.3390/cancers13081943

Academic Editor: Stephane Dedieu

Received: 28 March 2021
Accepted: 12 April 2021
Published: 17 April 2021

Simple Summary: Colorectal cancer (CRC) is a very heterogeneous disease. Efforts to characterize and search for biomarkers for these patients are currently ongoing in the hope of establishing a more targeted therapeutic approach. The role of PD-1 ligand (PD-L1) expression as a biomarker has not yet been fully elucidated. The Consensus Molecular Subtype classification has been delineated, but although already acknowledged in the most recent international guidelines, it has yet to be implemented in clinical practice. We investigate PD-L1 expression as a biomarker of prognosis in the early-stage setting and integrate it with the Consensus Molecular Subtype (CMS), in an effort to differentiate those patients with a worse prognosis who could potentially benefit from an early, more aggressive treatment. Our results suggest PD-L1 as an independent prognostic factor in early stage setting when assessed by immunohistochemistry. Additionally, PD-L1 expression appears to be a viable biomarker to differentiate patients in the CMS (CMS2/CMS3) who lack a clear prognosis.

Abstract: Background. There is a patent need to better characterize early-stage colorectal cancer (CRC) patients. PD-1 ligand (PD-L1) expression has been proposed as a prognostic factor but yields mixed results in different settings. The Consensus Molecular Subtype (CMS) classification has yet to be integrated into clinical practice. We sought to evaluate the prognostic value of PD-L1 expression overall and within CMS in early-stage colon cancer patients, in the hope of assisting treatment choice in this setting. Methods. Tissue-microarrays were constructed from tumor samples of 162 stage II/III CRC patients. They underwent automatic immunohistochemical staining for PD-L1 and the proposed CMS panel. Primary endpoints were overall survival (OS) and disease-free survival (DFS). Results. PD-L1 expression was significantly and independently associated with better prognosis (HR = 0.46 (0.26–0.82), $p = 0.009$) and was mostly seen in immune cells of the tumor-related stroma. CMS4 five-folds the risk of mortalitycompared with CMS1 (HR = 5.58 (1.36, 22.0), $p = 0.034$). In the subgroup CMS2/CMS3 analysis, PD-L1 expression significantly differentiated individuals with better OS ($p = 0.004$) and DFS ($p < 0.001$). Conclusions. Our study suggests that PD-L1 expression is an independent prognostic factor in patients with stage II/III colon cancer. Additionally, it successfully differentiates patients with better prognosis in the CMS2/CMS3 group and may prove significant for the clinical relevance of the CMS classification.

Keywords: PD-L1; CMS; colon cancer

1. Introduction

Colorectal cancer (CRC) is currently the third most commonly diagnosed cancer and the second leading cause of cancer death worldwide [1], due in part to it being a highly heterogeneous disease. A classification of CRC patients is needed to provide the basis for better treatment decisions and targeted therapies, particularly in early-stage settings [2]. Through transcriptomics the Consensus Molecular Subtype (CMS) classification has been delineated and proposed as a prognostic tool with predictive capabilities and therapeutic implications [3]. However, these types of techniques require specialized resources that are not within the reach of most hospitals.

In a first effort to address this issue, a subrogated panel of four proteins (CDX2, FRMD6, HTR2B and ZEB1) has been validated through immunohistochemistry (IHC) [4]. This panel allows a classification of CRC patients that is reproducible and more easily accessible to hospitals and laboratories. The proposed panel classifies patients into CMS4 and CMS2/CMS3 subtypes, with CMS1 being defined by the mismatch repair (MMR) proteins panel: CDX2 is a homeobox transcription factor expressed in early intestinal development, where it regulates proliferation, differentiation, cell adhesion and migration of intestinal epithelial cells [5]. Pilati et al. reported that lack of CDX2 expression in the CMS classification is useful for identifying poor prognosis patients (CMS4/CDX2-negative), whereas CMS2 and CMS3 tumors rarely show total loss of CDX2 [6]. HTR2B is a G-protein coupled receptor subtype of the serotonin family that is overexpressed in various solid tumors [7,8] and has a higher level of expression in mesenchymal-like tumors [4]. FRMD6 is an Ezrin/Radixin/Moesin family protein that is part of the Hippo signaling pathway kinase cascade [9]. Its loss of expression contributes to the epithelial-to-mesenchymal transition (EMT) while its overexpression antagonizes the yes-associated protein 1 (YAP) activity [10]. ZEB1 is a transcription factor regulated by a variety of signaling pathways including WNT [11]. It promotes invasion and metastasis by inducing EMT and is frequently observed in mesenchymal-like carcinoma cells that confer resistance to cancer therapy [12].

MMR protein expression (MLH1, MSH2, MSH6, PMS2) is studied to determine microsatellite instability (MSI) or deficient MMR (dMMR), which accounts for 15–20% of CRCs [13–15]. In the current ESMO and NCCN colon cancer guidelines, dMMR status is acknowledged as being a valid prognostic biomarker of CRC in some settings, although other major and minor prognostic tools, especially TNM staging, must be used when deciding whether to offer adjuvant therapy [16,17]. Recently, anti-PD1 treatment has proven to be beneficial in MSI high patients [18].

The aforementioned four-biomarker IHC panel and the MMR panel can identify the CMS4 and CMS1 subtypes, which have the worst and best prognosis, respectively. CMS2 and CMS3 account for more than 50% of the population and are indistinguishable from each other by these panels. This CMS2/CMS3 group includes patients with very different molecular characteristics [19] and survival [20–23]. Therefore, there is a need for new biomarkers that can provide clear expectations about prognosis, particularly for this group of patients.

Programmed cell death protein 1 (PD-1, also known as PDCD1 and CD279) is an inhibitory receptor that is expressed by T cells during activation. It regulates T cell effector functions during various physiological responses, including acute and chronic infection, cancer and autoimmunity, and in immune homeostasis [24]. Some cancer cells can develop PD-1 ligand (PD-L1) expression, which potentially shields it from immune attack by inhibiting T cell effector functions [25]. Its expression has been associated with the serrated pathway of colorectal carcinogenesis, with the presence of BRAF mutation, dMMR and poor differentiation [26]. The potential activation of the WNT/β-catenin pathway by this receptor has also been linked to progression [27]. Additionally, a recent study has reported the regulation of PD-L1 by the Zinc finger E-box binding homeobox 1 (ZEB1), an EMT inhibitor [28].

PD-L1 expression has been proposed as being a biomarker of prognosis in early CRC, but has yet to be fully devised, probably due to the lack of standardization in IHC assessment and homogenization for the studied population [29].

PD-L1 was first studied as a predictive tool in CRC, although early published studies yielded some apparently contradictory results [26,30,31]. More recently, it has proved to be of predictive value for anti-PD-1 therapy for overall survival (OS), and especially for overall response rate (ORR) and progression-free survival (PFS) [18,32]. This has been achieved mainly by post hoc analysis, which proposes, among other things, a different cut-off value for PD-L1 expression and a different methodology for the pathological assessment (>1%, >5%, >50%) [25,29,33,34].

We hypothesize that PD-L1 expression, when assessed by IHC using a standardized methodology, is a strong candidate for assessment as a potential biomarker. Furthermore, when added to the panel for CMS classification, it could prove helpful for patients whose immune response is not as clear as in those of the CMS2/CMS3 subtypes. Therefore, the objective is to first corroborate the possibility of classifying a large cohort of early-stage CRC patients into CMS subtypes through the proposed IHC panel, and then to investigate the prognostic role of PD-L1 expression in addition to the previous panel, specifically for patients in the CMS2/CMS3 subgroup. We expect to obtain clearer expectations about the CMS2/CMS3 subgroup that might inform physicians' choice of treatment for early-stage patients in routine clinical practice.

2. Materials and Methods

This study was performed in accordance with the World Medical Association Declaration of Helsinki. The study was approved by the Regional Clinical Research Ethics Committee (CEIC) Pyto2017/51 Cod. MOL_CRC, 15 May 2018. Patient consent was waived due to the use of stored tumor samples for research purposes in compliance with the current Spanish and European Union legislation (resolution 1387/2017 (08/11) and resolution 193/2018 (06/03) of the Navarra Health Service—Osasunbidea).

2.1. Patients

The cohort of this retrospective study consists of 162 patients diagnosed with stage II/III CRC consecutively surgically resected with curative intention in the Hospital Complex of Navarra between 2009 and 2013. All patients were diagnosed by the Department of Pathology, following the standardized treatment protocol established by the Colorectal Committee. Participants were then followed until death or last medical consultation, with a cut-off date of 1 October 2018, when the clinical data were retrieved, anonymized and analyzed.

The clinical follow-up protocol included a medical visit and carcinoembryonic antigen monitoring every three months for two years and then every six months for three more years. Computed tomography (CT) was performed annually, at the same time in years one and five as a colonoscopy. A CT of the abdomen and chest x-ray or CT was performed preoperatively to rule out distant metastases in all patients. The data retrieved subsequently included age, gender, localization (the right and left colons were, respectively, defined as proximal and distal to the splenic angle [35]), differentiation grade (defined as exhibiting less than 50% or at least 50% of glandular formations), lymph node ratio, histological type, and lymphatic, blood vessel and perineural invasion. Tumors were classified according to the TNM Classification of Malignant Tumors (TNM), 7th edition [36].

Only patients with stage II/III CRC and a confirmed pathological diagnosis of adenocarcinoma were included in the study. Patients who had insufficient tumor material, were lost to follow-up for at least three years or had died, had fewer than two IHC-stained blocks for evaluation, or whose information about their baseline characteristics was missing were then excluded. To further homogenize the study population, we decided to include only cases with colon carcinoma (CC).

One hundred and forty-four patients from the original cohort were included in the statistical analyses.

2.2. Pathological Study

Hematoxylin-eosin sections representative of the invasive carcinoma from the Formalin-fixed paraffin-embedded (FFPE) tissue specimens were selected for each patient. Four spots/areas were annotated per case in the infiltrating tumor, two from the tumor periphery or invasion front, and two from the central tumor area to minimize the heterogeneity within the tumor. Areas of abscess or necrosis were avoided.

Corresponding donor tissue cores were then transferred to the tissue microarray (TMA) recipient blocks using a manual Tissue Arrayer MTA-1 (Beecher Instruments Silver Spring, MD, USA). Four representative 1-mm-diameter cores were obtained in sequence for each tumor after confirmation of each annotation in selected areas. Each TMA block consisted of two sections containing 10 × 5 cores, and four tissue cores from benign colon and ovary tissue selected as controls and for orientation (Figure S1). Each TMA block was divided into two halves: the first half contained a pair of consecutive TMA samples for each patient; the second half also contained the second pair of consecutive TMA samples but in a different order, with each half containing a control. This method was adopted to reduce possible evaluator bias when analyzing consecutive samples from the same patient.

The constructed TMAs blocks were then sectioned in 4μm slides, stained, scanned and finally scored as described below.

2.3. Immunohistochemical Analysis

The TMAs sections underwent immunohistochemical staining against CDX2, FRMD6, HTR2B, and ZEB1 for the CMS classification. This IHC-based screening panel was used as a surrogate for gene expression profiling [37].

The antibodies used were anti-FRMD6 (Clone HPA001297; 1:50; Sigma), anti-HTR2B (Clone HPA012867; 1:50; Sigma), anti-ZEB1 (Clone HPA027524; 1:50; Sigma), using Roche's BenchMark Ventana automatic immunostainer. The anti-CDX2 (Clone PA0535; RTU; Novocastra), anti-cytokeratin (Clone PA0909; RTU; Leica), and the MMR proteins MLH1 (Clone PA0610; RTU; Biocare), MSH2 (Clone FE-11; 1:100; Calbiochem), MSH6 (Clone PM265AA; RTU; Biocare) and PMS2 (Clone PM344AA; RTU; Biocare), were used to determine dMMR status using Leica Biosystems' Bond automatic immunostainer, and the BRAF V600E mutation was determined through anti-BRAF (Clone VE1, 1:1, Roche) by IHC.

Finally, to determine PD-L1 expression, TMAs underwent staining using the antibody anti-PD-L1 (SP142; RTU; Roche) following the specifically approved protocol. Staining was performed in each section, after antigenic recovery and endogen peroxidase blockage, by sequentially incubating the specific primary and secondary antibodies, and revealed with the Optiview Universal DAB Detection Kit using an automatic BenchMark XT (VENTANA/Roche).

Each stained TMA array was then scanned and digitalized using the VENTANA iScan HT Slide scanner. Images were processed using the integrated Virtuoso image and workflow management software (VENTANA/Roche).

2.4. IHC Scoring and Evaluation

Once digitized, each individual sample from the TMA slide was scored by two independent evaluators (a trained senior scientist and an expert pathologist), both of whom were blinded to the patients' clinical data. In the event of discordant results, a wash-out period of three weeks was imposed, after which the evaluators scored the samples again and, with the aid of reference images of each antibody, arrived at a consensus score.

The CMS assessment was assessed according to published methodology [37]. The scores obtained were then uploaded to the online IHC classifier (https://crcclassifier.shinyapps.io/appTesting, accessed on: 2 September 2020) and the CMS2/CMS3 and CMS4 subtypes were established.

The status of MMR was determined as proficient (pMMR) or deficient (dMMR). A case was considered to be pMMR when any focus of the tumor exhibited positive nuclear staining for all MMR proteins (MLH1, PMS2, MSH2 and MSH6). If the tumor showed a total loss of staining for any of these proteins in all tumor cells, it was considered to be dMMR. The latter were first used to define patients as belonging to the CMS1 subtype. Lymphocytes were used as an internal control for evidence of positive staining. BRAF V600E was classified dichotomously as mutated (pathological) or wild type, also with a known mutated colon adenocarcinoma control for positive staining.

The expression of PD-L1 was measured when at least 50 viable tumor cells were present. A sample of the amygdala was used as a control in each TMA. For the IHC assessment, the SP142 antibody guidelines state that the determination of PD-L1 status evaluation is based on the percentage area of positive immune cells within the total area of inflammation and tumor-related stroma (%) of any intensity and the percentage area of PD-L1 expressing tumor cells within the total tumor area (%) of any intensity [29,38].

In our study, the presence of discernible PD-L1 staining of any intensity was discernible in immune cells (lymphocytes, macrophages and dendritic cells) in the tumor-related stroma (Figure 1). The assessment was performed using a four-level score based on the percentage stained, as follows: 0 when <1%; 1 between 1% and <5%; 2 between 5% and <50%; and 3 when >50%. To maximize sensitivity and specificity, a score of 0 was considered as negative, 1 or more as low expression (PD-L1–L) and 2 or more as high expression (PD-L1–H).

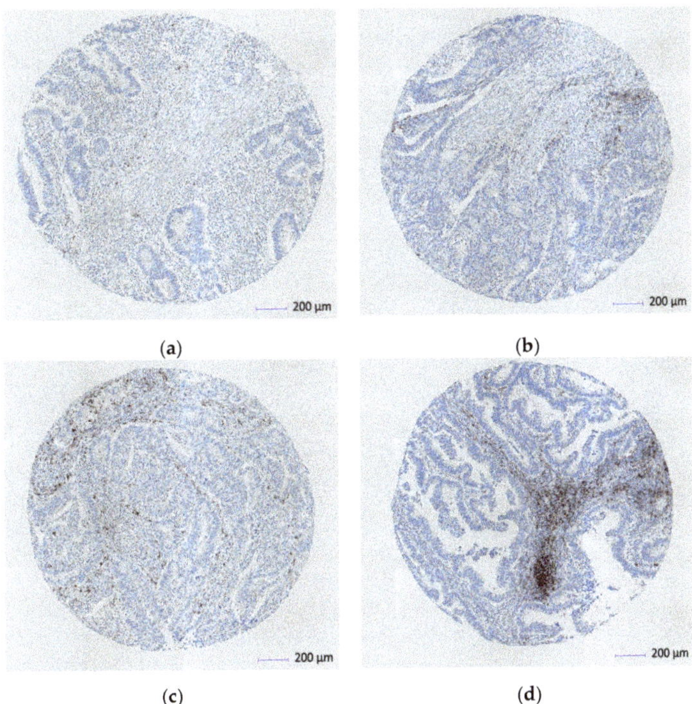

Figure 1. Immunohistochemistry (IHC) scan of PD-1 ligand (PD-L1) expression on immune cells at the tumor-stroma interface (Scanned images core X4). (**a**) PD-L1 < 1% (score 0); (**b**) PD-L1 of > 1–5% (score 1); (**c**) PD-L1 > 5%–< 50% (score 2); (**d**) PD-L1 > 50% (score 3).

2.5. Statistical Analysis

The primary endpoint OS was defined as time from surgery to death due to any cause and disease-free survival (DFS) was defined as time from surgery to relapse or death due to any cause. A predetermined subgroup analysis of OS and DFS for the expression of PD-L1 in the CMS2/CMS3 population was performed.

The statistical analysis was performed by using the SPSS 24.0 software (IBM, New York, USA). Associations between variables among groups were determined using the *t*-test, Mann–Whitney U test or ANOVA for quantitative variables and using Fisher's exact test and the χ^2 test for categorical ones. Univariate and multivariate Cox proportional hazard regression models were used for OS and DFS analysis. The multivariate model was adjusted for the factors that proved significant with survival in the univariate analysis, which included baseline clinical variables such as age and sex. The survival curves were calculated using the Kaplan–Maier method and the log-rank test. Statistical significance was set at two tailed *p*-value of <0.05.

3. Results

To assess the prognostic value of PD-L1 expression in addition to the CMS classification in early-stage CC patients, 144 patients were analyzed. Only patients with colon adenocarcinomas were included in the analysis. The majority of patients were men (68.1%), with stage II (55.6%), well or moderately differentiated (81.9%) and right-sided (54.9%) tumors. Patient baseline characteristics and main clinical parameters are summarized in Table 1.

Table 1. Demographic and pathological characteristics.

Variable		N (%) $n = 144$
Age (years) *		72.2 (9.6)
	Range	48–93
Gender		
	Female	46 (31.9)
	Male	98 (68.1)
Localization		
	Right	79 (54.9)
	Left	65 (45.1)
Differentiation grade		
	<50%	26 (18.1)
	≥50%	118 (81.9)
Lymph node ratio		
	Mean * (SD)	6.7 (12.1)
	Median (Q1–Q3)	0.0 (0–9.3)
Histologic type		
	Colloid	18 (12.5)
	Adenocarcinoma	125 (86.8)
	Signet ring cell carcinoma	1 (0.7)
TNM Stage		
	II	80 (55.6)
	III	64 (44.4)
Lymphatic vascular invasion		
	Negative	108 (75.0)
	Positive	36 (25.0)
Blood vessel invasion		
	Negative	102 (70.8)
	Positive	42 (29.2)
Perineural invasion		
	Negative	112 (77.8)
	Positive	32 (22.2)

* Values are means. SD: Standard deviation. Q1–Q3: Quartiles 1–3.

In the IHC analysis CMS2/CMS3 represents the largest subgroup (81.3%), whereas the CMS1/dMMR subgroup represented 12.5% and CMS4 made up 6.3% of the cohort. The expression of PD-L1–L (more than 1% of immune cells) was present in 55.5% of patients, and PD-L1–H (more than 5% of immune cells) was present in 20.1% of patients. Distributions of CMS subtypes, PD-L1 expression, BRAF expression and MMR protein deficiency, according to IHC analysis, are presented in Table 2.

Table 2. Prevalence of study variables.

Variable		N (%) *n* = 144
MMR status		
	pMMR	126 (87.5)
	dMMR	18 (12.5)
IHC BRAF (V600 mutation) *		
	Negative	125 (89.3)
	Positive	15 (10.7)
CMS classification		
	CMS1	18 (12.5)
	CMS2/CMS3	117 (81.3)
	CMS4	9 (6.3)
IHC PD-L1 expression		
	Negative	64 (44.4)
	≥1–<5%	51 (35.4)
	≥5%, >50%	29 (20.1)

* Not assessable in 4 tumor samples. p/dMMR: Mismatch repair proficient or deficient. CMS: Consensus Molecular Subtype. IHC: Immunohistochemical.

Patients were followed for a median of 65.0 months (95% CI (62.2–67.7)), during which time 27 patients (18.8%) relapsed and 51 patients (35.4%) died. The number of outcome events (relapse and death, respectively) in the CMS subgroups were as follows: CMS1 subgroup 1 (5.6%) and 4 (22.2%), CMS2/3 subgroup 22 (18.8%) and 42 (35.9%), CMS4 subgroup 4 (44.4%) and 5 (55.6%).

3.1. Comparative Analysis

Mismatch repair deficiency (dMMR) was significantly more frequent in the right-sided tumors ($p = 0.002$) and was also significantly more frequently associated with PD-L1 expression ($p < 0.001$).

The CMS classification showed a significant difference in the expression of both PD-L1–L ($p = 0.038$) and PD-L1–H ($p < 0.001$). The differences were mainly due to the overexpression in CMS1 and under expression in CMS4. For PD-L1–L the expression was found in 77.8% of the CMS1 group, in 22.2% of the CMS4 group and in almost half (49.6%) of the CMS2/CMS3 group.

A statistically significant difference was found for TNM stage and CMS ($p = 0.016$). Stage II and III patients were more frequently classified into CMS1 and CMS4, respectively, but similar numbers of stage II and III patients were classified as CMS2/CMS3. Differences were also found with respect to localization and CMS; right-sided tumors were more often classified into CMS1, whereas there were no differences for CMS2/CMS3.

PD-L1 expression was more often expressed in stage II tumors ($p = 0.014$) and was found concomitantly with BRAF mutation ($p = 0.002$). A full comparative analysis is presented in Table 3.

Table 3. Comparative analysis.

Variable		MMR pMMR *n* (%) dMMR *n* (%)	*p*	CMS1 *n* (%)	CMS2/CMS3 *n* (%)	CMS4 *n* (%)	*p*	PD-L1–L Neg (%) Pos (%)	*p*	PD-L1–H Neg (%) Pos (%)	*p*
Age	Mean (SD)	72.2 (9.5) 72.3 (10.6)	0.948 [1]	72.3 (10.6)	72.6 (9.2)	66.0 (11.2)	0.134 [1]	72.7 (10.4) 71.8 (9.0)	0.583 [3]	71.8 (9.6) 73.9 (9.4)	0.296 [3]
Gender	Men	86 (68.3) 12 (66.7)		12 (66.7)	79 (67.5)	7 (77.8)		41 (64.1) 57 (71.3)		77 (67.0) 21 (72.4)	
	Women	40 (31.7) 6 (33.3)	0.893 [2]	6 (33.3)	38 (32.5)	2 (22.2)	0.884 [4]	23 (35.9) 23 (28.7)	0.358 [2]	38 (33.0) 8 (27.6)	0.573 [2]
Localization	Right	63 (50.0) 16 (88.9)		16 (88.9)	57 (48.7)	6 (66.7)		29 (45.3) 50 (62.5)		61 (53.0) 18 (62.1)	
	Left	63 (50.0) 2 (11.1)	**0.002** [2]	2 (11.1)	60 (51.3)	3 (33.3)	**0.003** [4]	35 (54.7) 30 (37.5)	**0.039** [2]	54 (47.0) 11 (37.9)	0.412 [2]
TNM Stage	II	68 (54.0) 12 (66.7)		12 (66.7)	67 (57.3)	1 (11.1)		30 (46.9) 50 (53.1)		58 (50.4) 22 (75.9)	
	III	58 (46.0) 6 (33.3)	0.310 [2]	6 (33.3)	50 (42.7)	8 (88.9)	**0.016** [4]	34 (62.5) 30 (37.5)	0.061 [2]	57 (49.6) 7 (24.1)	**0.014** [2]
Lymphatic vascular invasion	No	92 (73.0) 16 (55.9)		16 (88.9)	87 (74.4)	5 (55.6)		47 (73.4) 61 (76.3)		83 (72.2) 25 (86.2)	
	Yes	34 (27.0) 2 (11.1)	0.243 [4]	2 (11.1)	30 (25.6)	4 (44.4)	0.169 [4]	17 (26.6) 19 (23.8)	0.699 [2]	32 (27.8) 4 (13.8)	0.119 [2]
Blood vessel invasion	No	88 (69.8) 14 (77.8)		14 (77.8)	90 (68.4)	8 (88.9)		40 (62.5) 62 (77.5)		79 (68.7) 23 (79.3)	
	Yes	38 (30.2) 4 (22.2)	0.488 [2]	4 (22.2)	37 (31.6)	1 (11.1)	0.373 [4]	24 (37.5) 18 (22.5)	**0.049** [2]	36 (31.3) 6 (20.7)	0.261 [2]
Perineural invasion	No	96 (76.2) 16 (88.9)		16 (88.9)	91 (77.8)	5 (55.6)		47 (73.4) 65 (81.3)		85 (73.9) 27 (93.1)	
	Yes	30 (23.8) 2 (11.1)	0.363 [4]	2 (11.1)	26 (22.2)	4 (44.4)	0.146 [4]	17 (26.6) 15 (18.8)	0.262 [2]	30 (26.1) 2 (6.9)	**0.026** [2]
BRAF IHC *	Wild	118 (94.4) 7 (46.7)		7 (46.7)	109 (94.0)	9 (100)		56 (90.3) 69 (88.5)		105 (93.8) 20 (71.4)	
	Mutant	7 (5.6) 8 (53.3)	**<0.001** [4]	8 (53.3)	7 (6.0)	0 (0.0)	**<0.001** [4]	6 (9.7) 9 (11.5)	0.724 [2]	7 (6.2) 8 (28.6)	**0.002** [2]
MMR status	pMMR	-		0 (0.0)	117 (100)	9 (100)		60 (93.8) 66 (82.5)		107 (93.0) 19 (65.5)	
	dMMR	-	-	18 (100)	0 (0.0)	0 (0.0)	**<0.001** [4]	4 (6.3) 14 (17.5)	**0.043** [2]	8 (7.0) 10 (34.5)	**0.001** [2]
PD-L1–L	Neg	60 (47.6) 4 (22.2)		4 (22.2)	58 (49.6)	2 (22.2)		-		-	
	Pos	66 (52.4) 14 (77.8)	**0.043** [2]	14 (77.8)	59 (50.4)	7 (77.8)	**0.038** [4]	-	-	-	-
PD-L1–H	Neg	107 (84.9) 8 (44.4)		8 (44.4)	99 (84.6)	8 (88.9)		-		-	
	Pos	19 (15.1) 10 (55.6)	**<0.001** [4]	10 (55.6)	18 (15.4)	1 (11.1)	**<0.001** [4]	-	-	-	-

[1] ANOVA, [2] Chi-square test, [3] t-test, [4] Fisher's exact test, two-tailed, [5] *t*-test. * Not assessable in 4 tumor samples. p/dMMR: Mismatch-repair proficient or deficient. CMS: Consensus Molecular Subtype. IHC: Immunohistochemical. Neg/Pos: Negative/Positive. SD: Standard deviation. Statistically significant *p* values are presented in **bold**.

3.2. Univariate Analysis

The univariate analysis showed that the risk of mortality increased with age by about 9% per year (HR = 1.09 95% CI (1.04, 1.14)), $p < 0.001$), and, as expected, a relapse event significantly increased the risk of mortality (HR = 7.93 (95% CI 3.05, 20.6), $p < 0.001$), as seen in Table 4. Finally, perineural invasion also showed a tendency towards poor prognosis (HR = 2.20 (95% CI 0.99, 4.90), $p = 0.050$). The expression of PD-L1 was related to a reduced risk of death, especially for PD-L1–L (HR = 0.40 (95% CI 0.20, 0.81), $p = 0.010$). No significant difference was found in mortality between stage II and stage III patients (HR = 1.51 (95% CI 0.76, 2.99), $p = 0.242$).

Table 4. Univariate analysis of overall survival.

Variable (Reference)	Hazard Ratio	95% CI	*p* Value
Age (Mean)	1.09	1.04–1.14	**<0.001**
Gender (Male/Female)	0.96	0.46–2.00	0.913
Localization (Right/Left)	0.78	0.39–1.55	0.479
Stage (II/III)	1.51	0.76–2.99	0.242
Lymphatic vascular invasion	0.88	0.40–1.96	0.763
Blood vessel invasion	1.81	0.86–3.78	0.114
Perineural invasion	2.20	0.99–4.90	**0.050**
BRAF IHC * (wt/mutant)	0.25	0.05–1.14	0.056
MMR status (p/d)	0.48	0.15–1.54	0.211

Table 4. *Cont.*

Variable (Reference)	Hazard Ratio	95% CI	*p* Value
CMS1–CMS2/CMS3	1.96	0.61–6.34	
CMSCMS1–CMS4	4.38	0.78–24.5	0.224
PD-L1–L	0.40	0.20–0.81	**0.010**
PD-L1–H	0.41	0.15–1.07	0.064

* Not assessable in four tumor samples. p/dMMR: Mismatch repair proficient or deficient. CMS: Consensus Molecular Subtype. IHC: Immunohistochemical. Statistically significant *p* values are presented in **bold**.

3.3. Multivariate Analysis

The expression of PD-L1–L was associated with good prognosis in the univariate analysis and was confirmed as being independently associated with better OS in the multivariate analysis (HR = 0.46 (95% CI 0.26–0.82), $p = 0.009$) and DFS (HR = 0.48 (95% CI 0.28–0.83), $p = 0.012$). A high expression of PD-L1–H also showed a tendency towards statistical significance for better OS (HR = 0.42 (95% CI 0.17–1.02), $p = 0.054$) and DFS (HR = 0.46 (95% CI 0.20–1.05), $p = 0.064$).

CMS4 patients had five times greater risk of mortality and six times the risk of DFS compared to the CMS1 group (HR = 5.58 (95% CI 1.36, 22.0), $p = 0.034$ and HR = 6.33 (95% CI 1.68, 23.8), $p = 0.012$, respectively). CMS2/CM3 exhibited an intermediate prognosis with no statistically significant difference.

Independent variables associated with worse prognosis of mortality were age (HR = 1.09 (95% CI 1.05–1.13), $p < 0.001$) and perineural invasion (HR = 2.25 (95% CI 1.19–4.26), $p = 0.012$). Similar results for age and perineural invasion were found for DFS, but no significant differences were noted between the sexes.

3.4. Survival

The Kaplan–Meier curves (Figure 2) for OS and DFS are consistent with previous results. With respect to OS and DFS, the CMS1 group displayed the longest survival, followed by the CMS2/CMS3 and finally the CMS4, which had the poorest outcome. In the subgroup analysis of CMS2/CMS3, PD-L1 expression significantly differentiated patients with good and poor prognosis for OS and time to relapse or death ($p = 0.004$ and $p < 0.001$, respectively).

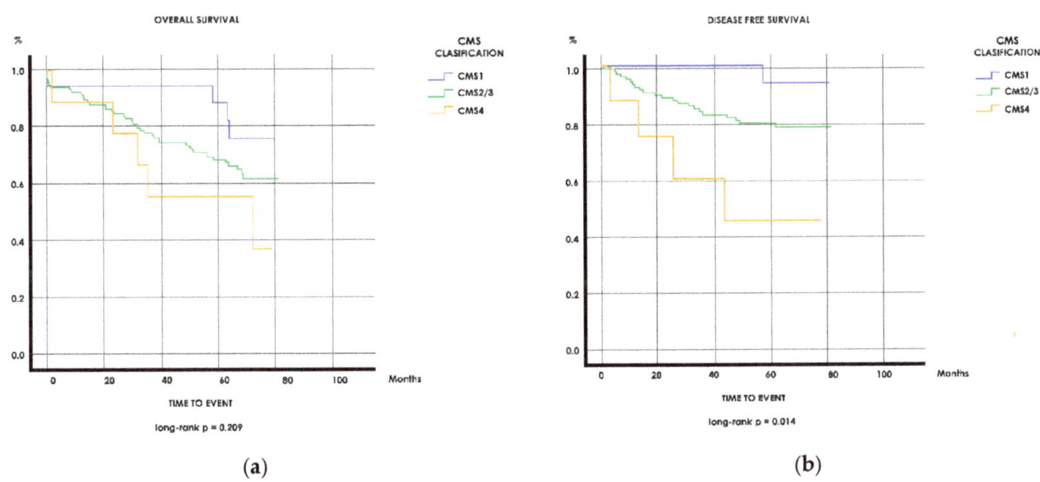

Figure 2. *Cont.*

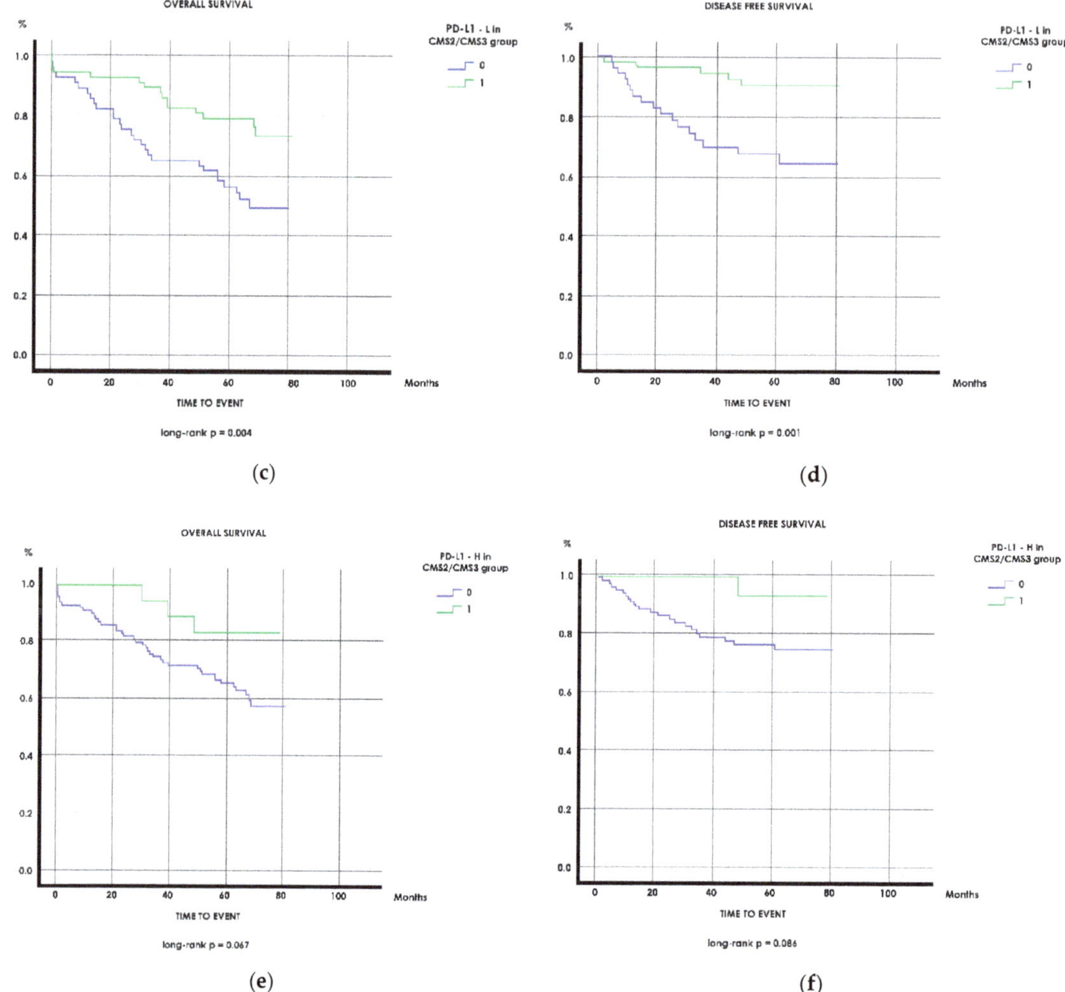

Figure 2. Survival analysis. Kaplan–Meier curves for (**a**) overall survival (OS) by CMS in the overall population; (**b**) disease-free survival (DFS) by CMS in the overall population; (**c**) OS for PD-L1–L in the CMS2/CMS3 group; (**d**) DFS for PD-L1–L in the CMS2/CMS3 group; (**e**) OS for PD-L1 – H in the CMS2/CMS3 group; (**f**) DFS for PD-L1–L in the CMS2/CMS3 group.

4. Discussion

Our findings suggest that PD-L1 expression is an independent prognostic factor in patients with cancer in the CMS2/CMS3 group. Patients in this group with positive expression of PD-L1–L ($\geq$1%) and of PD-L1–H ($\geq$5%) in immune cells in tumor-related stroma had longer OS and DFS than patients with a lower or null level of expression. After adjustment for known clinical prognostic factors, the prognostic effect of PD-L1 remained significant in the multivariate analysis for both OS and DFS. The CMS1 group provided the best prognosis, whereas the CMS4 group exhibited the worst outcome.

Consistent with the findings of similar studies, patients with a diagnosis of rectal cancer were excluded from our analysis in an effort to homogenize the patient population, since rectal cancer differs from colon cancer with respect to the therapeutic approach, tumor biology and prognosis [29,39–41]. Furthermore, through the use of IHC, some studies have revealed elevated PD-L1 expression in rectal cancer after chemo-radiotherapy in the perioperative setting [42,43].

PD-L1 expression depends on various factors and their possible interactions, for example the type of tumor, pathological assessment, tumor stage, and technical issues related to IHC (e.g., the type of clone, scoring method, cut-off values for positivity, etc.). CRC is considered to be a cold tumor with a low PD-L1 expression compared with other solid tumors such as lung cancer, renal cell carcinoma and urothelial carcinoma. PD-L1 expression in CRC is not frequently observed in tumor cells [29,38,44–46], although this may not be the case for all clones. Accordingly, the PD-L1 expression in our study with the SP142 clone mostly occurred in the immune cells of the tumor-related stroma, and not in any tumor cells (Figure S2a,b). In a few cases, the expression was initially thought to occur in the tumor epithelium, but, on closer assessment, it was found to be due to infiltration of intratumoral lymphocytes [38]. In these few cases of intertumoral expression, they all co-existed with positivity at the tumor–stroma interface.

It has been suggested that overexpression of PD-L1 in CRC is fundamentally related to an extrinsic/adaptive mechanism that drives PD-L1 expression in immune cells, highlighting the role of the tumor microenvironment, rather than being associated with an intrinsic gene alteration [44,47–49]. One example is the MSI in CRC, where an "extrinsic" immune cell-mediated PD-L1 upregulation mechanism has been hypothesized to be exerted by the induction of an active immune microenvironment by this instability on two fronts: an immune-stimulatory effect by increased cytotoxic effector T lymphocytes on one side, and immune inhibitory effect that includes PD-1/PD-L1 checkpoint on the other [38,44,50]. Likewise, our results showed that dMMR tumors were significantly associated with PD-L1 expression in immune cells. Furthermore, the level of expression of PD-L1 was also significantly related to dMMR tumors since more cases were assessed as being PD-L1–H than PD-L1–L (34.5% vs. 17.5%). The exosomes are another example supporting the "extrinsic/adaptive" mechanism. As recently reported by Tang et al., exosomes may play a role in immunosuppression and avoiding an anti-tumor immune response [51]. Overall, it has been suggested that there is a lack of evidence supporting "intrinsic" mechanisms in CRC, unlike other solid tumors [38].

We used the SP142 clone because it has proved useful in other tumor types with clinical implications and with a particular sensitivity of expression in immune cells (e.g., breast, urothelial and non-small cell lung cancer [52–55]). Special attention is required with the scoring method and the cut-off values defining positivity when comparing results, since there is no established consensus. Contradictory results can be found in other studies using different cut-off levels to determine the scoring method and PD-L1 positivity [49,56,57]. However, similar studies concur in setting the low level of expression of PD-L1 (PD-L1–L) as $\geq$1% and the high level of expression of PD-L1 (PD-L1–H) as $\geq$5%, since very few cases occur with >50% overexpression [26,29,47]. These studies also reported a similar overall incidence of PD-L1 for patients in stage II/III CRC as in our study.

Although some studies suggest that PD-L1 expression is a negative prognostic factor, this is mainly due to the assessment of expression in tumor cells [33,58,59] and tumor staging. The contradictory results from the metastatic setting and from the early-stages [60] are probably due to temporal and spatial differences in the microenvironment and PD-L1 expression [61–64].

Patients with dMMR express significantly higher levels of PD-L1 in the early stages [26,47,48,65,66], which is consistent with the findings of our study (p = 0.043 for PD-L1–L and $p < 0.001$ for PD-L1–H). With respect to survival, patients in the CMS1 group, defined by dMMR, have the best prognosis in early-stage CRC [2,29,67,68] independent of the degree of PD-L1 expression. Further, the value of PD-L1 expression as an immunohistochemical biomarker of good prognosis when assessed in immune cells has been suggested by several studies and meta-analyses [33,47,63,69]. It is independent of MMR status [28,29,67,70,71]. Thus, patients with positive PD-L1 expression in the CMS2/CMS3 or CMS4 groups might also be expected to have a better prognosis.

This probably explains why our findings suggest that PD-L1 can separate those patients in the CMS2/CMS3 group with good and bad prognoses, since positive PD-L1 expression is significantly associated with better prognosis, as illustrated by the Kaplan–Meier curves for OS and DFS. This analysis was not carried out in the CMS4 group given the small number of statistical events upon which to draw relevant conclusions.

These results seem to be valid for other advanced GI tumors in general. Some recently published data suggest that PD-L1 expression has prognostic and predictive value and patients are being considered for anti-PD-1/PD-L1 therapy in CRC and other solid tumors [18,52,53,72–74].

As mentioned previously, CMS1 and CMS4 show very different intrinsic biological characteristics that translate into better and worse patient prognosis, respectively, in early-stage CRC [2,67,68]; this is not so clear for CMS2 and CMS3. As noted above, CMS2 displays epithelial differentiation and strong upregulation of WNT and MYC downstream targets, and CMS3 is characterized by multiple metabolism signatures. However, they sometimes share these characteristics with CMS4 or with CMS1 without distinction and this may be the reason for their unclear or intermediate prognosis [68]. For example, CMS2 shares with CMS4 a high frequency of somatic copy-number alterations and WNT/MYC pathways, and shares with CMS1 PD-1 activation and immune cell infiltration, whereas CMS3 shares with CMS4 higher KRAS mutation rates and sugar metabolic signatures, and shares with CMS1 a hypermutated profile and caspase pathways [3,19,75]. The results of our study could be a first step towards integrating the use of biomarkers like PD-L1 expression to differentiate the prognosis in CMS2 and CMS3. As such, it may significantly help with the clinical relevance of this classification.

Adjuvant chemotherapy in stage II and stage III CC patients remains controversial. For stage II despite several randomized trials [76,77], there is still a need for robust evidence concerning the addition of adjuvant chemotherapy for all patients [68]. For stage III, some studies have been able to establish a basis for treatment decisions [56,78,79]. Overall, some early-stage CRC patients benefit from adjuvant chemotherapy, although their long-term response rate is still suboptimal, particularly in the elderly population [68,80]. According to the ESMO and NCCN guidelines, the TNM is the main factor when deciding between observation or chemotherapy treatment. Nevertheless, other histopathological and clinical factors are sometimes taken into consideration, even though their prognostic value has not been fully validated [16,17].

Therefore, further characterization of patients with clinical implications is urgently needed in the context of early-stage settings. Our results, with PD-L1 expression used as a biomarker in combination with the CMS classification, could be a response to this need and possibly help with the decision to provide adjuvant therapy in the early setting.

Certain limitations of this study should be acknowledged when interpreting our results. Firstly, there were relatively few mortality events during the follow-up period, as expected during the design of the study. We used DFS because it is a good indicator in the Kaplan–Meier curves when mortality events are limited. Secondly, we assumed treatment to be homogeneous among all patients during the full course of their disease, since it was established and monitored by the same cross-functional committee of the same hospital. However, the lack of consistent data across patient records regarding the full details of the treatments received, treatment dosage, treatment duration, and/or any modifications, meant that the design of the study could not accommodate treatment stratification. Finally, the known limitations for a single center and retrospective study should also be acknowledged.

As mentioned previously, there is a clear need for better tools and characterization strategies for early-stage CRC patients. The early-stage setting has been less widely studied than the metastatic setting, probably due to its complexity and variability, even though the overall benefits to patients could be greater. With current emerging data and newly available targeted therapies, we call for a continuation of efforts towards devising validated prognostic biomarkers. Furthermore, a multi-center prospective study should follow our findings to confirm a hypothesized predictive value of PD-L1 expression.

5. Conclusions

In conclusion, our study demonstrates that PD-L1 expression is an independent prognostic factor in patients with stage II/III colon cancer in the CMS2/CMS3 group. The PD-L1 expression of stromal-related immune cells (tumor microenvironment) in colon cancer (CC) provides valuable information of prognostic value. The CMS classification itself is also of prognostic utility for early-stage CC patients. The assessment of CMS and PD-L1 expression through IHC, when performed in early-stage CC patients, may also have predictive value, with the potential to guide physicians concerning the addition of adjuvant treatment.

We expect this study to be a first step towards integrating the use of biomarkers like PD-L1 expression into a unified IHC panel, which may significantly help with the clinical relevance and implementation of the CMS classification.

Supplementary Materials: The following are available online at https://www.mdpi.com/article/10.3390/cancers13081943/s1, Figure S1: Scanned image of TMA (hematoxylin-eosin stain), Figure S2: IHC scan detail on high power field of PD-L1 staining in immune cells (core X20).

Author Contributions: Conceptualization, P.A. and M.L.G.-D.; methodology, P.A., M.L.G.-D. and M.M.; software, A.G.; validation, P.A., D.G.S., J.S.A. and M.L.G.-D.; formal analysis, P.A., M.M. and R.V.; investigation, P.A., M.L.G.-D. and I.E.; resources, I.E., D.G.S. and M.M.; data curation, A.G. and J.S.A.; writing—original draft preparation, P.A., M.L.G.-D. and I.E.; writing—review and editing, P.A., I.E., D.G.S. and R.V.; visualization, R.V. and J.S.A.; supervision, M.L.G.-D.; project administration, M.L.G.-D. and I.E.; funding acquisition, D.G.S. and M.L.G.-D. All authors have read and agreed to the published version of the manuscript.

Funding: This research was partly funded by European Union Regional Development Fund ((17-20, RefBioII, Trans-Pyrenean cooperation network program 2016–2018, INTERREG-POCTEFA) https://ec.europa.eu/regional_policy/en/funding/erdf/ (accessed on 19 May 2020).

Institutional Review Board Statement: The study was approved by the Regional Clinical Research Ethics Committee (CEIC) Pyto2017/51 Cod. MOL_CRC, 15/05/2018.

Informed Consent Statement: Patient consent was waived due to usage of stored tumor samples for research purposes in compliance with the current Spanish and European Union legislation (resolution 1387/2017 (08/11) and resolution 193/2018 (06/03) of the Navarra Health Service—Osasunbidea).

Data Availability Statement: Full data for this study are available from the Corresponding Authors upon request.

Conflicts of Interest: P.A. is an employee of Servier Medical Affairs, although there were no affiliations or financial involvement with this or any other organization or entity, with financial or scientific interest, with the subject matter or materials discussed in the manuscript. All other authors declare no conflicts of interest. The funders had no role in the design of the study, in the collection, analyses, or interpretation of data, in the writing of the manuscript, or in the decision to publish the results.

References

1. Bray, F.; Ferlay, J.; Soerjomataram, I.; Siegel, R.L.; Torre, L.A.; Jemal, A. Global cancer statistics 2018: GLOBOCAN estimates of incidence and mortality worldwide for 36 cancers in 185 countries. *CA Cancer J. Clin.* **2018**, *68*, 394–424. [CrossRef] [PubMed]
2. Puccini, A.; Berger, M.D.; Zhang, W.; Lenz, H.-J. What We Know About Stage II and III Colon Cancer: It's Still Not Enough. *Target. Oncol.* **2017**, *12*, 265–275. [CrossRef]
3. Guinney, J.; Dienstmann, R.; Wang, X.; de Reyniès, A.; Schlicker, A.; Soneson, C.; Marisa, L.; Roepman, P.; Nyamundanda, G.; Angelino, P.; et al. The consensus molecular subtypes of colorectal cancer. *Nat. Med.* **2015**, *21*, 1350–1356. [CrossRef] [PubMed]
4. Trinh, A.; Trumpi, K.; De Sousa, F.; Wang, X.; de Jong, J.H.; Fessler, E.; Kuppen, P.; Reimers, M.; Swets, M.; Koopman, M.; et al. Practical and Robust Identification of Molecular Subtypes in Colorectal Cancer by Immunohistochemistry. *Clin Cancer Res.* **2017**, *23*, 387–398. [CrossRef] [PubMed]
5. Olsen, J.; Eiholm, S.; Kirkeby, L.; Espersen, M.; Jess, P.; Gögenür, I.; Troelsen, J.; Espersen, M.L.M. CDX2 downregulation is associated with poor differentiation and MMR deficiency in colon cancer. *Exp. Mol. Pathol.* **2016**, *100*, 59–66. [CrossRef] [PubMed]

6. Pilati, C.; Taieb, J.; Balogoun, R.; Marisa, L.; De Reyniès, A.; Laurent-Puig, P. CDX2 prognostic value in stage II/III resected colon cancer is related to CMS classification. *Ann. Oncol.* **2017**, *28*, 1032–1035. [CrossRef] [PubMed]
7. Soll, C.; Riener, M.-O.; Oberkofler, C.E.; Hellerbrand, C.; Wild, P.J.; DeOliveira, M.L.; Clavien, P.A. Expression of Serotonin Receptors in Human Hepatocellular Cancer. *Clin. Cancer Res.* **2012**, *1*, 5902–5910. [CrossRef] [PubMed]
8. Henriksen, R.; Dizeyi, N.; Abrahamsson, P.-A. Expression of Serotonin Receptors 5-HT1A.; 5-HT1B.; 5-HT2B and 5-HT4 in Ovary and in Ovarian Tumours. *Anticancer Res.* **2012**, *32*, 1361–1366.
9. Xu, Y.; Wang, K.; Yu, Q. FRMD6 inhibits human glioblastoma growth and progression by negatively regulating activity of receptor tyrosine kinases. *Oncotarget* **2016**, *25*, 70080–70091. [CrossRef]
10. Angus, L.; Moleirinho, S.; Herron, L.; Sinha, A.; Zhang, X.; Niestrata, M.; Dholakia, K.; Prystowsky, M.B.; Harvey, K.F.; Reynolds, P.; et al. Willin/FRMD6 expression activates the Hippo signaling pathway kinases in mammals and antagonizes oncogenic YAP. *Oncogene* **2011**, *31*, 238–250. [CrossRef]
11. Zhang, P.; Wei, Y.; Wang, L.; Debeb, B.G.; Yuan, Y.; Zhang, J.; Yuan, J.; Wang, M.; Chen, D.; Sun, Y.; et al. ATM-mediated stabilization of ZEB1 promotes DNA damage response and radioresistance through CHK1. *Nat. Cell Biol.* **2014**, *16*, 864–875. [CrossRef]
12. Zhang, P.; Sun, Y.; Ma, L. ZEB1: At the crossroads of epithelial-mesenchymal transition, metastasis and therapy resistance. *Cell Cycle* **2015**, *14*, 481–487. [CrossRef]
13. Zhang, B.Y.; Jones, J.C.; Briggler, A.M.; Hubbard, J.M.; Kipp, B.R.; Sargent, D.J.; Dixon, J.; Grothey, A. Lack of Caudal-Type Homeobox Transcription Factor 2 Expression as a Prognostic Biomarker in Metastatic Colorectal Cancer. *Clin. Colorectal Cancer* **2017**, *16*, 124–128. [CrossRef]
14. Gatalica, Z.; Vranic, S.; Xiu, J.; Swensen, J.; Reddy, S. High microsatellite instability (MSI-H) colorectal carcinoma: A brief review of predictive biomarkers in the era of personalized medicine. *Fam. Cancer* **2016**, *15*, 405–412. [CrossRef] [PubMed]
15. Vilar, E.; Gruber, S.B. Microsatellite instability in colorectal cancer—The stable evidence. *Nat. Rev. Clin. Oncol.* **2010**, *7*, 153–162. [CrossRef]
16. National Comprehensive Cancer Network (NCCN). NCCN Clinical Practice Guidelines in Oncology. Colon Cancer Version 4.2019 [Internet]. National Comprehensive Cancer Network. 2019. Available online: https://www.nccn.org/professionals/physician_gls/pdf/colon.pdf (accessed on 14 July 2020).
17. Argilés, G.; Tabernero, J.; Labianca, R.; Hochhauser, D.; Salazar, R.; Iveson, T.; Laurent-Puig, P.; Quirke, P.; Yoshino, T.; Taieb, J.; et al. Localised colon cancer: ESMO Clinical Practice Guidelines for diagnosis, treatment and follow-up. *Ann. Oncol.* **2020**, *31*, 1291–1305. [CrossRef]
18. Andre, T.; Shiu, K.-K.; Kim, T.W.; Jensen, B.V.; Jensen, L.H.; Punt, C.J.A.; Smith, M.; Garcia-Carbonero, R.; Benavides, M.; Gibbs, P.; et al. Pembrolizumab versus chemotherapy for microsatellite instability-high/mismatch repair deficient metastatic colorectal cancer: The phase 3 KEYNOTE-177 Study. *J. Clin. Oncol.* **2020**, *20* (Suppl. 18), LBA4. [CrossRef]
19. Dienstmann, R.; Vermeulen, L.; Guinney, J.; Kopetz, S.; Tejpar, S.; Tabernero, J. Consensus molecular subtypes and the evolution of precision medicine in colorectal cancer. *Nat. Rev. Cancer* **2017**, *17*, 79–92. [CrossRef]
20. De Sousa, E.; Melo, F.; Wang, X.; Jansen, M.; Fessler, E.; Trinh, A.; de Rooij, L.P.M.H.; Jong, H.; de Boer, O.; van Leersum, R.; et al. Poor-prognosis colon cancer is defined by a molecularly distinct subtype and develops from serrated precursor lesions. *Nat. Med.* **2013**, *19*, 614–618. [CrossRef]
21. Roepman, P.; Schlicker, A.; Tabernero, J.; Majewski, I.; Tian, S.; Moreno, V.; Snel, M.H.; Chresta, C.M.; Rosenberg, R.; Nitsche, U.; et al. Colorectal cancer intrinsic subtypes predict chemotherapy benefit, deficient mismatch repair and epithelial-to-mesenchymal transition. *Int. J. Cancer* **2014**, *134*, 552–562. [CrossRef]
22. Song, N.; Pogue-Geile, K.L.; Gavin, P.G.; Yothers, G.; Kim, S.R.; Johnson, N.L.; Lipchik, C.; Allegra, C.; Petrelli, J.; O'Connell, M.; et al. Clinical Outcome From Oxaliplatin Treatment in Stage II/III Colon Cancer According to Intrinsic Subtypes: Secondary Analysis of NSABP C-07/NRG Oncology Randomized Clinical Trial. *JAMA Oncol* **2016**, *1*, 1162. [CrossRef]
23. Fontana, E.; Eason, K.; Cervantes, A.; Salazar, R.; Sadanandam, A. Context matters—consensus molecular subtypes of colorectal cancer as biomarkers for clinical trials. *Ann. Oncol.* **2019**, *30*, 520–527. [CrossRef]
24. Sharpe, A.H.; Pauken, K.E. The diverse functions of the PD1 inhibitory pathway. *Nat. Rev. Immunol.* **2018**, *18*, 153–167. [CrossRef]
25. Oliveira, A.F.; Bretes, L.; Furtado, I. Review of PD-1/PD-L1 Inhibitors in Metastatic dMMR/MSI-H Colorectal Cancer. *Front. Oncol.* **2019**, *14*, 396. [CrossRef]
26. Rosenbaum, M.W.; Bledsoe, J.R.; Morales-Oyarvide, V.; Huynh, T.G.; Mino-Kenudson, M. PD-L1 expression in colorectal cancer is associated with microsatellite instability, BRAF mutation, medullary morphology and cytotoxic tumor-infiltrating lymphocytes. *Mod. Pathol.* **2016**, *29*, 1104–1112. [CrossRef]
27. Ruan, Z.; Liang, M.; Lai, M.; Shang, L.; Deng, X.; Su, X. KYA1797K down-regulates PD-L1 in colon cancer stem cells to block immune evasion by suppressing the β-catenin/STT3 signaling pathway. *Int. Immunopharmacol.* **2020**, *78*, 106003. [CrossRef]
28. Shen, Z.; Gu, L.; Mao, D.; Chen, M.; Jin, R. Clinicopathological and prognostic significance of PD-L1 expression in colorectal cancer: A systematic review and meta-analysis. *World J. Surg. Oncol.* **2019**, *17*, 1–9. [CrossRef]
29. Wyss, J.; Dislich, B.; Koelzer, V.H.; Galván, J.A.; Dawson, H.; Hädrich, M.; Inderbitzin, D.; Lugli, A.; Zlobec, I.; Berger, M.D. Stromal PD-1/PD-L1 Expression Predicts Outcome in Colon Cancer Patients. *Clin. Color. Cancer* **2019**, *18*, e20–e38. [CrossRef]

30. Droeser, R.A.; Hirt, C.; Viehl, C.T.; Frey, D.M.; Nebiker, C.; Huber, X.; Zlobecc, I.; Eppenberger-Castorid, S.; Tzankovd, A.; Rosso, R.; et al. Clinical impact of programmed cell death ligand 1 expression in colorectal cancer. *Eur. J. Cancer* **2013**, *49*, 2233–2242. [CrossRef]
31. Song, M.; Chen, D.; Lu, B.; Wang, C.; Zhang, J.; Huang, L.; Wang, X.; Timmons, C.L.; Hu, J.; Liu, B.; et al. PTEN Loss Increases PD-L1 Protein Expression and Affects the Correlation between PD-L1 Expression and Clinical Parameters in Colorectal Cancer. *PLoS ONE* **2013**, *8*, e65821. [CrossRef]

32. Mettu, N.B.; Twohy, E.; Ou, F.-S.; Halfdanarson, T.R.; Lenz, H.J.; Breakstone, R.; Boland, P.M.; Crysler, O.; Wu, C.; Grothey, A.; et al. BACCI: A phase II randomized, double-blind, multicenter, placebo-controlled study of capecitabine (C) bevacizumab (B) plus atezolizumab (A) or placebo (P) in refractory metastatic colorectal cancer (mCRC): An ACCRU network study. *Ann. Oncol.* **2019**, *30*, 203. [CrossRef]
33. Li, Y.; He, M.; Zhou, Y.; Yang, C.; Wei, S.; Bian, X.; Christopher, O.; Xie, L. The Prognostic and Clinicopathological Roles of PD-L1 Expression in Colorectal Cancer: A Systematic Review and Meta-Analysis. *Front. Pharm.* **2019**, *10*, 139. [CrossRef] [PubMed]
34. Bae, S.U.; Jeong, W.K.; Baek, S.K.; Kim, N.K.; Hwang, I. Prognostic impact of programmed cell death ligand 1 expression on long-term oncologic outcomes in colorectal cancer. *Oncol. Lett.* **2018**, *16*, 5214–5222. [CrossRef]
35. Lee, M.S.; Menter, D.G.; Kopetz, S. Right Versus Left Colon Cancer Biology: Integrating the Consensus Molecular Subtypes. *J. Natl. Compr. Canc. Netw.* **2017**, *15*, 411–419. [CrossRef] [PubMed]
36. Sobin, L.H.; Gospodarowicz, M.K.; Wittekind, C. *TNM Classification of Malignant Tumours*, 7th ed.; Wiley-Blackwell: Hoboken, NJ, USA, 2011; p. 336.
37. Hoorn, S.; Trinh, A.; de Jong, J.; Koens, L.; Vermeulen, L. Classification of Colorectal Cancer in Molecular Subtypes by Immunohistochemistry. In *Colorectal Cancer*; Beaulieu, J.-F., Ed.; Humana Press: New York, NY, USA, 2018; pp. 179–191.
38. Liu, S.; Gönen, M.; Stadler, Z.K.; Weiser, M.R.; Hechtman, J.F.; Vakiani, E.; Wang, T.; Vyas, M.; Joneja, U.; Al-Bayati, M.; et al. Cellular localization of PD-L1 expression in mismatch-repair-deficient and proficient colorectal carcinomas. *Mod. Pathol.* **2019**, *32*, 110–121. [CrossRef]
39. Berger, M.D.; Yang, D.; Sunakawa, Y.; Zhang, W.; Ning, Y.; Matsusaka, S.; Okazaki, S.; Miyamoto, Y.; Suenaga, M.; Schirripa, M.; et al. Impact of sex, age, and ethnicity/race on the survival of patients with rectal cancer in the United States from 1988 to 2012. *Oncotarget* **2016**, *7*, 53668–53678. [CrossRef] [PubMed]
40. Salem, M.E.; Yin, J.; Weinberg, B.A.; Renfro, L.A.; Pederson, L.D.; Maughan, T.S.; Adams, R.A.; Van Custem, E.; Falcone, A.; Tebbutt, N.C.; et al. Clinicopathological differences and survival outcomes with first-line therapy in patients with left-sided colon cancer and rectal cancer: Pooled analysis of 2879 patients from AGITG (MAX), COIN.; FOCUS2, OPUS.; CRYSTAL and COIN-B trials in the ARCAD database. *Eur. J. Cancer* **2018**, *103*, 205–213.
41. Tamas, K.; Walenkamp, A.M.E.; de Vries, E.G.E.; van Vugt, M.A.T.M.; Beets-Tan, R.G.; van Etten, B.; de Groota, J.A.; Hospersa, G.A.P. Rectal and colon cancer: Not just a different anatomic site. *Cancer Treat. Rev.* **2015**, *41*, 671–679. [CrossRef]
42. Tominaga, T.; Akiyoshi, T.; Yamamoto, N.; Taguchi, S.; Mori, S.; Nagasaki, T.; Fukunaga, Y.; Ueno, M. Clinical significance of soluble programmed cell death-1 and soluble programmed cell death-ligand 1 in patients with locally advanced rectal cancer treated with neoadjuvant chemoradiotherapy. *PLoS ONE* **2019**, *26*, e0212978. [CrossRef]
43. Shao, L.; Peng, Q.; Du, K.; He, J.; Dong, Y.; Lin, X.; Li, J.; Wu, J. Tumor cell PD-L1 predicts poor local control for rectal cancer patients following neoadjuvant radiotherapy. *Cancer Manag. Res.* **2017**, *9*, 249–258. [CrossRef]
44. Llosa, N.J.; Cruise, M.; Tam, A.; Wicks, E.C.; Hechenbleikner, E.M.; Taube, J.M.; Blosser, R.L.; Fan, H.; Wang, H.; Luber, B.S.; et al. The Vigorous Immune Microenvironment of Microsatellite Instable Colon Cancer Is Balanced by Multiple Counter-Inhibitory Checkpoints. *Cancer Discov.* **2015**, *5*, 43–51. [CrossRef]
45. Marginean, E.C.; Melosky, B. Is There a Role for Programmed Death Ligand-1 Testing and Immunotherapy in Colorectal Cancer With Microsatellite Instability? Part II—The Challenge of Programmed Death Ligand-1 Testing and Its Role in Microsatellite Instability-High Colorectal Cancer. *Arch. Pathol. Lab. Med.* **2018**, *1*, 26–34. [CrossRef]
46. Elfishawy, M.; Abd-ELaziz, S.A.; Hegazy, A.; El-Yasergy, D.F. Immunohistochemical Expression of Programmed Death Ligand-1 (PDL-1) in Colorectal carcinoma and Its Correlation with Stromal Tumor Infiltrating Lymphocytes. *Asian Pac. J. Cancer Prev.* **2020**, *21*, 225–232. [CrossRef] [PubMed]
47. El Jabbour, T.; Ross, J.S.; Sheehan, C.; Affolter, K.; Geiersbach, K.B.; Boguniewicz, A.; Ainechi, S.; Bronner, M.P.; Jones, D.M.; Lee, H. PD-L1 protein expression in tumour cells and immune cells in mismatch repair protein-deficient and -proficient colorectal cancer: The foundation study using the SP142 antibody and whole section immunohistochemistry. *J. Clin. Pathol.* **2018**, *71*, 46–51. [CrossRef] [PubMed]
48. Kim, J.H.; Park, H.E.; Cho, N.-Y.; Lee, H.S.; Kang, G.H. Characterisation of PD-L1-positive subsets of microsatellite-unstable colorectal cancers. *Br. J. Cancer* **2016**, *115*, 490–496. [CrossRef]
49. Masugi, Y.; Nishihara, R.; Mingyang, S.; Mima, K.; Da Silva, A.; Shi, Y.; Inamura, K.; Cao, Y.; Song, M.; Nowak, J.A.; et al. Tumour CD274 (PD-L1) expression and T cells in colorectal cancer. *Gut* **2017**, *66*, 1463–1473. [CrossRef] [PubMed]
50. Le, D.T.; Uram, J.N.; Wang, H.; Bartlett, B.R.; Kemberling, H.; Eyring, A.D.; Skora, A.D.; Luber, B.S.; Azad, D.; Laheru, D.; et al. PD-1 Blockade in Tumors with Mismatch-Repair Deficiency. *N. Engl. J. Med.* **2015**, *372*, 2509–2520. [CrossRef]
51. Tang, Y.; Zhang, P.; Wang, Y.; Wang, J.; Su, M.; Wang, Y.; Zhou, L.; Xiong, W.; Zeng, Z.; Zhou, Y.; et al. The Biogenesis, Biology, and Clinical Significance of Exosomal PD-L1 in Cancer. *Front. Immunol.* **2020**, *11*, 604. [CrossRef]
52. Fehrenbacher, L.; Spira, A.; Ballinger, M.; Kowanetz, M.; Vansteenkiste, J.; Mazieres, J.; Park, K.; Smith, D.; Artal-Cortes, A.; Lewanski, C.; et al. Atezolizumab versus docetaxel for patients with previously treated non-small-cell lung cancer (POPLAR): A multicentre, open-label, phase 2 randomised controlled trial. *Lancet* **2016**, *387*, 1837–1846. [CrossRef]
53. Rittmeyer, A.; Barlesi, F.; Waterkamp, D.; Park, K.; Ciardiello, F.; von Pawel, J.; Gadgeel, S.M.; Hida, T.; Kowalski, D.M.; Dols, M.C.; et al. Atezolizumab versus docetaxel in patients with previously treated non-small-cell lung cancer (OAK): A phase 3, open-label, multicentre randomised controlled trial. *Lancet* **2017**, *389*, 255–265. [CrossRef]

54. Schmid, P.; Adams, S.; Rugo, H.S.; Schneeweiss, A.; Barrios, C.H.; Iwata, H.; Diéras, V.; Hegg, R.; Im, S.A.; Shaw Wright, G.; et al. Atezolizumab and Nab-Paclitaxel in Advanced Triple-Negative Breast Cancer. *N. Engl. J. Med.* **2018**, *379*, 2108–2121. [CrossRef] [PubMed]
55. Eckstein, M.; Erben, P.; Kriegmair, M.C.; Worst, T.S.; Weiß, C.-A.; Wirtz, R.M.; Wach, S.; Stoehr, R.; Sikic, D.; Geppert, C.I.; et al. Performance of the Food and Drug Administration/EMA-approved programmed cell death ligand-1 assays in urothelial carcinoma with emphasis on therapy stratification for first-line use of atezolizumab and pembrolizumab. *Eur. J. Cancer* **2019**, *106*, 234–243. [CrossRef]
56. Shi, Q.; Sobrero, A.F.; Shields, A.F.; Yoshino, T.; Paul, J.; Taieb, J.; Sougklakos, I.; Kerr, R.; Labianca, R.; Meyerhardt, J.A.; et al. Prospective pooled analysis of six phase III trials investigating duration of adjuvant (adjuv) oxaliplatin-based therapy (3 vs. 6 months) for patients (pts) with stage III colon cancer (CC): The IDEA (International Duration Evaluation of Adjuvant chemotherapy) collaboration. *J. Clin. Oncol.* **2017**, *35* (Suppl. 18), LBA1.
57. Zhu, J.; Chen, L.; Zou, L.; Yang, P.; Wu, R.; Mao, Y.; Zhou, H.; Li, R.; Wang, K.; Wang, W.; et al. MiR-20b, -21, and -130b inhibit PTEN expression resulting in B7-H1 over-expression in advanced colorectal cancer. *Hum. Immunol.* **2014**, *75*, 348–353. [CrossRef] [PubMed]
58. Lee, L.H.; Cavalcanti, M.S.; Segal, N.H.; Hechtman, J.F.; Weiser, M.R.; Smith, J.J.; Garcia-Aguilar, J.; Sadot, E.; Ntiamoah, P.; Markowitz, A.J.; et al. Patterns and prognostic relevance of PD-1 and PD-L1 expression in colorectal carcinoma. *Mod. Pathol.* **2016**, *29*, 1433–1442. [CrossRef]
59. Shan, T.; Chen, S.; Wu, T.; Yang, Y.; Li, S.; Chen, X. PD-L1 expression in colon cancer and its relationship with clinical prognosis. *Int. J. Clinc. Exp. Pathol.* **2019**, *12*, 1764–1769.
60. Ni, X.; Sun, X.; Wang, D.; Chen, Y.; Zhang, Y.; Li, W.; Wang, L.; Suo, J. The clinicopathological and prognostic value of programmed death-ligand 1 in colorectal cancer: A meta-analysis. *Clin. Transl. Oncol.* **2019**, *21*, 674–686. [CrossRef]
61. Zhou, K.I.; Peterson, B.; Serritella, A.; Thomas, J.; Reizine, N.; Moya, S.; Tan, C.; Wang, Y.; Catenacci, D.V.T. Spatial and Temporal Heterogeneity of PD-L1 Expression and Tumor Mutational Burden in Gastroesophageal Adenocarcinoma at Baseline Diagnosis and after Chemotherapy. *Clin. Cancer Res.* **2020**, *26*, 6453–6463. [CrossRef] [PubMed]
62. Yamashita, K.; Iwatsuki, M.; Ajani, J.A.; Baba, H. Programmed death ligand-1 expression in gastrointestinal cancer: Clinical significance and future challenges. *Ann. Gastroenterol. Surg.* **2020**, *4*, 369–378. [CrossRef]
63. Shklovskaya, E.; Rizos, H. Spatial and Temporal Changes in PD-L1 Expression in Cancer: The Role of Genetic Drivers, Tumor Microenvironment and Resistance to Therapy. *Int. J. Mol. Sci.* **2020**, *21*, 7139. [CrossRef]
64. Chen, B.J.; Dashnamoorthy, R.; Galera, P.; Makarenko, V.; Chang, H.; Ghosh, S.; Evens, A.M. The immune checkpoint molecules PD-1, PD-L1, TIM-3 and LAG-3 in diffuse large B-cell lymphoma. *Oncotarget* **2019**, *10*, 2030–2040. [CrossRef]
65. Le, D.T.; Durham, J.N.; Smith, K.N.; Wang, H.; Bartlett, B.R.; Aulakh, L.K.; Lu, S.; Kemberling, H.; Wilt, C.; Luber, B.S.; et al. Mismatch repair deficiency predicts response of solid tumors to PD-1 blockade. *Science* **2017**, *357*, 409–413. [CrossRef] [PubMed]
66. Jung, D.H.; Park, H.J.; Jang, H.H.; Kim, S.-H.; Jung, Y.; Lee, W.-S. Clinical Impact of PD-L1 Expression for Survival in Curatively Resected Colon Cancer. *Cancer Investig.* **2020**, *38*, 406–414. [CrossRef] [PubMed]
67. Huang, C.-Y.; Chiang, S.-F.; Ke, T.-W.; Chen, T.-W.; You, Y.-S.; Chen, W.T.-L.; Chao, K.S.C. Clinical significance of programmed death 1 ligand-1 (CD274/PD-L1) and intra-tumoral CD8+ T-cell infiltration in stage II–III colorectal cancer. *Sci. Rep.* **2018**, *8*, 15658. [CrossRef]
68. Chakrabarti, S.; Peterson, C.Y.; Sriram, D.; Mahipal, A. Early stage colon cancer: Current treatment standards, evolving paradigms, and future directions. *World J. Gastrointest. Oncol.* **2020**, *12*, 808–832. [CrossRef] [PubMed]
69. Zhao, T.; Li, C.; Wu, Y.; Li, B.; Zhang, B. Prognostic value of PD-L1 expression in tumor infiltrating immune cells in cancers: A meta-analysis. *PLoS ONE* **2017**, *12*, e0176822. [CrossRef]
70. Yang, L.; Xue, R.; Pan, C. Prognostic and clinicopathological value of PD-L1 in colorectal cancer: A systematic review and meta-analysis. *Oncotargets* **2019**, *12*, 3671–3682. [CrossRef] [PubMed]
71. Wang, L.; Liu, Z.; Fisher, K.W.; Ren, F.; Lv, J.; Davidson, D.D.; Baldridge, L.A.; Du, X.; Cheng, L. Prognostic value of programmed death ligand 1, p53, and Ki-67 in patients with advanced-stage colorectal cancer. *Hum. Pathol.* **2018**, *71*, 20–29. [CrossRef] [PubMed]
72. Yaghoubi, N.; Soltani, A.; Ghazvini, K.; Hassanian, S.M.; Hashemy, S.I. PD-1/ PD-L1 blockade as a novel treatment for colorectal cancer. *Biomed. Pharm.* **2019**, *110*, 312–318. [CrossRef] [PubMed]
73. Overman, M.J.; McDermott, R.; Leach, J.L.; Lonardi, S.; Lenz, H.-J.; Morse, M.; Desai, J.; Hill, A.; Axelson, M.; Moss, R.; et al. Nivolumab in patients with metastatic DNA mismatch repair-deficient or microsatellite instability-high colorectal cancer (CheckMate 142): An open-label, multicentre, phase 2 study. *Lancet Oncol.* **2017**, *18*, 1182–1191. [CrossRef]
74. Kluger, H.M.; Zito, C.R.; Turcu, G.; Baine, M.K.; Zhang, H.; Adeniran, A.; Sznol, M.; Rimm, D.L.; Kluger, Y.; Chen, L.; et al. PD-L1 Studies Across Tumor Types, Its Differential Expression and Predictive Value in Patients Treated with Immune Checkpoint Inhibitors. *Clin. Cancer Res.* **2017**, *23*, 4270–4279. [CrossRef]
75. Müller, M.F.; Ibrahim, A.E.K.; Arends, M.J. Molecular pathological classification of colorectal cancer. *Virchows Arch.* **2016**, *469*, 125–134. [CrossRef] [PubMed]
76. Iveson, T.; Sobrero, A.F.; Yoshino, T.; Sougklakos, I.; Ou, F.-S.; Meyers, J.P.; Shi, Q.; Saunders, M.P.; Labianca, R.; Yamanaka, T.; et al. Prospective pooled analysis of four randomized trials investigating duration of adjuvant oxaliplatin-based therapy (3 vs 6 months) for patients with high-risk stage II colorectal cancer. *J. Clin. Oncol.* **2019**, *37* (Suppl. 15), 3501. [CrossRef]

77. Tournigand, C.; André, T.; Bonnetain, F.; Chibaudel, B.; Lledo, G.; Hickish, T.; Tabernero, J.; Boni, C.; Bachet, J.-B.; Teixeira, L.; et al. Adjuvant Therapy With Fluorouracil and Oxaliplatin in Stage II and Elderly Patients (between ages 70 and 75 years) With Colon Cancer: Subgroup Analyses of the Multicenter International Study of Oxaliplatin, Fluorouracil, and Leucovorin in the Adjuvant Treatment of Colon Cancer Trial. *J. Clin. Oncol.* **2012**, *30*, 3353–3360. [PubMed]
78. Grothey, A.; Sobrero, A.F.; Shields, A.F.; Yoshino, T.; Paul, J.; Taieb, J.; Souglakos, J.; Shi, Q.; Kerr, R.; Labianca, R.; et al. Duration of Adjuvant Chemotherapy for Stage III Colon Cancer. *N. Engl. J. Med.* **2018**, *378*, 1177–1188. [CrossRef] [PubMed]
79. André, T.; De Gramont, A.A.; Vernerey, D.; Chibaudel, B.B.; Bonnetain, F.; Tijeras-Raballand, A.A.; Scriva, A.A.; Hickish, T.T.; Tabernero, J.; Van Laethem, J.L.; et al. Adjuvant Fluorouracil, Leucovorin, and Oxaliplatin in Stage II to III Colon Cancer: Updated 10-Year Survival and Outcomes According to BRAF Mutation and Mismatch Repair Status of the MOSAIC Study. *J. Clin. Oncol.* **2015**, *33*, 4176–4187. [CrossRef] [PubMed]
80. Sargent, D.J.; Marsoni, S.; Monges, G.; Thibodeau, S.N.; Labianca, R.; Hamilton, S.R.; French, A.J.; Kabat, B.; Foster, N.R.; Torri, V.; et al. Defective Mismatch Repair As a Predictive Marker for Lack of Efficacy of Fluorouracil-Based Adjuvant Therapy in Colon Cancer. *J. Clin. Oncol.* **2010**, *28*, 3219–3226. [CrossRef] [PubMed]

 *cancers*

Article

High Dual Expression of the Biomarkers CD44v6/$\alpha 2\beta 1$ and CD44v6/PD-L1 Indicate Early Recurrence after Colorectal Hepatic Metastasectomy

Friederike Wrana [1], Katharina Dötzer [1], Martin Prüfer [1], Jens Werner [1,2] and Barbara Mayer [1,2,*]

[1] Department of General, Visceral, and Transplant Surgery, Ludwig-Maximilians-University, Marchioninistraße 15, 81377 Munich, Germany; friederike.schlueter@med.uni.muenchen.de (F.W.); katharina.doetzer@muenchen-klinik.de (K.D.); martin.pruefer@dritter-orden.de (M.P.); jens.werner@med.uni-muenchen.de (J.W.)

[2] German Cancer Consortium (DKTK), Partner Site Munich, Pettenkoferstraße 8a, 80336 Munich, Germany

* Correspondence: barbara.mayer@med.uni-muenchen.de; Tel.: +49-89-4400-76438; Fax: +49-89-4400-76433

Simple Summary: Distant metastasis in colorectal cancer still correlates with poor prognosis, emphasizing the high need for new diagnostic and therapeutic strategies. In the present study, liver and lung metastases revealed profound differences in the expression pattern of metastasis-driving protein biomarkers. This suggests the adaption of the therapy to the biology of the metastatic organ site. High expression of the cell adhesion molecule CD44v6 and high dual expression of CD44v6, combined with the cell adhesion molecules integrin $\alpha 2\beta 1$, as well as the checkpoint inhibitor molecule PD-L1, correlated significantly with early recurrence after hepatectomy, in a substantial number of liver metastatic patients. These findings suggest the need for the implementation of biological risk factors into clinical risk scores, aiming to make the prognosis of the individual patient more precise. Further, dual expression of protein biomarkers that are druggable, such as CD44v6/$\alpha 2\beta 1$ and CD44v6/PD-L1, can identify high-risk patients for targeted therapy that might provide a survival benefit.

Abstract: Considering the biology of CRC, distant metastases might support the identification of high-risk patients for early recurrence and targeted therapy. Expression of a panel of druggable, metastasis-related biomarkers was immunohistochemically analyzed in 53 liver (LM) and 15 lung metastases (LuM) and correlated with survival. Differential expression between LM and LuM was observed for the growth factor receptors IGF1R (LuM 92.3% vs. LM 75.8%, $p = 0.013$), EGFR (LuM 68% vs. LM 41.5%, $p = 0.004$), the cell adhesion molecules CD44v6 (LuM 55.7% vs. LM 34.9%, $p = 0.019$) and $\alpha 2\beta 1$ (LuM 88.3% vs. LM 58.5%, $p = 0.001$) and the check point molecule PD-L1 (LuM 6.1% vs. LM 3.3%, $p = 0.005$). Contrary, expression of HGFR, Hsp90, Muc1, Her2/neu, ERα and PR was comparable in LuM and LM. In the LM cohort ($n = 52$), a high CD44v6 expression was identified as an independent factor of poor prognosis (PFS: HR 2.37, 95% CI 1.18–4.78, $p = 0.016$). High co-expression of CD44v6/$\alpha 2\beta 1$ (HR 4.14, 95% CI 1.65–10.38, $p = 0.002$) and CD44v6/PD-L1 (HR 2.88, 95% CI 1.21–6.85, $p = 0.017$) indicated early recurrence after hepatectomy, in a substantial number of patients (CD44v6/$\alpha 2\beta 1$: 11 (21.15%) patients; CD44v6/PD-L1: 12 (23.1%) patients). Dual expression of druggable protein biomarkers may refine prognostic prediction and stratify high-risk patients for new therapeutic concepts, depending on the metastatic location.

Keywords: colorectal cancer; liver metastases; lung metastases; protein biomarker; dual expression; early recurrence; poor prognosis

Citation: Wrana, F.; Dötzer, K.; Prüfer, M.; Werner, J.; Mayer, B. High Dual Expression of the Biomarkers CD44v6/$\alpha 2\beta 1$ and CD44v6/PD-L1 Indicate Early Recurrence after Colorectal Hepatic Metastasectomy. *Cancers* **2022**, *14*, 1939. https://doi.org/10.3390/cancers14081939

Academic Editor: Stephane Dedieu

Received: 9 March 2022
Accepted: 7 April 2022
Published: 12 April 2022

1. Introduction

According to international guidelines [1–3], metastasectomy currently offers the best chance for long-term survival for selected colorectal cancer patients. Additional standard chemotherapy for patients with resectable liver metastases resulted in the prolongation of disease-free survival (DFS) and progression-free survival (PFS) but revealed no significant improvement in overall survival (OS) [4,5]. In patients with resectable pulmonary metastases, the outcome of peri-operative chemotherapy is inconclusive [6,7]. However, despite curative-intent metastasectomy, more than half of the patients suffer recurrence [8,9]. This highlights the urgent need for the implementation of new strategies to identify high-risk patients suitable for personalized therapy, aiming to improve treatment outcome and survival [10].

Colorectal cancer preferentially metastasizes to the liver, followed by the lung and the peritoneum and, more rarely, in bone, ovary and the brain [11–13]. The metastatic pattern depends on the sidedness of the primary colorectal tumor. Elucidating the underlying mechanisms of the metastatic organotropism, profound molecular differences were observed between right-sided and left-sided CRC cancers. Similarly, the tumor microenvironment seems to have a deep impact on the metastatic site [14]. Indeed, for primary metastatic colorectal cancer, a growing body of molecular data is available, resulting in the continuous development of targeted therapies and improvement in survival [15,16].

Comparative analysis of primary CRC and corresponding metastatic sites revealed maintenance of the main driver mutations in both liver and lung metastases, some of which are approved for CRC therapy, such as RAS, BRAF and MSI [17–19]. In contrast, genomic [20–22], transcriptomic [23] and proteomic [24] profiling identified molecular differences between primary tumor, liver and lung metastases that might have potential therapeutic implications for specific metastatic sites. Moreover, distant metastases in different organs revealed discordant responses to standard chemotherapy [25], all together, supporting the concept of inter- and intratumor heterogeneity, which is one of the key factors in tumor progression, therapeutic resistance, and poor patient outcome.

In the present study, a panel of protein biomarkers was selected, which drive the complex metastatic process of primary colorectal cancer and lead to poor prognosis. In contrast, little information is available on the expression pattern of these prognostic factors in liver and lung metastases. The protein biomarker panel encompassed the growth factor receptors epidermal growth factor receptor (EGF-R) and hepatocyte growth factor receptor (HGF-R) [26], human epidermal growth factor receptor (Her2/neu) [27], insulin-like growth factor 1 receptor (IGF-1R) [28], estrogen receptor alpha (Erα) [29] and progesterone receptor (PR) [30], the cell adhesion molecules CD44v6 [31], Muc1 [32] and integrin $\alpha 2\beta 1$ [33], the chaperone heat shock protein 90 (Hsp90) [34], and the immune checkpoint molecule programmed death ligand 1 (PD-L1) [35]. Interestingly, the protein biomarkers selected are drug targets, for which drugs are already approved or for which clinical trials are ongoing, in primary colorectal cancer or other cancer types. This could open up new options for second and further line treatments in colorectal cancer.

The present study aimed (1) to identify the phenotypic heterogeneity in tumor biology between colorectal liver and lung metastases and (2) to stratify patients with a high risk for early recurrence after hepatic metastasectomy.

2. Materials and Methods

2.1. Patient Cohort

The patient cohort consists of 68 patients with metastatic colorectal cancer, receiving metastasectomy with curative intent at the Department of General, Visceral, and Transplant Surgery, Ludwig-Maximilians-University, Munich, Germany. A liver metastasis (LM, $n = 53$) or a lung metastasis (LuM, $n = 15$) was analyzed from each patient. Double-coded tissues and the corresponding data used in this study were provided by the Biobank of the Department of General, Visceral, and Transplant Surgery, Ludwig-Maximilians-University Munich, Munich, Germany. This Biobank operates under the administration of the Human Tissue and Cell Research (HTCR) Foundation. The framework of HTCR Foundation, which includes obtaining written informed consent from all donors, has been approved by the ethics commission of the Faculty of Medicine at the LMU (approval number 025-12) as well as the Bavarian State Medical Association (approval number 11142) in

Germany. All liver metastases were diagnosed as the first relapse of the individual patient. Lung metastases represented first (n = 3), second (n = 8) and later stage relapse (n = 4). Survival analysis was performed for 52 patients diagnosed with liver metastases. One patient was lost to follow up. Follow-up period of the patient cohort was from December 2010 until February 2018.

2.2. Immunohistochemistry and Evaluation of Biomarker Expression

Fresh tumor samples including adjacent benign reference tissue were collected according to international biobanking standards. After surgery the tumor samples were immediately snap frozen in liquid nitrogen. Serial cryosections (5 µm) were performed and air dried over night at room temperature. Sections were either fixed in acetone, or for the ERα und PR staining in formalin solution (10%). Immunohistochemistry was performed using the standard avidin-biotin-peroxidase complex method [36–38]. Briefly, unspecific Fc receptors were blocked with 10% AB-serum in D-PBS, pH 7.4 for 20 min. Endogenous biotin was blocked using the Avidin-/Biotin-blocking Kit for 15 min. The primary antibodies (Table 1) were incubated for one hour. Some antibodies were detected with the secondary biotinylated antibody (111-065-114; wc 7.0 µg/mL; JacksonImmunoResearch, West Grove, PA, USA for anti-rabbit and 315-065-048; wc 0.75 µg/mL; JacksonImmunoResearch for anti-mouse) for 30 min, followed by the peroxidase-conjugated streptavidin (016-030-084; wc 1.0 µg/mL; affymetrix eBiosciences, Santa Clara, CA, USA) for another 30 min. Other primary antibodies were detected with the amplification Kit ZytoChem Plus (HRP060; Zytomed Systems, Bargteheide, Germany) according to the instructions of the manufacturer (marked in Table 1 with Kit: +). For visualization of the antigen–antibody reaction all slides were developed in a 3-Amino-9-ethylcarbazole solution containing 35% hydrogen peroxide (AEC staining) for eight minutes in darkness. Counterstaining was performed with Mayer's hemalum solution. All incubation steps were performed in a humid chamber at room temperature. Specificity of the staining was controlled by the corresponding isotype controls (Table 1). Cancer cells were visualized by EpCAM and pan-cytokeratin expression.

For the evaluation of biomarker expression, the size of the measurement field was standardized using a normalized grid at 100× magnification (Olympus microscope BX50, Olympus, Hamburg, Germany). The biomarker-positive tumor area was determined in relation to the total tumor area. The percentage of biomarker-positive tumor cells was expressed by semiquantitative estimation in 10% increments. Staining results were evaluated by two independent observers (FW, BM). External monitoring was performed by local pathologists (Institute of Pathology, LMU Munich, Munich, Germany, T. Kirchner) and for Her2/neu expression by J. Rüschoff (Institute of Pathology Nordhessen, Kassel, Germany, Rüschoff) [39].

For some biomarkers standardized cut-off values are given, namely ERα and PR [40], Her2/neu [39,41], Muc1 [42,43], and PD-L1 [36,44]. In the absence of standardized cut-offs for other biomarkers, cut-offs were assessed using the biphasic distribution, which was statistically defined using the mean antigen expression in liver or lung metastases. Biomarker expression below the calculated cut-off was defined as low expression, and biomarker expression above the calculated cut-off was defined as high expression. The same cut-off values were used for single biomarker analysis and the evaluation of dual biomarker expression. In addition to the tumor tissue, antigen expression was evaluated on the adjacent benign liver and lung tissues.

Table 1. Antibody Panel for Immunophenotyping of Colorectal Liver and Lung Metastases.

Biomarker	Antibody/Clone	Species	Isotype	Working Concentration (µg/mL)	Kit	Source
HGF-R	Sp44	rabbit	IgG1	2.12	-	Spring Bioscience/Biomol, Pleasanton, CA, USA
IGF1-R	24–31	mouse	IgG1	4.0	+	Invitrogen, Carlsbad, CA, USA
EGR-R	H11	mouse	IgG1	2.94	-	Dako, Santa Clara, CA, USA
Her2/neu	4B5	rabbit	IgG1	1.5	-	Ventana, Roche, Basel, Switzerland
Erα	ID5	mouse	IgG1	2.5	+	Dako
PR	PgR 636	mouse	IgG1	2.5	+	Dako
Muc1	Ma55.2	mouse	IgG1	0.5	-	Monosan, Uden, The Netherlands
CD44v6	VFF-18	mouse	IgG1	1.0	-	eBioscience Affymetrix
α2β1	BHA2.1	mouse	IgG1	2.5	-	Millipore, Burlington, MA, USA
Hsp90	AC88	mouse	IgG1	10.0	+	Abcam, Cambridge, UK
PD-L1	MIH1	mouse	IgG1	10.0	+	Affymetrix
Positive controls						
Epcam	Ber-EP4	mouse	IgG1	5.0	-	Dako
Pan Cytokeratin	KL-1	mouse	IgG1	0.32	-	Zytomed Systems
isotype controls						
MOPC-21	MOPC-21	mouse	IgG1	5.0	-	Sigma-Aldrich, St. Louis, MO, USA
MOPC-21	MOPC-21	mouse	IgG1	4.0	+	Sigma-Aldrich
MOPC-21	MOPC-21	mouse	IgG1	10.0	+	Sigma-Aldrich
Rabbit mAb	DA1E	rabbit	IgG1	2.12	-	Cell Signaling, Danvers, MA, USA

2.3. Statistical Analysis

All statistical analyses were performed with IBM SPSS v. 23. Mean biomarker expression between liver and lung metastases was compared using the Mann–Whitney U-test. The prognostic impact of single and dual biomarker expression was evaluated using Kaplan–Meier analysis (log rank test, 'pairwise over strata') and multivariate Cox regression analysis (biomarker expression used as 'categorical covariate', 'First' as reference category). OS was defined as the time from metastasectomy until the last follow-up or death of the patient. PFS was defined as the time from metastasectomy until the next progression. A p-value of ≤ 0.05 was considered as significant.

3. Results

3.1. Patient Characteristics

In the present study, 53 liver metastases and 15 lung metastases surgically resected from colorectal cancer patients were analyzed. Men were more frequently affected than women (LM: ratio 1.79:1; LuM: ratio 4:1). Most (66.04%) liver metastases were detected at primary diagnosis (synchronous), whereas all lung metastases were documented at a later time (metachronous). Liver and lung metastases were diagnosed as single organ metastases. However, at the organ site, tumor disease was frequently extensive (number of nodules within the metastatic organ >1; LM: 64.15%, LuM: 53.33%; multilobular involvement; LM: 56.6%, LuM: 66.67%). Still, most patients were resected with curative intent (R0; LM: 73.58%, LuM: 80%). Further, 32 of 53 (60.38%) patients diagnosed with liver metastases received first-line chemotherapy (5-FU as single agent: 34.38%, oxaliplatin-based: 43.75%, irinotecan-based: 15.63%, others: 6.25%) and 23 of 53 (43.40%) received neoadjuvant chemotherapy before liver metastasectomy. Of these, 10 of 15 (66.67%) patients were treated with front line chemotherapy (5-FU as single agent: 10%, oxaliplatin-based: 80%, others: 10%) and 8 of 15 (53.33%) patients received neoadjuvant chemotherapy, right before surgery of the lung metastasis studied. Complete treatment records were not available for all patients with lung metastases.

Patient characteristics are summarized in detail in Table 2.

Table 2. Patient Characteristics.

Parameters	Liver Metastases		Lung Metastases	
	n	%	*n*	%
patient related				
sex				
male	34	64.15	12	80.00
female	19	35.85	3	20.00
age (years)				
median	64		62	
mean	64		59	
range	30–89		37–74	
metastasis related				
grading				
G1/G2	39	81.25	11	73.33
G3	9	18.75	4	26.67
missing	5		0	
number of metastases *				
1	19	35.85	7	46.67
>1	34	64.15	8	53.33
diameter of the largest metastases (cm)				
median	3.5		1.8	
mean	4.29		2.25	
range	1.3–21.7		0.9–3.3	
type of metastasis				
synchronous	35	66.04	0	0.00
metachronous	18	33.6	15	100.00
R-status				
R0	39	73.58	12	80.00
R1	14	26.42	3	20.00
distinction of metastasis				
unilobular	23	43.4	5	33.33
multilobular	30	56.6	10	66.67
anatomical site				
left sided	7	13.21	7	46.67
right sided	15	28.30	8	53.33
both sided	31	58.49		
neoadjuvant chemotherapy #				
yes	23	43.40	8	53.33
no	30	56.60	7	46.67
therapy options				
oxaliplatin-based	11	47.83	1	12.5
irinotecan-based	7	30.43	5	62.5
others	5	21.74	2	25.0

n, number of patients; **R-status**, residual status after surgery; *, nodules within the metastatic organ; #, administered directly before metastasectomy.

Survival analysis was performed in the patient cohort with liver metastases but was omitted in patients with lung metastases because of small sample size. Patients diagnosed with multiple (>1) LM had a significantly shorter PFS compared to patients diagnosed with a single liver metastasis (multiple metastases, PFS: 6.5 months; single metastasis, PFS: 10 months; log-rank, $p = 0.014$). Patients with synchronous LM relapsed much faster compared to patients with metachronous LM (synchronous, PFS: 7 months; metachronous, PFS: 16 months; log rank, $p = 0.001$). None of the patient characteristics revealed an impact on OS.

3.2. Differential Biomarker Expression in Colorectal Liver and Lung Metastases

Liver and lung metastases were comparatively analyzed with a panel of metastasis-related protein biomarkers. A differential expression pattern between liver and lung metastases was observed for the growth factor receptors IGF-1R (LuM 92.3% vs. LM 75.8%, $p = 0.013$) and EGF-R (LuM 68% vs. LM 41.5%, $p = 0.004$), showing a significantly higher fraction of positive cancer cells in the lung metastases, respectively. Similar results were obtained for the cell adhesion molecules CD44v6 (LuM 55.7% vs. LM 34.9%, $p = 0.019$) and integrin α2β1 (LuM 88.3% vs. LM 58.5%, $p = 0.001$), as well as for the check point molecule PD-L1 (LuM 6.1% vs. LM 3.3%, $p = 0.005$). In contrast, no significant difference was observed for the growth factor receptor HGF-R and the chaperon molecule Hsp90, both showing a high fraction of positive cancer cells in almost all distant metastases. Conversely, all but one metastatic lesion were found negative for the hormone receptors ERα and PR. One individual liver metastasis demonstrated 30% ERα positive cancer cells. Moreover, in colorectal liver and lung metastases, a minor fraction of the cancer cells were found positive for the cell adhesion molecule Muc1 and growth factor receptor Her2/neu. In fact, only one liver metastasis (60% Her2/neu positive cancer cells) qualified for anti-Her2/neu therapy. The number of biomarker-positive lesions and the means of biomarker expression are given in Table 3. The distribution of biomarker expression is shown for liver and lung metastases (Figure 1).

Table 3. Positivity and Distribution of Biomarkers in Liver and Lung Metastases.

Biomarker	Number of Positive Lesions Liver		Number of Positive Lesions Lung		Number of Positive Cancer Cells (%) Median			Number of Positive Cancer Cells (%) Mean			Number of Positive Lesions above Cut-Offs Liver		Number of Positive Lesions above Cut-Offs Lung	
	$n = 53$	%	$n = 15$	%	Liver	Lung	p-Value	Liver	Lung	Cut Off *	$n = 53$	%	$n = 15$	%
HGF-R	52	98.1	15	100	95	95	0.166	87.7	95.3			n.t.		
IGF-1R	50	94.3	15	100	90	100	**0.013**	75.8	92.3	>80	29	54.7	12	80
EGF-R	45	84.9	15	100	40	70	**0.004**	41.5	68.0	>50	25	47.2	12	80
Her2/neu	19	35.8	8	53.3	0	1	0.575	5.7	1.7	>50	1	1.9	0	0
ERα	1	1.9	0	0	0	0	n.t.	0.6	0	≥1		n.t.		
PR	0	0	0	0	0	0	n.t.	0	0	≥1		n.t.		
Muc1	26	49.1	9	60	0	1	0.614	6.8	5.9	+/−	26	49.1	9	60
CD44v6	45	84.9	15	100	30	60	**0.019**	34.9	55.7	>30	23	43.4	10	66.7
α2β1	46	86.8	15	100	70	90	**0.001**	58.5	88.3	>80	20	37.7	11	73.3
Hsp90	51	96.2	15	100	75	80	0.475	68.7	73.9	>70	26	49.1	9	60
PD-L1	24	45.3	13	86.7	0	1	**0.005**	6.1	3.25	>1	24	45.3	11	73.3

n, number of patients; **n.t.**, not tested; *, calculation of the cut-offs is given in the Materials and Methods Section.

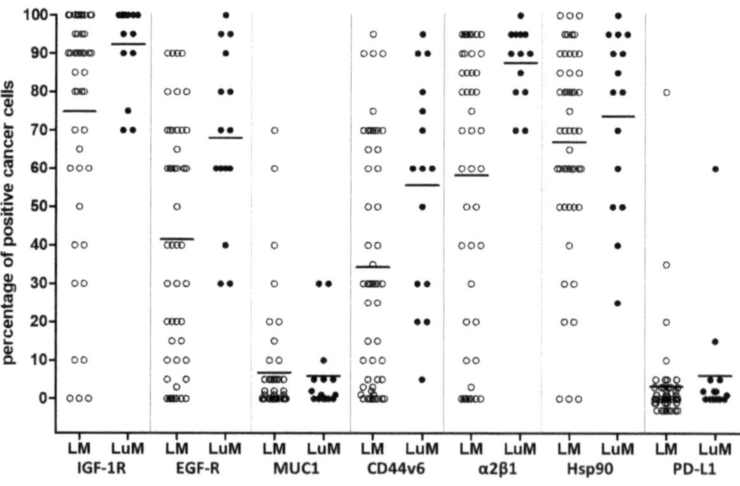

Figure 1. Biomarker Expression Pattern of Liver and Lung Metastases. horizontal bars, Means. Each dot represents a metastatic lesion; empty dots represent liver metastases (**LM**); filled dots represent lung metastases (**LuM**).

Biomarker analysis showed most of the benign liver tissues positive for HGF-R, EGF-R, and Hsp90. IGF-1R and PD-L1 were detected in a fraction of benign liver samples (IGF-1R: 11 out of 52, 21.2%; PD-L1: 10 out of 52, 19.2%). Interestingly, benign liver tissue was negative for Muc1, CD44v6 and the integrin α2β1. In contrast, all biomarkers tested were detected on benign lung tissue, although the integrin α2β1 (10 out of 15, 66.6%) and Muc1 (8 out of 15, 53%) were observed on a reduced number of adjacent lung tissues. Data obtained in benign tissue samples are summarized in Table S1. Figure 2 demonstrates the significantly different staining patterns by each biomarker of liver and lung metastases.

3.3. Prognostic Impact of Biomarker Expression in Colorectal Liver Metastases

The prognostic impact of the biomarkers was analyzed in patients with liver metastases. CD44v6, but none of the other biomarkers tested, was identified as an indicator for early recurrence. Liver metastases with a high fraction (>30%, n = 22) of CD44v6+ tumor cells significantly correlated with a shorter (median 7.0 months) PFS compared to LM with a low CD44v6 expression (≤30% CD44v6+ cells, n = 30; median 15.5 months; log rank p = 0.01). Recurrent liver metastases with a high proportion of CD44v6+ cancer cells showed more frequent multi-organ metastases (6 out of 19, 31.58%), compared to liver metastases with a low proportion of CD44v6+ cancer cells (3 out of 22, 13.65%). Almost all multi-organ metastases involved liver and lung, regardless of the extent of CD44v6 expression. Cox regression analysis confirmed the independent prognostic impact of CD44v6 on PFS (Table 4). No significant correlation was found between CD44v6 expression in LM and OS.

Table 4. Multivariate Survival Analysis of CD44v6 Expression in Colorectal liver Metastases.

Variable	Groups	HR	Cox Regression p-Value	95% CI
age (median in years)	>64/≤64	1.424	0.357	0.671–3.021
number of metastases *	>1/≤1	1.221	0.572	0.610–2.454
type of metastases	synchronous/metachronous	4.206	**0.004**	1.572–11.254
CD44v6 expression	>30%/≤30%	2.369	**0.016**	1.175–4.777

HR, Hazard ratio; p-value was calculated for progression free survival; **CI**, confidence interval; *, nodules within the metastatic organ.

3.4. CD44v6-Related Dual Biomarker Expression in Colorectal Liver Metastases

Co-expression analysis was performed on CD44v6 and the metastasis-related biomarkers. Univariate analysis identified three pairs of highly expressed biomarkers associated with short PFS. Patients with liver metastases with strong expression of CD44v6 and integrin α2β1 showed a shorter mean PFS (3 months) compared to the group with only high expression of CD44v6 (7 months) (Table 5, Figure 3). Multivariate Cox regression analysis identified the combination of a high CD44v6 and a high integrin α2β1 expression (HR: 4.135, 95% CI: 1.648–10.375, p = 0.002) and the combination of a high CD44v6 and a high PD-L1 expression (HR: 2.882, 95% CI: 1.213–6.848, p = 0.017), as independent prognostic factors for short progression-free survival (Table 6). High co-expression was detected in a substantial number of patients; i.e., CD44v6 high (>30% positive tumors cells) combined with integrin α2β1 high (>80% positive tumor cells) in 11 out of 52 (21.15%) patients, CD44v6 high combined with Hsp90 high (>70% positive tumor cells) in 14 out of 52 (26.92%) patients and CD44v6 high combined with PD-L1 high (>1% positive cells) in 12 out of 52 (23.1%) patients.

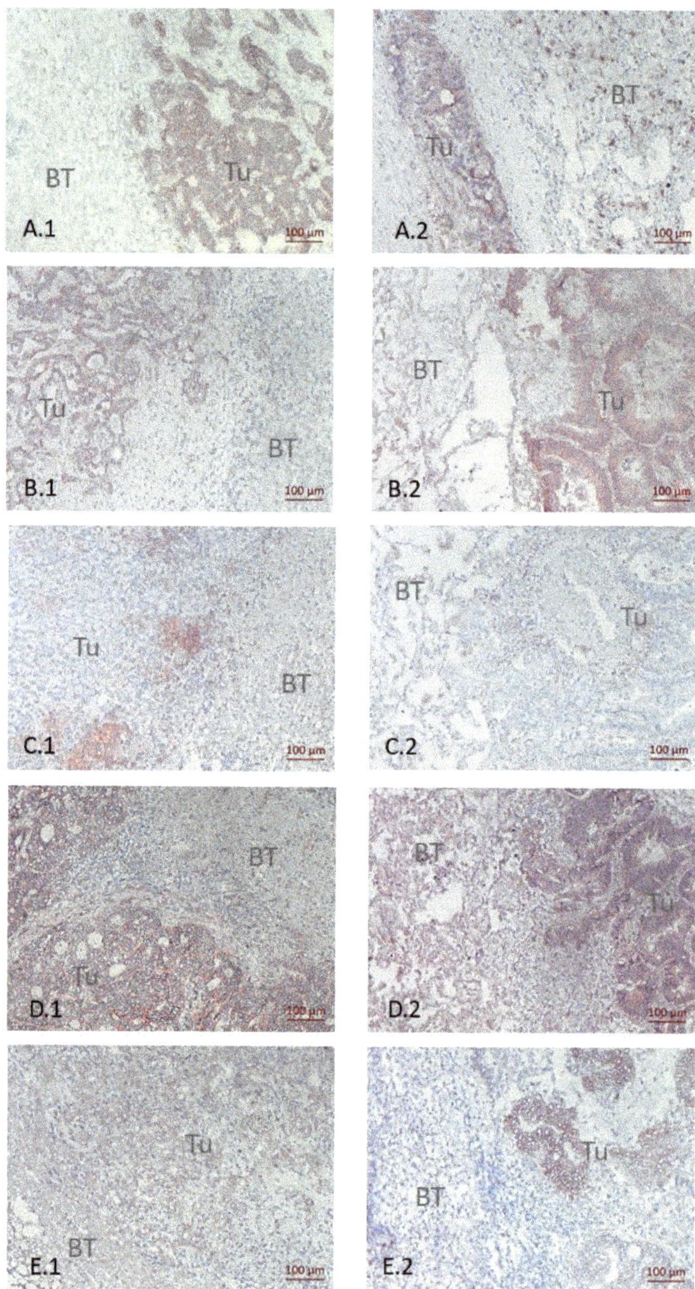

Figure 2. Immunohistochemical Staining of Different Biomarkers. Differential biomarker expression between liver (**1**) and lung (**2**) metastases demonstrated by immuno-histochemistry. (**A**), CD44v6; (**B**), $\alpha 2\beta 1$; (**C**), PD-L1; (**D**), IGF-1R; (**E**), EGFR; **Tu**, tumor tissue; **BT**, Benign tissue.

Table 5. Univariate Survival Analysis of CD44v6-Related Dual Biomarker Expression in Colorectal Liver Metastases.

Combination	Number of Patients (n)	Log Rank p-Value	Median PFS (month)
CD44v6 high *	22	**0.01**	7
CD44v6 low	30		15.5
CD44v6 high/IGF1-R high	15	0.142	7
CD44v6 high/IGF1-R low or CD44v6 low/IGF1-R high	20		9
CD44v6 low/IGF1-R low	17		17
CD44v6 high/EGF-R high	11	0.217	6
CD44v6 high/EGF-R low or CD44v6 low/EGF-R high	24		11.5
CD44v6 low/EGF-R low	17		9
CD44v6 high/Muc1 high	11	0.574	8
CD44v6 high/Muc1 low or CD44v6 low/Muc1 high	23		11
CD44v6 high/Muc1 low	18		7.5
CD44v6 high/ α2β1 high	11	**0.002**	3
CD44v6 high/ α2β1 low or CD44v6 low/ α2β1 high	18		9
CD44v6 low/ α2β1 low	23		24
CD44v6 high/Hsp90 high	14	**0.022**	7
CD44v6 high/Hsp90 low or CD44v6 low/Hsp90 high	21		9
CD44v6 low/Hsp90 low	17		17
CD44v6 high/PD-L1 high	12	**0.023**	7
CD44v6 high/PD-L1 low or CD44v6 low/PD-L1 high	21		14
CD44v6 low/PD-L1 low	19		11

PFS, progression-free survival; cut-off values defining high and low for the individual biomarker are given in Table 3; *, calculation of the cut-offs is given in the Materials and Methods Section.

Table 6. Multivariate Survival Analysis of CD44v6-Related Dual Biomarker Expression in Colorectal Liver Metastases.

Variable	Groups	HR	Cox Regression (PFS) p-Value	95% CI
age (median in years)	>64/≤64	1.561	0.256	0.724–3.366
number of metastases *	>1/≤1	1.398	0.358	0.684–2.855
type of metastases	synchronous/metachronous	3.813	**0.008**	1.407–10.332
CD44v6/α2β1 expression	high/high vs. low/low	4.135	**0.002**	1.648–10.375
	high/low and low/high vs. low/low	1.784	0.145	0.819–3.886
age (median in years)	>64/≤64	1.129	0.773	0.496–2.568
number of metastases	>1/≤1	1.321	0.460	0.632–2.762
type of metastases	synchronous/metachronous	3.345	**0.013**	1.289–8.680
CD44v6/Hsp90 expression	high/high vs. low/low	2.039	0.085	0.906–4.586
	high/low and low/high vs. low/low	1.412	0.443	0.585–3.404
age (median in years)	>64/≤64	1.290	0.493	0.623–2.675
number of metastases	>1/≤1	1.341	0.418	0.659–2.728
type of metastases	synchronous/metachronous	4.154	**0.004**	1.584–10.893
CD44v6/PD-L1 expression	high/high vs. low/low	2.882	**0.017**	1.213–6.848
	high/low and low/high vs. low/low	0.872	0.723	0.409–1.860

HR, Hazard ratio; **PFS**, progression free survival; **CI**, confidence interval; *, nodules within the metastatic organ; cut-off values defining high- and low-level expression for the individual biomarker are given in Table 3.

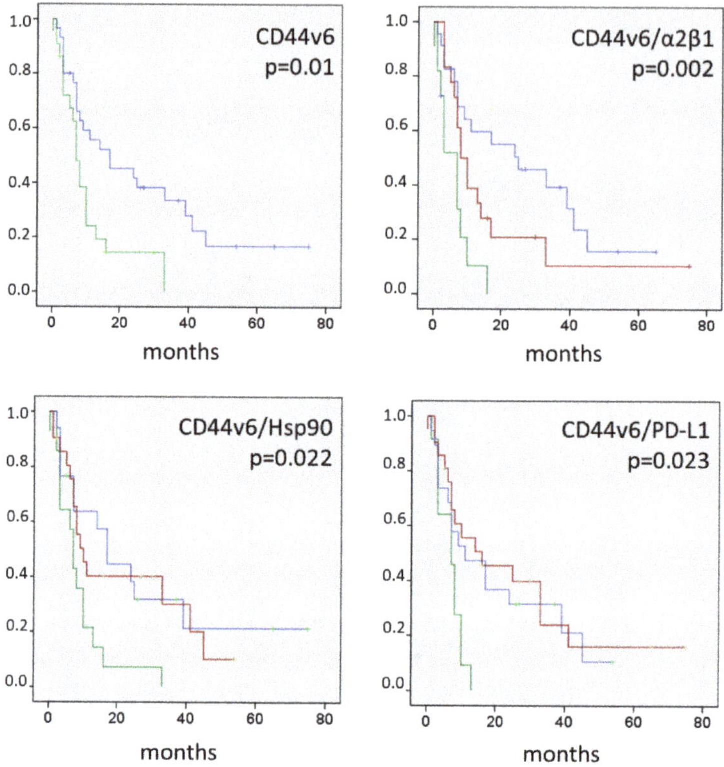

Figure 3. Kaplan–Meier Curves of CD44v6-Related Biomarker Expression in Colorectal Liver Metastases. **Blue lines**, low/low expression; **green lines**, high/high expression; **red lines**, high/low and low/high expression; log-rank *p*-values are given; cut-off values defining high- and low-level expression for the individual biomarker are given in Table 3.

4. Discussion

It is well published that primary colorectal cancer differs in its biology, depending on sidedness [45]. This also includes treatment-relevant characteristics, such as the RAS [46,47], MSI [48] and BRAF status [49]. In the present study, biomarker heterogeneity was identified between colorectal liver and lung metastases, namely for the cell adhesion molecules α2β1, CD44v6, the growth factor receptors IGF-1R, EGF-R and the immune checkpoint biomarker PD-L1. These site-specific differences in biomarker expression might reflect the complex multifactorial interactions between disseminated cancer cells and the target organ microenvironment [50]. Cancer cells with a unique tumor biology are homing to metastatic niches with a microenvironment promoting colonization, survival, and proliferation [51,52]. Liver and lung metastases reveal biological differences; for example, in the cellular composition of the microenvironment [36,52–54], the ECM signature [52,55,56] and the secretome profile [52,57]. Quantitative differences in protein biomarker expression were found between liver and lung metastases, showing a significantly higher proportion of IGF-1R-, EGR-R-, CD44v6-, α2β1-, and PD-L1-positive cancer cells in the lung. This observation confirms published data, showing a higher frequency of genetic drivers, such as KRAS alterations and MET amplification in lung metastases [20,58]. At the same time, lung metastases exhibit an increased immunosuppressive microenvironment and prometastatic inflammation [36,59]. These findings suggest distinct colonization mechanisms, involving both specific cancer cells with a higher propensity to metastasize to the lung and a lung-specific environment that facilitates metastasis of specific cancer cells.

Targeting metastasis-relevant biomarker expression will open up new therapeutic opportunities, adjusted to specific metastatic localizations. This is in deep contrast to the current guideline, which recommends the concept of treating distant metastasis with the same therapy, independent from the metastatic organ site.

The protein biomarker expression pattern in liver metastases was tested for prognostic relevance. A high (>30%) fraction of CD44v6+ liver metastatic cells was identified as an independent prognostic factor mediating short progression-free survival. This finding supports CD44v6 as a metastatic driver. Multiple underlying molecular mechanisms have been described for CD44v6-mediated progression in colorectal cancer. Examples are interactions with the extracellular matrix components osteopontin and hyaluronic acid and the binding of different cytokines, such as HGF, EGF and VEGF [31,60]. Co-expression analysis identified two new independent risk factors associated with poor prognosis of CRC patients with liver metastases. Most interesting, high dual expression of CD44v6 and integrin $\alpha 2\beta 1$ represents an indicator of early recurrence, defined as tumor relapse within six months after liver resection for colorectal metastases [61,62]. Direct and extracellular matrix-mediated molecular crosstalk between CD44v6 and various integrins, including $\alpha 2\beta 1$, was found to promote cancer cell proliferation and invasion, tumor angiogenesis and chemoresistance, all involved in a considerable shortening of progression-free survival compared to the single CD44v6 expression [63–65]. In addition, dual expression of CD44v6 and PD-L1, indicating the crosstalk between tumor cells and the tumor microenvironment, significantly correlated with short survival. The subset of CD44v6+ colorectal cancers simultaneously expressing PD-L1 might represent stem-like properties and contributes to immune evasion mediating poor prognosis [66,67]. Similarly, co-mutations in RAS, TP53 and SMAD4, as well as in APC and PIK3CA, resulted in a worse outcome after hepatectomy compared to single mutations [19]. Therefore, our findings support the strategy of combining prognostic protein biomarkers to render the prediction of outcome more precise [68,69]. Further, these new factors might be included in clinical risk scores, similar as reported for the KRAS status in the GAME score [70] and the KRAS/NRAS/BRAF status in the CERR score [71], which resulted in the refinement to predict recurrence after resection of CRC liver metastases. In contrast to some of the most investigated therapeutic biomarkers, namely BRAF, MSI-high, and Her2/neu, all detected in a very small patient cohort [19,20], dual expression of the druggable targets CD44v6/$\alpha 2\beta 1$ and CD44v6/PD-L1 was identified in about 20% of the liver metastatic patients.

In addition, these novel findings might have an impact on the development of new therapeutic strategies for liver metastatic CRC patients. Currently, new anti-CD44v6 treatment strategies, such as half antibodies conjugated nanoparticles [72], peptides (NCT03009214) and CD44v6-specific CAR gene-engineered T cells (NCT04427449, [73]) are under investigation and might also become a treatment option for CRC patients with CD44v6-positive liver metastases. Combination of two biomarkers might help to stratify patients more precisely for targeted therapy compared to single biomarker expression. For example, Shek et al., 2021, reported that only a subgroup of PD-L1-positive mCRCs responded to checkpoint inhibitor therapy [74]. In addition, dual expression of druggable biomarkers will further promote the promising concept of multiple target inhibition, aiming to improve treatment outcome and reduce the risk of drug resistance. Recently, the combination of the BRAF inhibitor Encorafenib with the EGF-R inhibitor Cetuximab has been reported as the new standard for the treatment of metastatic BRAF-mutated colorectal cancer [75]. Currently, a number of clinical trials are ongoing in advanced colorectal cancer, simultaneously inhibiting different targets. This includes combination therapy of the EGF-R inhibitor Panitumumab with the multi-kinase inhibitor Cabozantinib [76]. Further, anti-PD-L1 checkpoint inhibitors have been combined with targeted therapies, aiming to improve the response to immunotherapy [77]. In the present study, dual expression of PD-L1 and CD44v6 was found to correlate with poor prognosis and might represent a new therapeutic option for combination therapy. The second interesting pair of therapeutic targets identified in the present study was the co-expression of CD44v6 and the integrin $\alpha 2\beta 1$. Both cell adhesion molecules were found to mediate chemoresistance [65,78]. Simultaneous inhibition of both targets might result in the circumvention of chemoresistance and represent a new anti-metastatic strategy of targeted therapy. Consideration of metastasis-driving protein biomarkers that predict early recurrence after hepatectomy might play a critical role in the clinical management of patients diagnosed with liver metastases [79]. The findings in the present study need to be confirmed in a larger, prospective trial.

5. Conclusions

A differential expression pattern of the druggable protein biomarkers α2β1, CD44v6, IGF-1R, EGF-R and PD-L1 was identified between colorectal liver and lung metastases. High expression of CD44v6, CD44v6/α2β1, and CD44v6/PD-L1 correlated significantly with early recurrence after hepatic metastasectomy. Dual biomarker expression may render the prognostic prediction more precise and stratify high-risk patients for new therapeutic concepts, depending on the metastatic organ site.

Supplementary Materials: The following supporting information can be downloaded at: https://www.mdpi.com/article/10.3390/cancers14081939/s1, Table S1: Positivity and Distribution of Biomarkers in Benign Tissue of Liver and Lung Metastases.

Author Contributions: Conceptualization, B.M.; methodology, K.D. and F.W.; validation, B.M. and K.D.; investigation, F.W. and B.M.; resources, B.M.; data curation, F.W. and M.P.; writing—original draft preparation, F.W. and B.M.; writing—review and editing, K.D., J.W. and M.P.; visualization, F.W.; supervision, B.M.; project administration, B.M.; funding acquisition, B.M. All authors have read and agreed to the published version of the manuscript.

Funding: This research was funded by the German Federal Ministry of Education and Research, Leading Edge Cluster m4 (B.M.) under Grant FKZ 16EX1021N.

Institutional Review Board Statement: The study was conducted according to the guidelines of the Biobank of the Department of General, Visceral, and Transplant Surgery, Ludwig-Maximilians-University Munich, Munich, Germany. This Biobank operates under the administration of the Human Tissue and Cell Research (HTCR) Foundation. The framework of the HTCR Foundation has been approved by the ethics commission of the Faculty of Medicine at the LMU (approval number 025-12), as well as the Bavarian State Medical Association (approval number 11142) in Germany.

Informed Consent Statement: Informed consent from all patients was obtained by the HTCR Foundation.

Data Availability Statement: Data corresponding to the analyzed tissues were delivered in anonymized form by the HTCR Foundation.

Acknowledgments: We thank the staff members of the Biobank, especially Maresa Demmel, Nadine Gese, Ute Bossmanns and Beatrice Rauter for tissue sample/data organization, Michael Pohr for technical support and Robin v. Holzschuher for providing language help.

Conflicts of Interest: The authors declare no conflict of interest. The funders had no role in the design of the study; in the collection, analyses, or interpretation of data; in the writing of the manuscript, or in the decision to publish the results.

References

1. Benson, A.B.; Venook, A.P.; Al-Hawary, M.M.; Arain, M.A.; Chen, Y.J.; Ciombor, K.K.; Cohen, S.; Cooper, H.S.; Deming, D.; Farkas, L.; et al. Colon Cancer, Version 2.2021, NCCN Clinical Practice Guidelines in Oncology. *J. Natl. Compr. Canc. Netw.* **2021**, *19*, 329–359. [CrossRef] [PubMed]
2. *2020 Exceptional Surveillance of Colorectal Cancer (NICE Guideline NG151)*; Copyright © NICE 2020; National Institute for Health and Care Excellence: London, UK, 2020.
3. Van Cutsem, E.; Cervantes, A.; Adam, R.; Sobrero, A.; Van Krieken, J.H.; Aderka, D.; Aranda Aguilar, E.; Bardelli, A.; Benson, A.; Bodoky, G.; et al. ESMO consensus guidelines for the management of patients with metastatic colorectal cancer. *Ann. Oncol. Off. J. Eur. Soc. Med. Oncol.* **2016**, *27*, 1386–1422. [CrossRef] [PubMed]

4. Nordlinger, B.; Sorbye, H.; Glimelius, B.; Poston, G.J.; Schlag, P.M.; Rougier, P.; Bechstein, W.O.; Primrose, J.N.; Walpole, E.T.; Finch-Jones, M.; et al. Perioperative FOLFOX4 chemotherapy and surgery versus surgery alone for resectable liver metastases from colorectal cancer (EORTC 40983): Long-term results of a randomised, controlled, phase 3 trial. *Lancet Oncol.* **2013**, *14*, 1208–1215. [CrossRef]
5. Kanemitsu, Y.; Shimizu, Y.; Mizusawa, J.; Inaba, Y.; Hamaguchi, T.; Shida, D.; Ohue, M.; Komori, K.; Shiomi, A.; Shiozawa, M.; et al. Hepatectomy Followed by mFOLFOX6 Versus Hepatectomy Alone for Liver-Only Metastatic Colorectal Cancer (JCOG0603): A Phase II or III Randomized Controlled Trial. *J. Clin. Oncol. Off. J. Am. Soc. Clin. Oncol.* **2021**, *39*, 3789–3799. [CrossRef] [PubMed]
6. Guerrera, F.; Falcoz, P.E.; Renaud, S.; Massard, G. Does perioperative chemotherapy improve survival in patients with resectable lung metastases of colorectal cancer? *Interact. Cardiovasc. Thorac. Surg.* **2017**, *24*, 789–791. [CrossRef]
7. Li, Y.; Qin, Y. Peri-operative chemotherapy for resectable colorectal lung metastasis: A systematic review and meta-analysis. *J. Cancer. Res. Clin. Oncol.* **2020**, *146*, 545–553. [CrossRef]
8. Martin, J.; Petrillo, A.; Smyth, E.C.; Shaida, N.; Khwaja, S.; Cheow, H.K.; Duckworth, A.; Heister, P.; Praseedom, R.; Jah, A.; et al. Colorectal liver metastases: Current management and future perspectives. *World J. Clin. Oncol.* **2020**, *11*, 761–808. [CrossRef]
9. Lang, H.; Baumgart, J.; Roth, W.; Moehler, M.; Kloth, M. Cancer gene related characterization of patterns and point of recurrence after resection of colorectal liver metastases. *Ann. Transl. Med.* **2021**, *9*, 1372. [CrossRef]
10. Filip, S.; Vymetalkova, V.; Petera, J.; Vodickova, L.; Kubecek, O.; John, S.; Cecka, F.; Krupova, M.; Manethova, M.; Cervena, K.; et al. Distant Metastasis in Colorectal Cancer Patients-Do We Have New Predicting Clinicopathological and Molecular Biomarkers? A Comprehensive Review. *Int. J. Mol. Sci.* **2020**, *21*, 5255. [CrossRef]
11. Riihimaki, M.; Hemminki, A.; Sundquist, J.; Hemminki, K. Patterns of metastasis in colon and rectal cancer. *Sci. Rep.* **2016**, *6*, 29765. [CrossRef]
12. Gao, Y.; Bado, I.; Wang, H.; Zhang, W.; Rosen, J.M.; Zhang, X.H. Metastasis Organotropism: Redefining the Congenial Soil. *Dev. Cell* **2019**, *49*, 375–391. [CrossRef] [PubMed]
13. Pretzsch, E.; Bösch, F.; Neumann, J.; Ganschow, P.; Bazhin, A.; Guba, M.; Werner, J.; Angele, M. Mechanisms of Metastasis in Colorectal Cancer and Metastatic Organotropism: Hematogenous versus Peritoneal Spread. *J. Oncol.* **2019**, *2019*, 7407190. [CrossRef] [PubMed]
14. Chandra, R.; Karalis, J.D.; Liu, C.; Murimwa, G.Z.; Voth Park, J.; Heid, C.A.; Reznik, S.I.; Huang, E.; Minna, J.D.; Brekken, R.A. The Colorectal Cancer Tumor Microenvironment and Its Impact on Liver and Lung Metastasis. *Cancers* **2021**, *13*, 6206. [CrossRef] [PubMed]
15. Dekker, E.; Tanis, P.J.; Vleugels, J.L.A.; Kasi, P.M.; Wallace, M.B. Colorectal cancer. *Lancet* **2019**, *394*, 1467–1480. [CrossRef]
16. Imyanitov, E.; Kuligina, E. Molecular testing for colorectal cancer: Clinical applications. *World J. Gastrointest. Oncol.* **2021**, *13*, 1288–1301. [CrossRef]
17. Bhullar, D.S.; Barriuso, J.; Mullamitha, S.; Saunders, M.P.; O'Dwyer, S.T.; Aziz, O. Biomarker concordance between primary colorectal cancer and its metastases. *eBioMedicine* **2019**, *40*, 363–374. [CrossRef]
18. Testa, U.; Castelli, G.; Pelosi, E. Genetic Alterations of Metastatic Colorectal Cancer. *Biomedicines* **2020**, *8*, 414. [CrossRef]
19. Diener, M.K.; Fichtner-Feigl, S. Biomarkers in colorectal liver metastases: Rising complexity and unknown clinical significance? *Ann. Gastroenterol. Surg.* **2021**, *5*, 477–483. [CrossRef]
20. Wang, Z.; Zheng, X.; Wang, X.; Chen, Y.; Li, Z.; Yu, J.; Yang, W.; Mao, B.; Zhang, H.; Li, J.; et al. Genetic differences between lung metastases and liver metastases from left-sided microsatellite stable colorectal cancer: Next generation sequencing and clinical implications. *Ann. Transl. Med.* **2021**, *9*, 967. [CrossRef]
21. Jiang, B.; Mu, Q.; Qiu, F.; Li, X.; Xu, W.; Yu, J.; Fu, W.; Cao, Y.; Wang, J. Machine learning of genomic features in organotropic metastases stratifies progression risk of primary tumors. *Nat. Commun.* **2021**, *12*, 6692. [CrossRef]
22. Puccini, A.; Seeber, A.; Xiu, J.; Goldberg, R.M.; Soldato, D.; Grothey, A.; Shields, A.F.; Salem, M.E.; Battaglin, F.; Berger, M.D.; et al. Molecular differences between lymph nodes and distant metastases compared with primaries in colorectal cancer patients. *NPJ Precis. Oncol.* **2021**, *5*, 95. [CrossRef] [PubMed]
23. Eide, P.W.; Moosavi, S.H.; Eilertsen, I.A.; Brunsell, T.H.; Langerud, J.; Berg, K.C.G.; Røsok, B.I.; Bjørnbeth, B.A.; Nesbakken, A.; Lothe, R.A.; et al. Metastatic heterogeneity of the consensus molecular subtypes of colorectal cancer. *NPJ Genom. Med.* **2021**, *6*, 59. [CrossRef] [PubMed]
24. Fahrner, M.; Bronsert, P.; Fichtner-Feigl, S.; Jud, A.; Schilling, O. Proteome biology of primary colorectal carcinoma and corresponding liver metastases. *Neoplasia* **2021**, *23*, 1240–1251. [CrossRef]
25. Vigano, L.; Corleone, P.; Darwish, S.S.; Turri, N.; Famularo, S.; Viggiani, L.; Rimassa, L.; Del Fabbro, D.; Di Tommaso, L.; Torzilli, G. Hepatic and Extrahepatic Colorectal Metastases Have Discordant Responses to Systemic Therapy. Pathology Data from Patients Undergoing Simultaneous Resection of Multiple Tumor Sites. *Cancers* **2021**, *13*, 464. [CrossRef] [PubMed]
26. Xie, Y.H.; Chen, Y.X.; Fang, J.Y. Comprehensive review of targeted therapy for colorectal cancer. *Signal. Transduct. Target.* **2020**, *5*, 22. [CrossRef] [PubMed]
27. Siena, S.; Sartore-Bianchi, A.; Marsoni, S.; Hurwitz, H.I.; McCall, S.J.; Penault-Llorca, F.; Srock, S.; Bardelli, A.; Trusolino, L. Targeting the human epidermal growth factor receptor 2 (HER2) oncogene in colorectal cancer. *Ann. Oncol. Off. J. Eur. Soc. Med. Oncol.* **2018**, *29*, 1108–1119. [CrossRef] [PubMed]
28. Oliveres, H.; Pesántez, D.; Maurel, J. Lessons to Learn for Adequate Targeted Therapy Development in Metastatic Colorectal Cancer Patients. *Int. J. Mol. Sci.* **2021**, *22*, 5019. [CrossRef]

29. Liang, R.; Lin, Y.; Yuan, C.L.; Liu, Z.H.; Li, Y.Q.; Luo, X.L.; Ye, J.Z.; Ye, H.H. High expression of estrogen-related receptor α is significantly associated with poor prognosis in patients with colorectal cancer. *Oncol. Lett.* **2018**, *15*, 5933–5939. [CrossRef]
30. Ye, S.B.; Cheng, Y.K.; Zhang, L.; Wang, X.P.; Wang, L.; Lan, P. Prognostic value of estrogen receptor-α and progesterone receptor in curatively resected colorectal cancer: A retrospective analysis with independent validations. *BMC Cancer* **2019**, *19*, 933. [CrossRef]
31. Ma, L.; Dong, L.; Chang, P. CD44v6 engages in colorectal cancer progression. *Cell Death Dis.* **2019**, *10*, 30. [CrossRef]
32. Li, C.; Zuo, D.; Liu, T.; Yin, L.; Li, C.; Wang, L. Prognostic and Clinicopathological Significance of MUC Family Members in Colorectal Cancer: A Systematic Review and Meta-Analysis. *Gastroenterol Res. Pr.* **2019**, *2019*, 2391670. [CrossRef] [PubMed]
33. Hou, S.; Wang, J.; Li, W.; Hao, X.; Hang, Q. Roles of Integrins in Gastrointestinal Cancer Metastasis. *Front Mol. Biosci.* **2021**, *8*, 708779. [CrossRef] [PubMed]
34. Zhang, S.; Guo, S.; Li, Z.; Li, D.; Zhan, Q. High expression of HSP90 is associated with poor prognosis in patients with colorectal cancer. *Peer J.* **2019**, *7*, e7946. [CrossRef]
35. Oliveira, A.F.; Bretes, L.; Furtado, I. Review of PD-1/PD-L1 Inhibitors in Metastatic dMMR/MSI-H Colorectal Cancer. *Front Oncol.* **2019**, *9*, 396. [CrossRef] [PubMed]
36. Schlueter, F.; Doetzer, K.; Pruefer, M.; Bazhin, A.V.; Werner, J.; Mayer, B. Integrating Routine Clinical Factors to Stratify Colorectal Cancer Patients with Liver and Lung Metastases for Immune Therapy. *J. Cancer Sci. Clin. Ther.* **2021**, *5*, 49–62. [CrossRef]
37. Dotzer, K.; Schluter, F.; Schoenberg, M.B.; Bazhin, A.V.; von Koch, F.E.; Schnelzer, A.; Anthuber, S.; Grab, D.; Czogalla, B.; Burges, A.; et al. Immune Heterogeneity Between Primary Tumors and Corresponding Metastatic Lesions and Response to Platinum Therapy in Primary Ovarian Cancer. *Cancers* **2019**, *11*, 1250. [CrossRef] [PubMed]
38. Mayer, B.; Lorenz, C.; Babic, R.; Jauch, K.W.; Schildberg, F.W.; Funke, I.; Johnson, J.P. Expression of leukocyte cell adhesion molecules on gastric carcinomas: Possible involvement of LFA-3 expression in the development of distant metastases. *Int. J. Cancer* **1995**, *64*, 415–423. [CrossRef]
39. Rüschoff, J.; Dietel, M.; Baretton, G.; Arbogast, S.; Walch, A.; Monges, G.; Chenard, M.P.; Penault-Llorca, F.; Nagelmeier, I.; Schlake, W.; et al. HER2 diagnostics in gastric cancer-guideline validation and development of standardized immunohistochemical testing. *Virchows Arch. Int. J. Pathol.* **2010**, *457*, 299–307. [CrossRef]
40. Sieh, W.; Kobel, M.; Longacre, T.A.; Bowtell, D.D.; deFazio, A.; Goodman, M.T.; Hogdall, E.; Deen, S.; Wentzensen, N.; Moysich, K.B.; et al. Hormone-receptor expression and ovarian cancer survival: An Ovarian Tumor Tissue Analysis consortium study. *Lancet Oncol.* **2013**, *14*, 853–862. [CrossRef]
41. Wolff, A.C.; Hammond, M.E.H.; Allison, K.H.; Harvey, B.E.; Mangu, P.B.; Bartlett, J.M.; Bilous, M.; Ellis, I.O.; Fitzgibbons, P.; Hanna, W. Human epidermal growth factor receptor 2 testing in breast cancer: American Society of Clinical Oncology/College of American Pathologists clinical practice guideline focused update. *Arch. Pathol. Lab. Med.* **2018**, *142*, 1364–1382. [CrossRef]
42. Nagai, S.; Takenaka, K.; Sonobe, M.; Ogawa, E.; Wada, H.; Tanaka, F. A novel classification of MUC1 expression is correlated with tumor differentiation and postoperative prognosis in non-small cell lung cancer. *J. Thorac. Oncol.* **2006**, *1*, 46–51.
43. Zeng, Y.; Zhang, Q.; Zhang, Y.; Lu, M.; Liu, Y.; Zheng, T.; Feng, S.; Hao, M.; Shi, H. MUC1 Predicts Colorectal Cancer Metastasis: A Systematic Review and Meta-Analysis of Case Controlled Studies. *PLoS ONE* **2015**, *10*, e0138049. [CrossRef] [PubMed]
44. Korehisa, S.; Oki, E.; Iimori, M.; Nakaji, Y.; Shimokawa, M.; Saeki, H.; Okano, S.; Oda, Y.; Maehara, Y. Clinical significance of programmed cell death-ligand 1 expression and the immune microenvironment at the invasive front of colorectal cancers with high microsatellite instability. *Int. J. Cancer* **2018**, *142*, 822–832. [CrossRef] [PubMed]
45. Baran, B.; Mert Ozupek, N.; Yerli Tetik, N.; Acar, E.; Bekcioglu, O.; Baskin, Y. Difference Between Left-Sided and Right-Sided Colorectal Cancer: A Focused Review of Literature. *Gastroenterol. Res.* **2018**, *11*, 264–273. [CrossRef] [PubMed]
46. Zihui Yong, Z.; Ching, G.T.H.; Ching, M.T.C. Metastatic Profile of Colorectal Cancer: Interplay Between Primary Tumor Location and KRAS Status. *J. Surg. Res.* **2020**, *246*, 325–334. [CrossRef] [PubMed]
47. Xue, X.; Li, X.; Pan, Z.; Zhao, L.; Ding, Y. Comparison of clinicopathological features and KRAS gene mutation of left-sided and right-sided colon cancers. *Int. J. Clin. Exp. Pathol.* **2017**, *10*, 11353–11359. [PubMed]
48. Takahashi, Y.; Sugai, T.; Habano, W.; Ishida, K.; Eizuka, M.; Otsuka, K.; Sasaki, A.; Takayuki, M.; Morikawa, T.; Unno, M.; et al. Molecular differences in the microsatellite stable phenotype between left-sided and right-sided colorectal cancer. *Int. J. Cancer* **2016**, *139*, 2493–2501. [CrossRef] [PubMed]
49. Brulé, S.Y.; Jonker, D.J.; Karapetis, C.S.; O'Callaghan, C.J.; Moore, M.J.; Wong, R.; Tebbutt, N.C.; Underhill, C.; Yip, D.; Zalcberg, J.R.; et al. Location of colon cancer (right-sided versus left-sided) as a prognostic factor and a predictor of benefit from cetuximab in NCIC CO.17. *Eur. J. Cancer* **2015**, *51*, 1405–1414. [CrossRef] [PubMed]
50. Welch, D.R.; Hurst, D.R. Defining the Hallmarks of Metastasis. *Cancer Res.* **2019**, *79*, 3011–3027. [CrossRef]
51. Dmello, R.S.; To, S.Q.; Chand, A.L. Therapeutic Targeting of the Tumour Microenvironment in Metastatic Colorectal Cancer. *Int. J. Mol. Sci.* **2021**, *22*, 2067. [CrossRef]
52. Altorki, N.K.; Markowitz, G.J.; Gao, D.; Port, J.L.; Saxena, A.; Stiles, B.; McGraw, T.; Mittal, V. The lung microenvironment: An important regulator of tumour growth and metastasis. *Nat. Rev. Cancer* **2019**, *19*, 9–31. [CrossRef] [PubMed]
53. Kong, J.; Tian, H.; Zhang, F.; Zhang, Z.; Li, J.; Liu, X.; Li, X.; Liu, J.; Li, X.; Jin, D.; et al. Extracellular vesicles of carcinoma-associated fibroblasts creates a pre-metastatic niche in the lung through activating fibroblasts. *Mol. Cancer* **2019**, *18*, 175. [CrossRef] [PubMed]
54. Gonzalez-Zubeldia, I.; Dotor, J.; Redrado, M.; Bleau, A.M.; Manrique, I.; de Aberasturi, A.L.; Villalba, M.; Calvo, A. Co-migration of colon cancer cells and CAFs induced by TGFβ_1 enhances liver metastasis. *Cell Tissue Res.* **2015**, *359*, 829–839. [CrossRef] [PubMed]

55. Huang, J.; Pan, C.; Hu, H.; Zheng, S.; Ding, L. Osteopontin-enhanced hepatic metastasis of colorectal cancer cells. *PLoS ONE* **2012**, *7*, e47901. [CrossRef]
56. Naba, A.; Clauser, K.R.; Whittaker, C.A.; Carr, S.A.; Tanabe, K.K.; Hynes, R.O. Extracellular matrix signatures of human primary metastatic colon cancers and their metastases to liver. *BMC Cancer* **2014**, *14*, 518. [CrossRef]
57. Peinado, H.; Zhang, H.; Matei, I.R.; Costa-Silva, B.; Hoshino, A.; Rodrigues, G.; Psaila, B.; Kaplan, R.N.; Bromberg, J.F.; Kang, Y.; et al. Pre-metastatic niches: Organ-specific homes for metastases. *Nat. Rev. Cancer* **2017**, *17*, 302–317. [CrossRef]
58. Passot, G.; Kim, B.J.; Glehen, O.; Mehran, R.J.; Kopetz, S.E.; Goere, D.; Overman, M.J.; Pocard, M.; Marchal, F.; Conrad, C.; et al. Impact of RAS Mutations in Metastatic Colorectal Cancer After Potentially Curative Resection: Does Site of Metastases Matter? *Ann. Surg. Oncol.* **2018**, *25*, 179–187. [CrossRef]
59. García-Mulero, S.; Alonso, M.H.; Pardo, J.; Santos, C.; Sanjuan, X.; Salazar, R.; Moreno, V.; Piulats, J.M.; Sanz-Pamplona, R. Lung metastases share common immune features regardless of primary tumor origin. *J. Immunother. Cancer* **2020**, *8*, 491. [CrossRef]
60. Wang, J.L.; Su, W.Y.; Lin, Y.W.; Xiong, H.; Chen, Y.X.; Xu, J.; Fang, J.Y. CD44v6 overexpression related to metastasis and poor prognosis of colorectal cancer: A meta-analysis. *Oncotarget* **2017**, *8*, 12866–12876. [CrossRef]
61. Viganò, L.; Capussotti, L.; Lapointe, R.; Barroso, E.; Hubert, C.; Giuliante, F.; Ijzermans, J.N.; Mirza, D.F.; Elias, D.; Adam, R. Early recurrence after liver resection for colorectal metastases: Risk factors, prognosis, and treatment. A LiverMetSurvey-based study of 6025 patients. *Ann. Surg. Oncol.* **2014**, *21*, 1276–1286. [CrossRef]
62. Dai, S.; Ye, Y.; Kong, X.; Li, J.; Ding, K. A predictive model for early recurrence of colorectal-cancer liver metastases based on clinical parameters. *Gastroenterol. Rep.* **2021**, *9*, 241–251. [CrossRef] [PubMed]
63. Gao, C.; Guo, H.; Downey, L.; Marroquin, C.; Wei, J.; Kuo, P.C. Osteopontin-dependent CD44v6 expression and cell adhesion in HepG2 cells. *Carcinogenesis* **2003**, *24*, 1871–1878. [CrossRef] [PubMed]
64. Jung, T.; Gross, W.; Zöller, M. CD44v6 coordinates tumor matrix-triggered motility and apoptosis resistance. *J. Biol. Chem.* **2011**, *286*, 15862–15874. [CrossRef] [PubMed]
65. Naci, D.; Vuori, K.; Aoudjit, F. Alpha2beta1 integrin in cancer development and chemoresistance. *Semin Cancer Biol.* **2015**, *35*, 145–153. [CrossRef]
66. Lee, Y.; Shin, J.H.; Longmire, M.; Wang, H.; Kohrt, H.E.; Chang, H.Y.; Sunwoo, J.B. CD44+ Cells in Head and Neck Squamous Cell Carcinoma Suppress T-Cell-Mediated Immunity by Selective Constitutive and Inducible Expression of PD-L1. *Clin. Cancer Res. Off. J. Am. Assoc. Cancer Res.* **2016**, *22*, 3571–3581. [CrossRef]
67. Lin, C.C.; Liao, T.T.; Yang, M.H. Immune Adaptation of Colorectal Cancer Stem Cells and Their Interaction With the Tumor Microenvironment. *Front. Oncol.* **2020**, *10*, 588542. [CrossRef]
68. Chen, K.; Li, Z.; Jiang, P.; Zhang, X.; Zhang, Y.; Jiang, Y.; He, Y.; Li, X. Co-expression of CD133, CD44v6 and human tissue factor is associated with metastasis and poor prognosis in pancreatic carcinoma. *Oncol. Rep.* **2014**, *32*, 755–763. [CrossRef]
69. Ueda, A.; Takasawa, A.; Akimoto, T.; Takasawa, K.; Aoyama, T.; Ino, Y.; Nojima, M.; Ono, Y.; Murata, M.; Osanai, M.; et al. Prognostic significance of the co-expression of EGFR and HER2 in adenocarcinoma of the uterine cervix. *PLoS ONE* **2017**, *12*, e0184123. [CrossRef]
70. Margonis, G.A.; Sasaki, K.; Gholami, S.; Kim, Y.; Andreatos, N.; Rezaee, N.; Deshwar, A.; Buettner, S.; Allen, P.J.; Kingham, T.P.; et al. Genetic And Morphological Evaluation (GAME) score for patients with colorectal liver metastases. *Br. J. Surg.* **2018**, *105*, 1210–1220. [CrossRef]
71. Chen, Y.; Chang, W.; Ren, L.; Chen, J.; Tang, W.; Liu, T.; Jian, M.; Liu, Y.; Wei, Y.; Xu, J. Comprehensive Evaluation of Relapse Risk (CERR) Score for Colorectal Liver Metastases: Development and Validation. *Oncologist* **2020**, *25*, e1031–e1041. [CrossRef]
72. Lourenço, B.N.; Pereira, R.F.; Barrias, C.C.; Fischbach, C.; Oliveira, C.; Granja, P.L. Engineering Modular Half-Antibody Conjugated Nanoparticles for Targeting CD44v6-Expressing Cancer Cells. *Nanomaterials* **2021**, *11*, 295. [CrossRef] [PubMed]
73. Porcellini, S.; Asperti, C.; Corna, S.; Cicoria, E.; Valtolina, V.; Stornaiuolo, A.; Valentinis, B.; Bordignon, C.; Traversari, C. CAR T Cells Redirected to CD44v6 Control Tumor Growth in Lung and Ovary Adenocarcinoma Bearing Mice. *Front. Immunol.* **2020**, *11*, 99. [CrossRef] [PubMed]
74. Shek, D.; Akhuba, L.; Carlino, M.S.; Nagrial, A.; Moujaber, T.; Read, S.A.; Gao, B.; Ahlenstiel, G. Immune-Checkpoint Inhibitors for Metastatic Colorectal Cancer: A Systematic Review of Clinical Outcomes. *Cancers* **2021**, *13*, 4345. [CrossRef] [PubMed]
75. Tabernero, J.; Grothey, A.; Van Cutsem, E.; Yaeger, R.; Wasan, H.; Yoshino, T.; Desai, J.; Ciardiello, F.; Loupakis, F.; Hong, Y.S.; et al. Encorafenib Plus Cetuximab as a New Standard of Care for Previously Treated BRAF V600E-Mutant Metastatic Colorectal Cancer: Updated Survival Results and Subgroup Analyses from the BEACON Study. *J. Clin. Oncol. Off. J. Am. Soc. Clin. Oncol.* **2021**, *39*, 273–284. [CrossRef]
76. Strickler, J.H.; Rushing, C.N.; Uronis, H.E.; Morse, M.A.; Niedzwiecki, D.; Blobe, G.C.; Moyer, A.N.; Bolch, E.; Webb, R.; Haley, S.; et al. Cabozantinib and Panitumumab for RAS Wild-Type Metastatic Colorectal Cancer. *Oncologist* **2021**, *26*, 465-e917. [CrossRef]
77. Ooki, A.; Shinozaki, E.; Yamaguchi, K. Immunotherapy in Colorectal Cancer: Current and Future Strategies. *J. Anus Rectum Colon* **2021**, *5*, 11–24. [CrossRef]
78. Pereira, C.; Ferreira, D.; Mendes, N.; Granja, P.L.; Almeida, G.M.; Oliveira, C. Expression of CD44v6-Containing Isoforms Influences Cisplatin Response in Gastric Cancer Cells. *Cancers* **2020**, *12*, 858. [CrossRef]
79. Kamphues, C.; Beyer, K.; Margonis, G.A. Prognostic and Therapeutic Implications of Tumor Biology in Colorectal Liver Metastases. *Cancers* **2021**, *14*, 88. [CrossRef]

)22, 14, 928

Article

Discoidin Domain Receptor 1 Expression in Colon Cancer: Roles and Prognosis Impact

Kouther Ben Arfi [1], Christophe Schneider [2], Amar Bennasroune [2], Nicole Bouland [2,3], Aurore Wolak-Thierry [4], Guillaume Collin [5], Cuong Cao Le [2], Kevin Toussaint [2], Cathy Hachet [2], Véronique Lehrter [5], Stéphane Dedieu [2], Olivier Bouché [5,6], Hamid Morjani [5], Camille Boulagnon-Rombi [1,2,3,†] and Aline Appert-Collin [2,*,†]

1 Laboratoire de Biopathologie, Centre Hospitalier Universitaire de Reims, 51090 Reims, France; kaouther.ben.arfi@gmail.com (K.B.A.); camille.boulagnon@gmail.com (C.B.-R.)
2 UMR 7369, Matrice Extracellulaire et Dynamique Cellulaire (MEDyC), Université de Reims Champagne Ardenne (URCA), 51097 Reims, France; christophe.schneider@univ-reims.fr (C.S.); amar.bennasroune@univ-reims.fr (A.B.); nicole.bouland@univ-reims.fr (N.B.); lecuongbi@gmail.com (C.C.L.); kevin.toussaint@univ-reims.fr (K.T.); cathy.hachet@univ-reims.fr (C.H.); stephane.dedieu@univ-reims.fr (S.D.)
3 Laboratoire d'Anatomie Pathologique, Faculté de Médecine, 51100 Reims, France
4 Unité d'Aide Méthodologique, Centre Hospitalier Universitaire, 51100 Reims, France; awolak-thierry@chu-reims.fr
5 Unité BioSpecT, EA7506, Université de Reims Champagne Ardenne (URCA), 51096 Reims, France; guillaume.collin@univ-reims.fr (G.C.); veronique.lehrter@univ-reims.fr (V.L.); obouche@chu-reims.fr (O.B.); hamid.morjani@univ-reims.fr (H.M.)
6 Service d'Hépato-Gastroentérologie, Centre Hospitalier Universitaire, 51100 Reims, France
* Correspondence: aline.bennasroune@univ-reims.fr; Tel.: +33-(0)3-26-91-83-61
† These authors contributed equally to this work.

Simple Summary: Colorectal cancer (CRC) is the third leading cause of cancer death in both sexes. Identification of the influencing factors and molecular mechanisms in CRC progression could improve patient survival. This study aimed first to characterize the expression of Discoidin Domain Receptor 1 (DDR1), a receptor tyrosine kinase for collagens in a large cohort of CRC patients, and second to establish in vitro whether DDR1 expression level is linked to CRC aggressiveness potential. Our immunohistochemical study indicated that DDR1 is highly expressed in colon cancer compared to normal colonic mucosa and its expression is associated with shorter event-free survival. In vitro, the invasive properties of several CRC cell lines seem to be correlated with the expression level of DDR1. Taken altogether, our results show that DDR1 is highly expressed in most colon adenocarcinomas and appears as an indicator of worse event free survival.

Abstract: Extracellular matrix components such as collagens are deposited within the tumor microenvironment at primary and metastatic sites and are recognized to be critical during tumor progression and metastasis development. This study aimed to evaluate the clinical and prognostic impact of Discoidin Domain Receptor 1 (DDR1) expression in colon cancers and its association with a particular molecular and/or morphological profile and to evaluate its potential role as a prognosis biomarker. Immunohistochemical expression of DDR1 was evaluated on 292 colonic adenocarcinomas. DDR1 was highly expressed in 240 (82.2%) adenocarcinomas. High DDR1 immunostaining score was significantly associated, on univariate analysis, with male sex, left tumor location, *BRAF* wild type status, *KRAS* mutated status, and Annexin A10 negativity. High DDR1 immunohistochemical expression was associated with shorter event free survival only. Laser capture microdissection analyses revealed that DDR1 mRNA expression was mainly attributable to adenocarcinoma compared to stromal cells. The impact of DDR1 expression on cell invasion was then evaluated by modified Boyden chamber assay using cell types with distinct mutational profiles. The invasion capacity of colon adenocarcinoma is supported by DDR1 expression. Thus, our results showed that DDR1 was highly expressed in most colon adenocarcinomas and appears as an indicator of worse event free survival.

Citation: Ben Arfi, K.; Schneider, C.; Bennasroune, A.; Bouland, N.; Wolak-Thierry, A.; Collin, G.; Le, C.C.; Toussaint, K.; Hachet, C.; Lehrter, V.; et al. Discoidin Domain Receptor 1 Expression in Colon Cancer: Roles and Prognosis Impact. *Cancers* **2022**, *14*, 928. https://doi.org/10.3390/cancers14040928

Academic Editor: Isabelle Van Seuningen

Received: 29 December 2021
Accepted: 10 February 2022
Published: 13 February 2022

Keywords: colon cancer; discoidin domain receptor; prognosis; event free survival; survival

1. Introduction

Colorectal cancer (CRC) is ranked among the most common cancers in the world and is a significant public health issue in developed countries. Recent data indicated that CRC is the third most common cancer and the second leading cause of cancer death in both sexes [1]. The important mortality in CRC patients is highly correlated to its potential of metastasis reported in 50% of patients after surgery [2]. Indeed, about 39% of CRC patients are diagnosed at early stage with localized-stage disease. For these patients, the 5-year survival rate is 90%, while for the patients diagnosed with stage IV CRC, the survival declines to 12% [3].

However, at the same stage, all CRC do not have the same prognosis. Some parameters set by the tumor stage could refine the prognosis prediction and some histoprognosis factors have been identified: lymphovascular invasion, perineural invasion, tumor differentiation, or molecular profiles [2]. Treatment decisions could be influenced by these factors. In fact, many studies have been recently conducted to find new molecularly based prognostic markers, which are complementary to the data obtained by pathological diagnosis and therefore may increase the patient's survival. However, new biomarkers able to stratify the prognosis groups of patients and improve treatment strategies remain necessary. For this purpose, several studies investigate the signaling pathways that promote the metastatic process in CRC in order to identify new key players in this process that could constitute potential targets [4].

Receptor tyrosine kinases (RTKs) play an important role in several cellular processes in tumors including growth, migration, invasion, and the response to therapies [5]. For instance, the mitogen-activated protein kinase (MAPK) pathway and the phosphoinositide-3-kinase–protein kinase B/Akt (PI3K-PKB/Akt) pathway, two main intracellular pathways activated by the epidermal growth factor receptor (EGFR), were the most used therapeutic targets in metastatic colon cancer [6].

Discoidin domain receptors (DDRs) are collagen receptors with tyrosine kinase activity. The expression of the two members of this family, DDR1 and DDR2, is different: DDR1 is preferentially located in epithelial cells whereas DDR2 is expressed more importantly in connective tissues of the embryonic mesoderm [7]. Both DDR1 and DDR2 are activated by fibrillar collagens such as type I collagen [8]. Several studies have suggested a pivotal role of DDRs in tumor progression [9–11]. DDR1 expression appears to be increased in a variety of tumors and is correlated to poor prognosis [9–11]. Indeed, high level of DDR1 expression has been observed in several tumors such as prostate [12], lungs [13], breast [14], and ovary [15], suggesting a potential role of DDR1 in tumorigenesis and tumor progression. Moreover, experimental models have demonstrated that DDR1 plays an important role in cell proliferation and the metastasis process [16–19].

However, its role appeared to be tumor dependent. DDR1 overexpression was associated with advanced tumor stages in esophageal cancer [20], brain tumors [21] and with poor survival, in lung adenocarcinoma [22] and serous ovarian cancer [15].

In colon carcinoma, the role of DDR1 remains incompletely elucidated. The prognosis impact of DDR1 in CRC had not been studied much until now. High DDR1 expression seemed to be associated with poor overall survival [23–25]. Moreover, Sirvent and co-workers have shown that DDR1 plays a key role in the invasion potential of CRC [26]. The pharmacological inhibition of DDR1-BCR signaling axis using nilotinib has indeed been reported to decrease invasion and metastatic processes in CRC. These results suggest that DDR1 could represent a potential target in CRC treatment [26].

In the present study, we evaluated the expression of DDR1 in a cohort of CRC that, to our knowledge, is the largest set of CRC specimens studied for this receptor up to date. Specifically, we assessed the association between DDR1 expression and associated clinicopathological and molecular characteristics and its potential value as a prognosis marker. Finally, we examined in vitro the role of DDR1 in cell invasion in several CRC cell lines to establish whether DDR1 expression level is linked to CRC aggressiveness potential.

2. Materials and Methods

2.1. Culture Cells

HCT116, SW480, SW620 and HT-29 colorectal carcinoma cell lines were purchased from American Type Culture Collection (ATCC, Rockville, MD, USA). HT-29DDR1-GFP and HT-29GFP were obtained as previously described [27]. All cell lines were grown in Dulbecco's Modified Eagle Medium (DMEM) with high glucose (4.5 g/L) (Thermo Fisher Scientific, Villebon sur Yvette, France) containing 10% (v/v) fetal bovine serum (FBS) (Dutscher, Bernolsheim, France) and 1% penicillin-streptomycin (v/v, Invitrogen). Cells were regularly controlled for the absence of mycoplasma by PCR methods.

2.2. RNA Isolation from Cell Culture

Total RNA from cells was extracted as described previously [28] and single-stranded cDNA was synthesized from 250 ng total mRNAs using VERSO cDNA kit (Thermo Fisher Scientific, Waltham, MA, USA) according to the manufacturer's instructions. Determination of the mRNA of DDR1 was carried out by real-time PCR as described [27].

2.3. Total Protein Extraction and Immunoblotting

Seventy-two hours after seeding, cells were washed with ice-cold phosphate-buffered saline (PBS) and harvested in lysis buffer (10 mM Tris, 150 mM NaCl, 5 mM EDTA, 1% triton, protease inhibitors (Roche Diagnostics, Indianapolis, IN), and 5 mM Na orthovanadate). Cell lysates were then centrifugated at 14,000 g for 10 min at 4 °C. Protein concentration was quantified by BCA Protein Assay Kit (Thermo Fisher Scientific, Waltham, MA, USA). Proteins were separated on acrylamide gels and electroblotted onto nitrocellulose membranes (Amersham Biosciences, Little Chalfont, UK). The blots were incubated with primary antibodies (DDR1 (D1G6), GFP (D5.1) and GAPDH (14C10)) and corresponding peroxidase conjugated secondary antibody as previously indicated [27].

2.4. Invasion Assay

Cell invasion was evaluated using type-I collagen-coated 24-well cell culture inserts with an 8 µm pore size (Dustscher, Bernolsheim, France). The Boyden chambers were coated with 25 µg cm^{-2} type-I collagen and then washed twice with PBS. A total of 5 × 104 cells were seeded into the upper chambers in a 200 µL DMEM culture medium, supplemented with 2% FBS, 1% penicillin–streptomycin. DMEM culture medium with 10% FBS and 1% penicillin–streptomycin was added in the lower chamber. After 24 h, the chambers were washed with PBS, fixed with methanol and stained with Di Aminido Phenyl lndol (DAPI, Santa Cruz Biotechnology). Cells remaining on the upper face of the membranes were suppressed by scraping, and those on the lower side were counted after being imaged on the EVOS® FL Auto Imaging System using a 40× objective (Thermofisher scientific, Waltham, MA, USA). Experiments were reproduced three times in triplicates.

Concerning experiments using Nilotinib and DDR1-IN-1 inhibitors, 7.5×10^4 cells were seeded into the upper chambers in a 200 µL DMEM culture medium, supplemented with 2% FBS, 1% penicillin–streptomycin in the presence or not of Nilotinib (100 nM, No.S1033) or DDR1-IN-1 (10 µM, No.S7498, Selleckchem). DMEM culture medium with 10% FBS and 1% penicillin–streptomycin was added in the lower chamber. After 24 h, the chambers were washed with PBS, fixed with methanol, and stained with crystal violet. Cells remaining on the upper face of the membranes were suppressed by scraping. Upon solubilization in acetic acid (10%), the amount of dye on the filter was quantified by spectrophotometry at 560 nm.

2.5. Patients

Patients and selection were clarified in paper from Boulagnon-Rombi et al. [29].

The study was conducted on adult patients who underwent surgery for sporadic colon cancer in the Digestive Surgery Department of the University Hospital of Reims between September 2006 and December 2012. Patients with rectal cancer were excluded.

Clinical data including age at the time of surgery, sex, performance status, surgical circumstances (tumor perforation, occlusion), tumor location, synchronous or metachronous metastases, tumor recurrence, treatment, death and pathological and molecular data including adenocarcinoma type, grade, and pTNM stage were collected. Patients were classified as having a right colonic cancer if the primary tumor was located in the caecum, ascending colon, hepatic flexure or transverse colon, and left colonic cancer if the tumor site was within the splenic flexure, descending colon, sigmoid colon, or rectosigmoid junction.

2.6. Pathology

All colon adenocarcinomas were classified and subtyped according to The World Health Organization criteria [30] and staged according to the International Union Against Cancer 2009 guidelines [31]. Tumor budding was assessed on Hematoxylin- Eosin-Saffron slides and classified as low budding rate if less than 5 buds were present in the 0.785 mm^2 hot spot [32].

2.7. Immunohistochemistry

Tissue samples were analyzed via tissue microarrays (TMA). For each tumor, 3 cores were punched in the central part and 3 cores at the invasive front of the tumor from the same original formalin-fixed paraffin-embedded tumor block. The cores were 2 mm in diameter and were precisely arrayed into a recipient paraffin block using the MiniCore Tissue Arrayer (Excilone, Elancourt, France). Sections of 4-µm thickness were cut and mounted on SuperFrost Plus Gold adhesive slides (Thermofisher Scientific, Waltham, MA, USA). Immunohistochemistry (IHC) was performed using DDR1 (D1G6) XP® Rabbit mAb, rabbit Monoclonal antibody (1/100, Cell Signaling ref: #5583) after heat-induced epitope retrieval in citrate pH 6 buffer (95 °C, 40 min) and overnight antibody incubation at 4 °C and then visualized using 3-Amino-9-Ethylcarbazole (AEC).

2.8. Scoring

Immunostaining intensity (SI) was graded independently by two pathologists (CBR, KBBA).

Immunopositivity was defined as a brown cytoplasmic color in the tumor cells. Staining intensity was scored as follows: 0, negative staining signal in >50% of tumor cells; 1+, weak staining signal detected in >50% of tumor cells; 2+, moderate staining signal in >50% of tumor cells; 3+, strong staining signal in >50% of tumor cells (Figure 1). The staining intensity was then divided into score 0/1+ for low DDR1 expression or score 2+/3+ for high DDR1 expression as previously described [23].

2.9. Molecular Analyses

Tumor DNA was extracted and the mutation profile (*BRAF*, *KRAS*, and MSI status) of the samples was determined as described earlier [33].

2.10. Laser Capture Microdissection

Laser capture microdissection was performed on fresh frozen colon cancer specimens cut into 12 µm serial sections and mounted on PALM membrane slides (Zeiss, Oberkochen, Germany) as previously noticed [29].

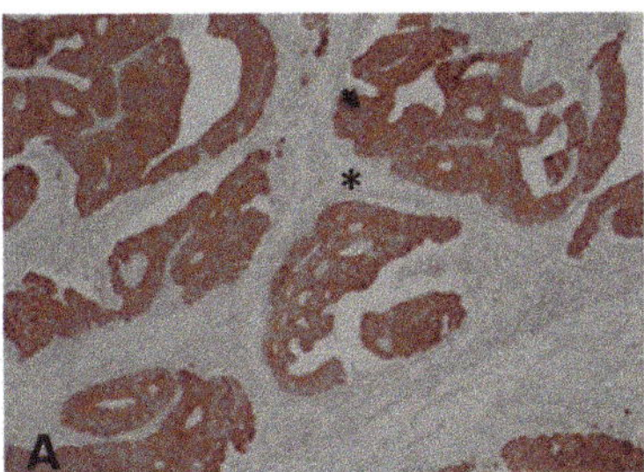

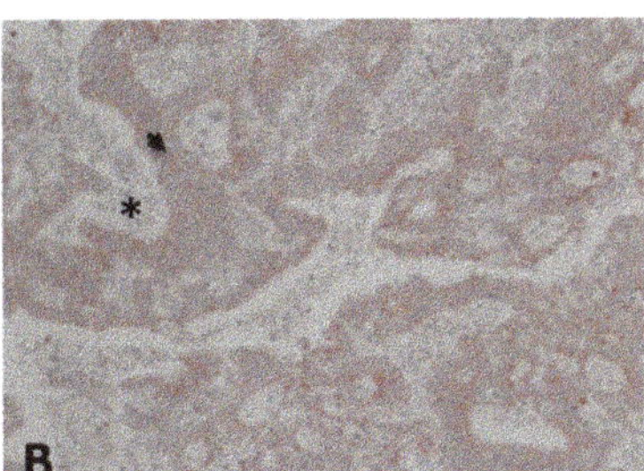

Figure 1. Representative images of DDR1 immunolabeling in colon adenocarcinoma. A. Strong and diffuse staining (red) in adenocarcinoma cells (arrow), (magnification ×10), scored 3+/high; B. Moderate and diffuse staining (red) in adenocarcinoma cells (arrow), (magnification ×20), scored 2+/high; C. Faint staining (red) in adenocarcinoma cells (arrow), (magnification ×10), scored 1+/low. Stromal cell highlighted by an asterisk (*) was weak (**A**) or negatively (**B**,**C**) stained in all cases.

RNA from tumor and stromal microdissected tissues were isolated and purified as indicated [29].

2.11. DDR1 mRNA Expression

Analysis of mRNA expression was performed as previously described [29]. Only RNAs with RQI values ≥5 were used for further analyses. Determination of the mRNA of DDR1 was carried out by real-time PCR as described [27].

2.12. Data Mining and Bioinformatic Analyses

Survival analyses were performed using publicly available data from TCGA, Martineau and SieberSmith gene expression dataset in the R2 microarray analysis and visualization platform (http://r2.amc.nl; last access date: 5 November 2021). The scan online algorithm was used to determine the cut-off values for separating high and low DDR1 expression groups.

2.13. Statistical and Survival Analyses

Statistical analyses and factors associated with immunohistochemical expression of DDR1 were clarified in paper from Boulagnon-Rombi et al. [29].

3. Results

3.1. Association of DDR1 Immunohistochemical Expression with Clinico-Pathological Features

The relationship between DDR1 expression and disease aggressiveness was investigated in a cohort of 292 colon cancer patients. The clinicopathological features are summarized in Table 1. The population consisted of 166 (57%) men and 126 (43%) women, whose mean age was 70.8 ± 10.8 years. Tumors were right-sided in 123 cases (42%), left-sided in 164 cases (56%), and multifocal in 5 cases (2%). The mean follow-up time was 43 months (±32 months).

Figure 1 illustrates representative IHC patterns of DDR1 expression. The immunostaining showed the localization of DDR1 mostly in the cytoplasm. The immunostaining intensity was strong in 144 (49.3%) samples, moderate in 96 (33%), and weak in 52 (17.8%), and no samples were found negative for DDR1 staining (score 0). DDR1 immunostaining was diffuse (>50% of positive tumor cells) in all cases. DDR1 immunolabeling in tumor stroma was weak or negative in all cases. For the statistical analysis, patients were divided into two groups: low expression of DDR1 for patients with immunostaining intensity scored 1 and high DDR1 expression for patients with immunostaining intensity scored 2 or 3. Thus, DDR1 expression by IHC was rated high in 240 (82.2%) cases. In case of samples presenting heterogeneity in immunostaining, the highest intensity was considered for scoring.

Table 1. Clinicopathological features of the cohort.

Clinicopathological Features	Total (%) n = 292
Gender	
Male	166(57)
Female	126 (43)
Age (Mean ± standard deviation) years	70.8 ± 10.8
UICC stage	
Stage I	34 (11.8)
Stage II	109 (37.8)
Stage III	72 (24.9)
Stage IV	74 (25.6)
Tumor location	
Left colon	164 (56)
Right colon	123 (42)
Multifocal	5 (2)
Occlusion	
Yes	34 (12)
No	258 (88)
Tumor perforation	
Yes	17 (6)
No	275 (94)
Differentiation grade	
Grade 1–2	245 (84)
Grade 3	47 (16)
Annexin A10	
Positive	36 (12)
Negative	255 (88)
***KRAS* status**	
Wild type	95 (67)
Mutant	46 (33)
***BRAF* status**	
Wild type	246 (86)
Mutant	40 (14)
Microsatellite status	
MSS	250 (87)
MSI	37 (13)
CIMP status	
No CIMP	20 (35.7)
CIMP-Low	30 (53.5)
CIMP-High	6 (10.7)

The relationship between DDR1 immunohistochemical expression and different clinicopathological and molecular characteristics was analyzed. Data are detailed in Table 2. In univariate analysis, a high DDR1 immunostaining score was significantly associated with male sex ($p = 0.0195$), left tumor location ($p = 0.0114$), BRAF wild-type status ($p < 0.0001$), KRAS mutated status ($p = 0.0041$), and absence of expression of the serrated markers Annexin A10 ($p = 0.0097$). In multivariate analysis, high DDR1 immunostaining score was independently associated with BRAF wild-type status only ($p < 0.0001$).

Table 2. Relationship between DDR1 expression and clinical and molecular characteristics.

Patients and Tumors Characteristics	*n*	DDR1		Univariate Analysis	Multivariate Analysis	
		High	Low	*p*-Value	OR [IC 95%]	*p*-Value
		n (%)	*n* (%)			
Age (Years)		70.23 ± 10.6	73.59 ± 11.2	0.052 *		
Gender				**0.0195 ‡**		n.s
Female	126	96 (40)	30 (57.7)			
Male	166	144 (60)	22 (42.3)			
Tumor location				**0.0114 ‡**		n.s
Left colon	164	143 (59.6)	21 (40.4)			
Right colon	128	97 (40.4)	31 (59.6)			
UICC stage				0.3240 ‡		
I	34	30 (12.5)	4 (8)			
II	109	90 (37.7)	19 (38)			
III	72	55 (23)	17 (34)			
IV	74	64 (26.8)	10 (20)			
Differentiation grade				0.0540 ‡		
1–2	245	206 (85.8)	39 (75)			
3	47	34 (14.2)	13 (25)			
Vascular invasion				0.2694 ‡		
Yes	115	90 (38.3)	15 (30)			
No	180	145 (61.7)	35 (70)			
Perineural invasion				0.6 ‡		
Yes	71	60 (25.5)	11 (22)			
No	214	175 (74.5)	39 (78)			
Budding score				1 †		
High	14	12 (5.4)	2 (4.2)			
Low	254	208 (94.5)	46 (95.8)			
CDX2				0.0565 †		
Positive	268	223 (94.9)	45 (86.5)			
Negative	19	12 (5.1)	7 (13.5)			
***KRAS* status**				**0.0041 ‡**		n.s
Wild type	95	69 (61.6)	26 (89.7)			
Mutant	46	43 (38.4)	3 (10.3)			
***BRAF* status**				**<0.0001 ‡**	**7.5 [4.11–13.67]**	**<0.0001**
Wild type	246	212 (90.2)	34 (66.7)			
Mutant	40	23 (9.8)	17 (33.3)			

Table 2. *Cont.*

Patients and Tumors Characteristics	*n*	DDR1		Univariate Analysis	Multivariate Analysis	
		High	Low	*p*-Value	OR [IC 95%]	*p*-Value
		n (%)	*n* (%)			
Microsatellite status				0.0909 †		
MSS	250	210 (89)	40 (78.4)			
MSI	36	25 (10.6)	11 (21.6)			
CIMP status				0.5488 †		
No CIMP	20	18 (39.1)	2 (20)			
CIMP-L	30	23 (50)	7 (70)			
CIMP-H	6	5 (10.88)	1 (10)			
Annexine A10				**0.0097 ‡**		n.s
Negative	255	215 (90)	40 (76.9)			
Positive	36	24 (10)	12 (23.1)			

n.s: not significant; ‡: khi-2; †: Fisher test; *: Satterthwaite.

3.2. Survival Analysis

We next investigated the relation between DDR1 expression and prognosis. Univariate analysis demonstrated that age, tumor stage, vascular invasion, and metastasis were predictors of overall survival (OS) in our cohort (Table 3).

High DDR1 immunostaining was not correlated with overall survival in all stages ($p = 0.5832$, Figure 2A) nor in metastatic (stage IV) patients ($p = 0.8376$, data not shown). Regarding event-free survival (EFS), univariate analysis revealed that occlusion, stage, vascular invasion, lymphatic invasion, differentiation grade, RAS status, CIMP status, and the level of DDR1 immunostaining scores were associated with shorter EFS (Table 3). High DDR1 expression was a predictor of shorter EFS in the entire cohort ($p = 0.0391$, Figure 2B). Stage specific analyses showed that DDR1 was not a predictor of EFS in stage II ($p = 0.1181$, Figure 3A), stage III ($p = 0.3389$, Figure 3B) and in metastatic patients ($p = 0.9102$, Figure 3C).

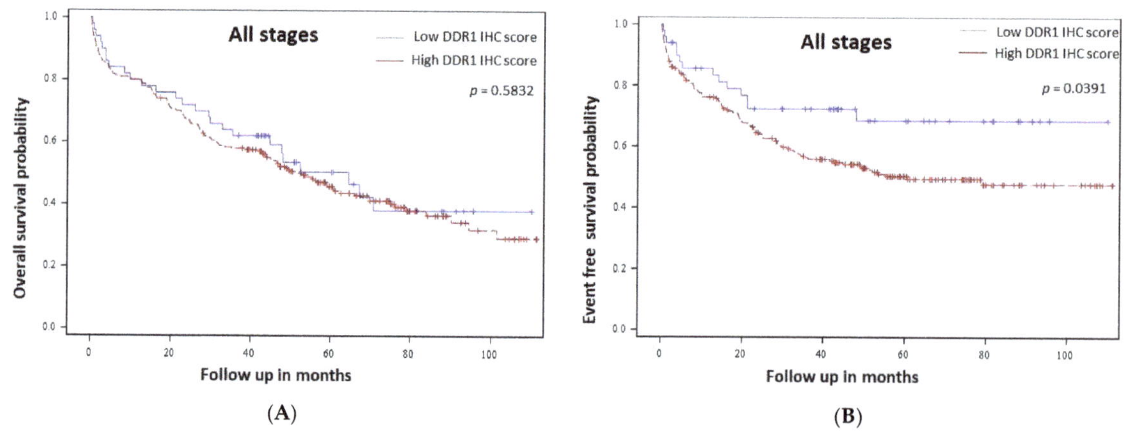

Figure 2. DDR1 value as a prognosis indicator in colon cancer patients from our cohort. Kaplan-Meier curves of overall survival (**A**) and event free-survival (**B**) probability for low (blue line) and high (red line) DDR1 immunohistochemical expression in adenocarcinoma cells from all tumor stages.

Table 3. Analysis of factors associated with overall and event-free survival.

Variables	*n*	Overall Survival	Event Free Survival
		p-value	*p*-value
Age	281	**0.0046**	0.3824
Perforation (yes vs. no)	281	0.0003	**<0.0001**
Occlusion (yes vs. no)	281	<0.0001	**<0.0001**
T4 (T4 vs. T1, T2, T3)	281	**<0.0001**	**<0.0001**
N (0, 1a vs. 1b and N2)	281	**<0.0001**	**<0.0001**
Vascular invasion (yes vs. no)	274	**0.0002**	**<0.0001**
Lymphatic invasion (yes vs. no)	273	0.0622	**0.0264**
Stage UICC	278	<0.0001	**<0.0001**
Differentiation grade (yes vs. no)	283	0.0032	**0.00018**
CDX2 IHC expression (yes vs. no)	276	0.0245	0.8486
Metastasis (M0 vs. M+)	276	**<0.0001**	**<0.0001**
KRAS mutation (yes vs. no)	135	0.0689	**0.0010**
BRAF mutation (yes vs.no)	276	0.7616	0.2882
CIMP status (low vs. High)	53	0.0644	**0.0003**
Microsatellite status (MSS vs. MSI)	72	0.4009	0.2294
DDR1 IHC tumor score (low vs. High)	281	0.5832	**0.0391**

In our cohort DDR1 mRNA expression levels successfully evaluated in 66 patients were not correlated with OS ($p = 0.86$) nor EFS ($p = 0.46$), whatever the CCR stage (data not shown).

To corroborate our previous results, we next performed survival analyses in Sieber-Smith ($n = 286$), Martineau ($n = 124$) [34] and TCGA cohorts ($n = 174$) obtained from R2 database [35,36]. In these cohorts, DDR1 mRNA expression was not correlated with overall nor relapse free survival (Figure 4).

3.3. *DDR1 Is More Expressed in Tumor Cells Compared with Stromal Cells*

DDR1 mRNA expression has been determined by RT-qPCR on 65 colonic adenocarcinoma samples and 78 colonic mucosa samples. Surprisingly, data showed a significant decrease in DDR1 expression within tumor samples when compared with normal colon samples (Figure 5A). Due to the difference observed in DDR1 expression between stromal and malignant cells when evaluated by IHC analysis, we used Laser Capture Microdissection (LCM) to thereafter quantify DDR1 mRNA expression in tumoral and stromal areas of each sample as previously described [29]. LCM was performed on 25 colon adenocarcinoma samples and RT-qPCR revealed that DDR1 mRNA expression was higher in the tumoral area than in the stroma (Figure 5B).

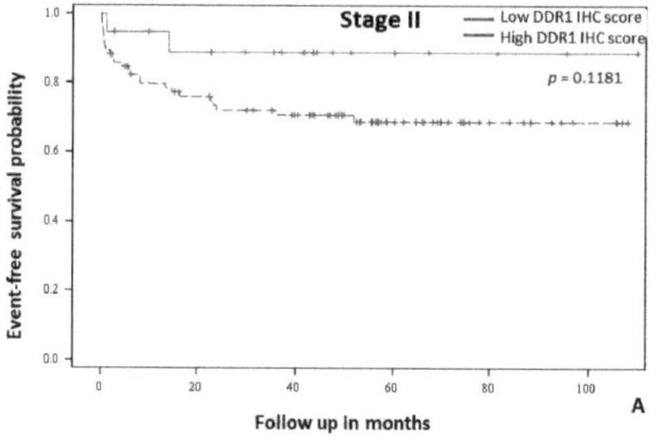

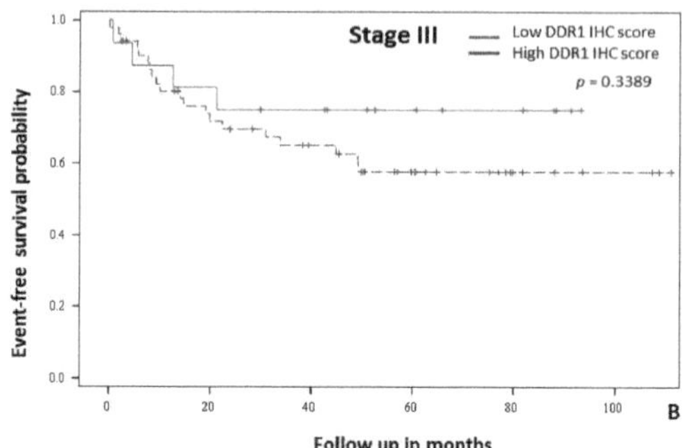

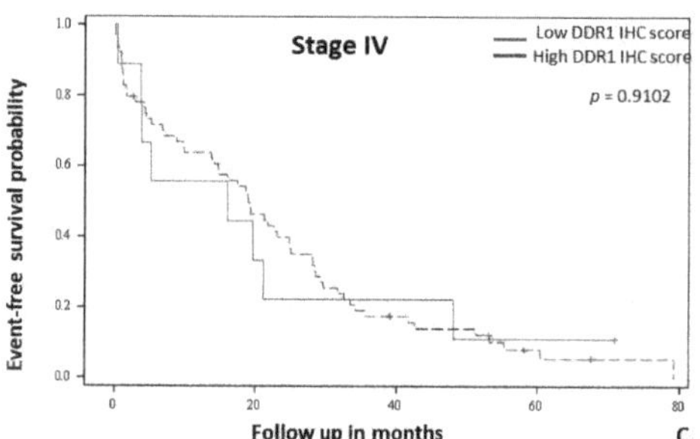

Figure 3. Stage specific event free survival analysis in colon cancer patients from our cohort according to DDR1 immunohistochemical expression. Kaplan-Meier curves of event free-survival probability for low (blue line) and high (red line) DDR1 immunohistochemical expression in cells in stage II (**A**), stage III (**B**) and stage IV patients (**C**).

3.4. DDR1 Mediates the Invasion of CRC Cells

We then investigated the possible role of DDR1 in CRC aggressiveness in vitro. We used HCT116, HT-29, SW480, and SW620 cell lines, which express different levels of DDR1 expression, and analyzed their ability to invade type I collagen as one of the main extracellular matrix components. These cell lines harbor different *KRAS/BRAF* statuses and their main characteristics are summarized in Supplementary Figure S2. The level of DDR1 expression was analyzed by both RT-qPCR and immunoblotting (Figure 6A,B, uncropped western blot images in Supplementary Figure S1). Data showed that the expression of DDR1 at the mRNA and protein levels was higher in HCT116 cells than in the other cell lines. In order to investigate deeply the impact of DDR1 on invasive properties of CRC cells, we used HT-29 cells expressing DDR1 at a basal level (HT-29^{GFP}) and overexpressing the receptor (HT-29$^{DDR1\text{-}GFP}$). As shown in Supplementary Figure S2, HT-29$^{DDR1\text{-}GFP}$ expressed a high level of DDR1 when compared to wild-type HT-29 or HT-29^{GFP} cells. The invasion potential of CRC cell lines was evaluated based on modified Boyden chamber assay using type I collagen coating. Data showed that HCT116 cells exhibited a higher invasion rate than SW480 and SW620 cells. When DDR1 was overexpressed in HT29 cells (HT-29$^{DDR1\text{-}GFP}$), the invasion rate was significantly increased compared to the control (HT-29^{GFP}) (Figure 6C). Interestingly, the invasion rate positively correlated with DDR1 expression level. To confirm the role of DDR1 in the invasion process of colorectal cells, nilotinib (100 nM) and DDR1-IN-1 (10 μM) have been used to inhibit specifically DDR1. As shown in Figure 6D, significant inhibition of cell invasiveness was observed when the cells were treated with nilotinib or DDR1-IN-1 compared with the control ones. Overall, these data suggest that DDR1 is involved in CRC invasion phenotype and could be associated in this way with the worse event free survival.

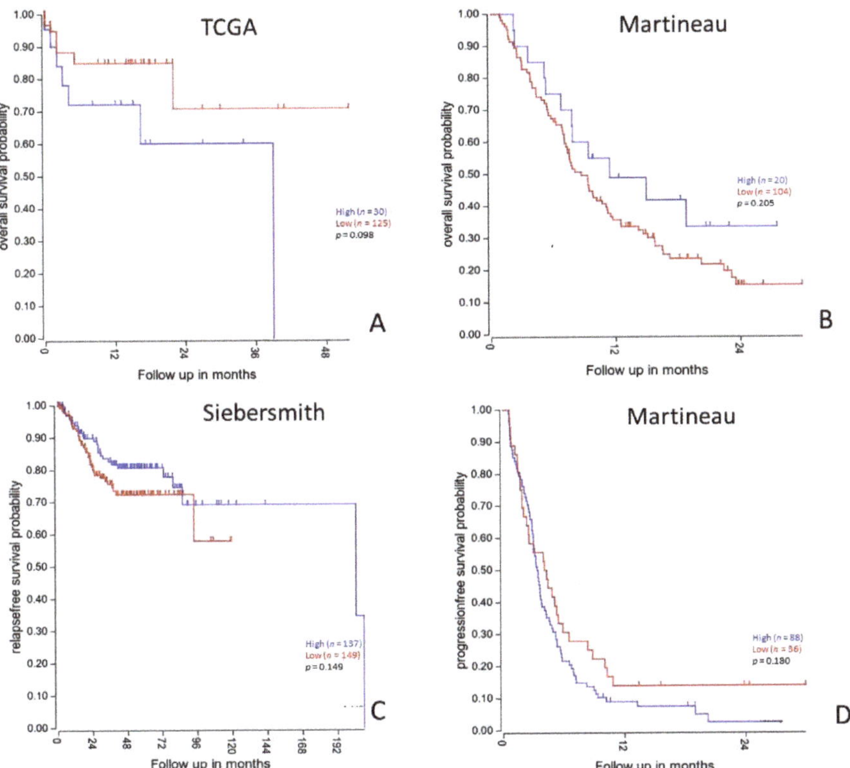

Figure 4. Survival analysis according to DDR1 mRNA expression profile in independent colorectal cancers patients' cohorts. Kaplan-Meier curves of overall survival (**A**,**B**) and relapse or progression free-survival (**C**,**D**) probability for low (red line) and high (blue line) DDR1 mRNA expression in all stages colorectal cancers patients and in stage IV (metastatic) patients (**C**,**D**). Survival analysis and Kaplan Meyier curves of the TCGA, Martineau and SieberSmith gene expression dataset were obtained from R2 platform (http://r2.amc.nl; last access date: 5 November 2021). All *p*-values were calculated using R2 online tools.

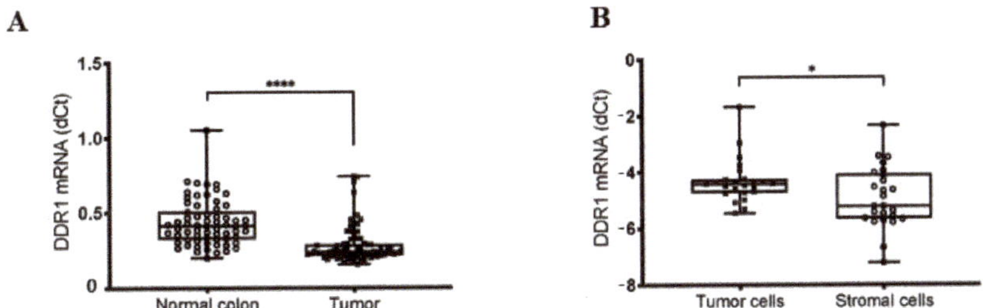

Figure 5. Comparison of DDR1 mRNA expression between tumor cells, normal colon and stromal cells. (**A**) Real-time PCR analysis of the DDR1 mRNA expression performed in colon adenocarcinoma and in normal colon mucosa fresh frozen samples. Values are represented as dCt normalized with RPL32. (**B**) Real-time PCR analysis of the DDR1 mRNA expression performed in adenocarcinoma cells and in stromal cells after laser capture microdissection. Values are represented as dCt normalized with RPL32. * $p < 0.05$, **** $p < 0.0001$, Mann Whitney test.

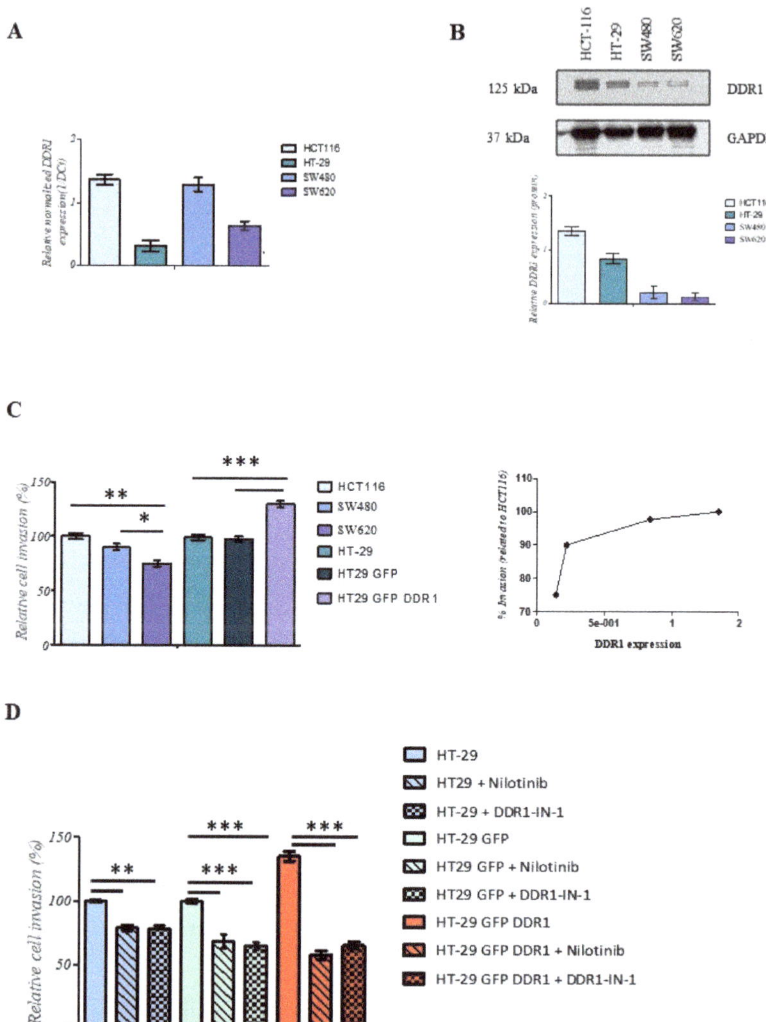

Figure 6. Human CRC cell invasion is modulated by DDR1 expression. (**A**) The relative mRNA expression of DDR1 was assessed using RT-qPCR. Values in HCT-116, HT-29, SW480, and SW620 were normalized with both RPL32 and RS18 mRNA expression. (**B**) The expression of DDR1 and GAPDH was assessed by western blotting using anti-DDR1 and anti-GAPDH antibodies in HCT-116, HT-29, SW480, and SW620 cells. Quantitative analysis of DDR1 protein was obtained by densitometry: the amount of DDR1 was normalized to GAPDH expression level (bottom panel). (**C**) HCT-116, HT-29, SW480, and SW620 were seeded into the collagen type I coated chambers for 24 h. Cells were then fixed with methanol and stained with DAPI. Results are expressed as mean ± SD of three independent experiments. Statistical significance was analyzed by a one-way ANOVA test using Dunnett's multiple comparisons. * $p = 0.05$, ** $p = 0.01$, *** $p = 0.001$ as compared to HCT-116 cells or HT-29 cells. Correlation between DDR1 expression and cell invasion (right panel). (**D**) HT-29, HT-29$^{DDR1\text{-}GFP}$, and HT-29^{GFP} cells were seeded into the collagen type I coated chambers for 24 h in absence or presence of nilotinib (100 nM) or DDR1-IN-1 (10 µM). Cells were then fixed with methanol and stained with crystal violet. Results are expressed as mean ± SD of three independent experiments. Statistical significance was analyzed by a one-way ANOVA test using Dunnett's multiple comparisons. ** $p = 0.01$, *** $p = 0.001$ as compared to HT-29, HT-29$^{DDR1\text{-}GFP}$, or HT-29^{GFP} cells.

4. Discussion

Many cancers are characterized by dysregulated expression of one or more RTKs. Such alteration has functional consequences at the cellular level which directly impact tumor progression, especially cell invasion and metastasis. DDRs play a key role in tumor progression, in part by regulating the reciprocal interplay between cancer cells and stromal collagens [37]. One of their major roles in the literature is their involvement in tumor invasion and metastasis [38].

In this study, we investigated the expression of DDR1 using immunohistochemistry in colon adenocarcinoma and studied the link between DDR1 expression with clinicopathologic and molecular parameters, including overall and event-free survival. Because DDR1 seems to play a role in CRC cell invasion and metastasis [5], we also investigated the impact of DDR1 on invasion properties of CRC cell lines in vitro using type I collagen as a main extracellular matrix component.

In this work, we showed that DDR1 expression was higher in adenocarcinoma cells than in normal colonic epithelium. DDR1 was highly expressed in a large majority (82.2%) of colon cancers. These results corroborate previous data showing a high DDR1 overexpression in 94% of colon cancer samples [23] and in tumor tissues from patients with primary CRC and hepatic CRC metastasis [24].

Our results demonstrated in univariate analysis that the clinico-pathological and molecular characteristics associated with DDR1 expression in colon adenocarcinoma were: male sex, left colon tumor localization, *BRAF* wild-type status, and absence of the expression of the serrated marker Annexin A10.

To our knowledge, no study had investigated these clinico-pathological and molecular characteristics in association with DDR1 expression in colon adenocarcinoma, especially the potential association with the serrated pathway highlighted by its markers Annexin A10. The molecular profile associated with DDR1 high expression could be integrated into the CMS4 molecular subtype of colorectal cancer. These tumors are characterized by strong stromal infiltration and show clear upregulation of genes playing a role in epithelial mesenchymal transition and associated to transforming growth factor β (TGF β) signaling pathway, angiogenesis, matrix remodeling pathways, and the complement-mediated inflammation. These CMS4 tumors presented worse overall survival and relapse-free survival [39]. Indeed, DDR1 mRNA expression was not associated with any CMS subtype [25]. Our bioinformatic analyses revealed that high DDR1 mRNA expression was independently associated with worse OS and PFS in stage IV patients. Moreover, any significant association between DDR1 mRNA expression and OS or EFS has been found in our cohort of patients undergoing surgery for colonic adenocarcinoma. However, divergent results showed that DDR1 high mRNA expression was associated with worse OS whatever the tumor stage [24].

The major limitations of our study were its retrospective and single-center design and that few patients had DDR1 mRNA expression data. However, our results were validated with bioinformatic analyses in three other patients' cohorts. In our patients' cohort, DDR1 immunohistochemical expression was only associated with worse EFS whatever the stage. DDR1 high protein expression was not associated with OS or with stage specific EFS. In a previous immunohistochemical study, high DDR1 immunoreactivity score was correlated with a shorter overall survival in a cohort of 100 patients with colorectal cancer [23]. In this study, EFS was not evaluated and stage specific survival analyses were not performed.

The molecular mechanisms underlying the roles of the DDRs in various steps of colon carcinoma progression are largely undefined. To fill this gap, we investigated the potential role of DDR1 in tumor cell invasion by using several colorectal cancer cell lines that differentially express DDR1. In addition, HT-29 cells overexpressing DDR1 were established and led to enhanced cell invasiveness. The data showed that the tumor cell invasion capacity is closely correlated to DDR1 expression. Moreover, specific pharmacological inhibition of DDR1 with nilotinib and DDR-IN-1 significantly reduced HT-29 cell invasion. These results ascertained previous reports indicating DDR1 pro-invasive role in several tumor cell lines and DDR1 metastatic function in many cancers [12,17,19,40], and demonstrate the importance of DDR1 in invasive tumors. For instance, DDR1 expression is increased by the microRNA MiR-199a-5p and promotes invasion in CRC by activating epithelial-mesenchymal transition [41]. In human A375 melanoma, HT29 colon carcinoma and SK-HEP hepatoma cells, chemical inhibition or silencing of DDR1 reduces cell adhesion to collagen I and MMP-dependent invasion [42]. Recently, Romayor and coworkers have demonstrated that DDR1 expressed by tumor cells promotes hepatic cell ability to alter the ECM structure by regulating collagen and MMPs expression, thus suggesting an impact of DDR1 in the desmoplastic response of hepatic tumor microenvironment during CRC tumorigenesis [24].

In addition, it has been recently demonstrated that DDR1 can have a great impact on the invasion function of metastatic colon carcinoma [26]. Indeed, invasion and metastatic processes were decreased by DDR1-BCR signaling axis inhibition in vivo in colon carcinoma suggesting that DDR1 could be an effective therapeutic target in this cancer. The authors concluded that the inhibition of DDR1 kinase activity with nilotinib may be a therapeutic benefit in patients with advanced CRC [26].

In other cancers, DDR1 expression could also have a prognostic implication. Indeed, high expression of DDR1 has also been identified in 52.2% of hepatocellular carcinoma samples [43], 61.0% of non-small cell lung cancer [13], and 69% of serous ovarian cancer tissues [15]. Moreover, high DDR1 expression was more frequently expressed in invasive carcinoma than in bronchioloalveolar carcinoma in lung cancers and was associated with shorter overall survival in non-small cell lung carcinomas [22]. On the contrary, DDR1 was not associated with survival in prostate cancer [12] and low DDR1 expression was associated with triple negative subtype of breast cancer and with shorter survival in this cancer type [44].

Thus, the overexpression of DDR1 in these malignant diseases, particularly in colorectal cancer, supports the hypothesis that DDR1 upregulation is widespread in cancer and can play an important role in tumorigenesis and/or tumor invasion and metastasis.

5. Conclusions

In summary, DDR1 is highly expressed in colon cancer compared to normal colonic mucosa. This overexpression of DDR1 is found in a large majority of colon cancers, suggesting a role of DDR1 in colorectal carcinogenesis. Although DDR1 was associated with shorter EFS, its role as a prognosis marker remains uncertain. However, frequent high expression of DDR1 in colon cancer could be further explored as a potential therapeutic target in this indication.

Supplementary Materials: The following are available online at https://www.mdpi.com/article/10.3390/cancers14040928/s1, Figure S1: (A) Summary table of the CRC cell lines characteristics. (B) DDR1 expression in HT-29, HT-29DDR1-GFP and HT-29GFP cell lines; Figure S2: Original, uncropped western blot scans.

Author Contributions: Conception or design of the work: A.A.-C., C.B.-R., H.M. Data collection: K.B.A., C.B.-R., N.B., A.A.-C., C.C.L., K.T., G.C., V.L., C.H. Data analysis and interpretation: K.B.A., C.B.-R., N.B., A.A.-C., C.C.L., C.S., A.W.-T. Drafting the article: K.B.A., C.B.-R., A.A.-C. Critical revision of the article: S.D., O.B., A.B., H.M. All authors have read and agreed to the published version of the manuscript.

Funding: This study was funded by grants from: 1. Ligue Nationale Contre le Cancer Conférence de Coordination Interrégionale du Grand Est, Appel d'Offre 2014 et 2015. Grant Numbers 61/2014/KM/ML/Recherche and 2015/KM/NM/PFRS/232 (SD). 2. Centre Hospitalier Universitaire de Reims, Appel d'Offre Local 2014 et 2015. Grant Numbers AU14-03 and AOL 2015-11 (CBR). 3. Structure Fédérative de Recherche Cap Santé (SFR Cap Santé) (A.A.-C., C.B.-R., H.M.). All these study sponsors have no roles in the study design, in the collection, analysis, and interpretation of data.

Institutional Review Board Statement: The study was performed in accordance with the ethical standards laid down in the Declaration of Helsinki. Written patients' consent for biospecimen use was obtained in all cases. Approval for the study was previously obtained from the local Institutional Review Board and the Tissue Bank Management Board [29]. The ethical committee of the Plateforme des Centre de Ressources Biologiques de Champagne Ardennes, protocol code: AC-2019-3408.

Informed Consent Statement: Informed consent was obtained from all subjects involved in the study.

Data Availability Statement: The data presented in this study are available on request from the corresponding author.

Acknowledgments: We thank Frederic Saltel (INSERM U1053, Bordeaux) for the gift of the lentivirus encoding DDR1-GFP. We also thank Damien Rioult from the technical platform MOBICYTE of the University of Reims Champagne-Ardenne for their excellent technical assistance.

Conflicts of Interest: S.D. serves as Chair of the Scientific and Clinical Advisory Board and has equity interest in Apmonia Therapeutics (Reims, France), a biotechnology company developing anticancer strategy. The other authors declare that they have no further financial or other conflicts of interest in relation to this research and its publication.

References

1. Sung, H.; Ferlay, J.; Siegel, R.L.; Laversanne, M.; Soerjomataram, I.; Jemal, A.; Bray, F. Global Cancer Statistics 2020: GLOBOCAN Estimates of Incidence and Mortality Worldwide for 36 Cancers in 185 Countries. *CA Cancer J. Clin.* **2021**, *71*, 209–249. [CrossRef]
2. Dienstmann, R.; Vermeulen, L.; Guinney, J.; Kopetz, S.; Tejpar, S.; Tabernero, J. Consensus molecular subtypes and the evolution of precision medicine in colorectal cancer. *Nat. Rev. Cancer* **2017**, *17*, 79–92. [CrossRef]
3. Miller, K.D.; Nogueira, L.; Mariotto, A.B.; Rowland, J.H.; Yabroff, K.R.; Alfano, C.M.; Jemal, A.; Kramer, J.; Siegel, R. Cancer treatment and survivorship statistics, 2019. *CA Cancer J. Clin.* **2019**, *69*, 363–385. [CrossRef] [PubMed]
4. Oliveres, H.; Pesántez, D.; Maurel, J. Lessons to Learn for Adequate Targeted Therapy Development in Metastatic Colorectal Cancer Patients. *Int. J. Mol. Sci.* **2021**, *22*, 5019. [CrossRef] [PubMed]
5. Lafitte, M.; Sirvent, A.; Roche, S. Collagen Kinase Receptors as Potential Therapeutic Targets in Metastatic Colon Cancer. *Front. Oncol.* **2020**, *10*, 125. [CrossRef] [PubMed]
6. Stefani, C.; Miricescu, D.; Stanescu-Spinu, I.-I.; Nica, R.I.; Greabu, M.; Totan, A.R.; Jinga, M. Growth Factors, PI3K/AKT/mTOR and, M.APK Signaling Pathways in Colorectal Cancer Pathogenesis: Where Are We Now? *Int. J. Mol. Sci.* **2021**, *22*, 10260. [CrossRef] [PubMed]
7. Hubbard, S.R.; Till, J.H. Protein tyrosine kinase structure and function. *Annu. Rev. Biochem.* **2000**, *69*, 373–398. [CrossRef]
8. Vogel, W.; Gish, G.D.; Alves, F.; Pawson, T. The discoidin domain receptor tyrosine kinases are activated by collagen. *Mol. Cell* **1997**, *1*, 13–23. [CrossRef]
9. Henriet, E.; Sala, M.; Abou Hammoud, A.; Tuariihionoa, A.; Di Martino, J.; Ros, M.; Saltel, F. Multitasking discoidin domain receptors are involved in several and specific hallmarks of cancer. *Cell Adhes. Migr.* **2018**, *12*, 363–377. [CrossRef] [PubMed]
10. Rammal, H.; Saby, C.; Magnien, K.; Van-Gulick, L.; Garnotel, R.; Buache, E.; El btaouri, H.; Jeannesson, P.; Morjani, H. Discoidin Domain Receptors: Potential Actors and Targets in Cancer. *Front. Pharmacol.* **2016**, *7*, 346. [CrossRef]
11. Gao, Y.; Zhou, J.; Li, J. Discoidin domain receptors orchestrate cancer progression: A focus on cancer therapies. *Cancer Sci.* **2021**, *112*, 962–969. [CrossRef] [PubMed]
12. Bonfil, R.D.; Chen, W.; Vranic, S.; Sohail, A.; Shi, D.; Jang, H.; Kim, H.R.; Prunotto, M.; Fridman, R. Expression and subcellular localization of Discoidin Domain Receptor 1 (DDR1) define prostate cancer aggressiveness. *Cancer Cell Int.* **2021**, *21*, 507. [CrossRef] [PubMed]
13. Zhu, S.; Wang, Y.; Li, Y.; Ding, J.; Dai, J.; Cai, H.; Zhang, D.; Song, Y. Discoidin domain receptor 1 is associated with poor prognosis of non-small cell lung cancer and promotes cell invasion via epithelial-to-mesenchymal transition. *Med. Oncol.* **2013**, *30*, 626.
14. Borza, C.M.; Pozzi, A. Discoidin domain receptors in disease. *Matrix Biol. J. Int. Soc. Matrix Biol.* **2014**, *34*, 185–192. [CrossRef] [PubMed]
15. Quan, J.; Yahata, T.; Adachi, S.; Yoshihara, K.; Tanaka, K. Identification of Receptor Tyrosine Kinase, Discoidin Domain Receptor 1 (DDR1), as a Potential Biomarker for Serous Ovarian Cancer. *Int. J. Mol. Sci.* **2011**, *12*, 971–982. [CrossRef] [PubMed]
16. Jin, H.; Ham, I.-H.; Oh, H.J.; Bae, C.A.; Lee, D.; Kim, Y.-B.; Son, S.-Y.; Chwae, Y.-J.; Han, S.-U.; Brekken, R.A.; et al. Inhibition of Discoidin Domain Receptor 1 Prevents Stroma-Induced Peritoneal Metastasis in Gastric Carcinoma. *Mol. Cancer Res.* **2018**, *16*, 1590–1600. [CrossRef] [PubMed]
17. Gao, H.; Chakraborty, G.; Zhang, Z.; Akalay, I.; Gadiya, M.; Gao, Y.; Sinha, S.; Hu, J.; Jiang, C.; Akram, M.; et al. Multi-organ Site Metastatic Reactivation Mediated by Non-canonical Discoidin Domain Receptor 1 Signaling. *Cell* **2016**, *166*, 47–62. [CrossRef] [PubMed]

18. Sun, X.; Gupta, K.; Wu, B.; Zhang, D.; Yuan, B.; Zhang, X.; Chiang, H.-C.; Zhang, C.; Curiel, T.J.; Bendeck, M.P.; et al. Tumor-extrinsic discoidin domain receptor 1 promotes mammary tumor growth by regulating adipose stromal interleukin 6 production in mice. *J. Biol. Chem.* **2018**, *293*, 2841–2849. [CrossRef]
19. Yuge, R.; Kitadai, Y.; Takigawa, H.; Naito, T.; Oue, N.; Yasui, W.; Tanaka, S.; Chayama, K. Silencing of Discoidin Domain Receptor-1 (DDR1) Concurrently Inhibits Multiple Steps of Metastasis Cascade in Gastric Cancer. *Transl. Oncol.* **2018**, *11*, 575–584. [CrossRef]
20. Nemoto, T.; Ohashi, K.; Akashi, T.; Johnson, J.D.; Hirokawa, K. Overexpression of protein tyrosine kinases in human esophageal cancer. *Pathobiol. J. Immunopathol. Mol Cell Biol.* **1997**, *65*, 195–203. [CrossRef]
21. Weiner, H.L.; Huang, H.; Zagzag, D.; Boyce, H.; Lichtenbaum, R.; Ziff, E.B. Consistent and Selective Expression of the Discoidin Domain Receptor-1 Tyrosine Kinase in Human Brain Tumors. *Neurosurgery* **2000**, *47*, 1400–1409. [CrossRef]
22. Yang, S.H.; Baek, H.A.; Lee, H.J.; Park, H.S.; Jang, K.Y.; Kang, M.J.; Lee, D.G.; Lee, Y.C.; Moon, W.S.; Chung, M.J. Discoidin domain receptor 1 is associated with poor prognosis of non-small cell lung carcinomas. *Oncol. Rep.* **2010**, *24*, 311–319. [PubMed]
23. Tao, Y.; Wang, R.; Lai, Q.; Wu, M.; Wang, Y.; Jiang, X.; Zeng, L.; Zhou, S.; Li, Z.; Yang, T.; et al. Targeting of DDR1 with antibody-drug conjugates has antitumor effects in a mouse model of colon carcinoma. *Mol. Oncol.* **2019**, *13*, 1855–1873. [CrossRef] [PubMed]
24. Romayor, I.; Márquez, J.; Benedicto, A.; Herrero, A.; Arteta, B.; Olaso, E. Tumor DDR1 deficiency reduces liver metastasis by colon carcinoma and impairs stromal reaction. *Am. J. Physiol. Gastrointest. Liver Physiol.* **2021**, *320*, G1002-13. [CrossRef] [PubMed]
25. Jeitany, M.; Leroy, C.; Tosti, P.; Lafitte, M.; Le Guet, J.; Simon, V.; Bonenfant, D.; Robert, B.; Grillet, F.; Mollevi, C.; et al. Inhibition of DDR1-BCR signalling by nilotinib as a new therapeutic strategy for metastatic colorectal cancer. *EMBO Mol. Med.* **2018**, *10*, e7918. [CrossRef]
26. Sirvent, A.; Lafitte, M.; Roche, S. DDR1 inhibition as a new therapeutic strategy for colorectal cancer. *Mol. Cell Oncol.* **2018**, *5*, e1465882. [CrossRef]
27. Le, C.C.; Bennasroune, A.; Collin, G.; Hachet, C.; Lehrter, V.; Rioult, D.; Dedieu, S.; Morjani, H.; Appert-Collin, A. LRP-1 Promotes Colon Cancer Cell Proliferation in 3D Collagen Matrices by Mediating DDR1 Endocytosis. *Front. Cell Dev. Biol.* **2020**, *8*, 412. [CrossRef] [PubMed]
28. Theret, L.; Jeanne, A.; Langlois, B.; Hachet, C.; David, M.; Khrestchatisky, M.; Devy, J.; Hervé, E.; Almagro, S.; Dedieu, S. Identification of LRP-1 as an endocytosis and recycling receptor for β1-integrin in thyroid cancer cells. *Oncotarget* **2017**, *8*, 78614–78632. [CrossRef]
29. Boulagnon-Rombi, C.; Schneider, C.; Leandri, C.; Jeanne, A.; Grybek, V.; Bressenot, A.M.; Barbe, C.; Marquet, B.; Nasri, S.; Coquelet, S.; et al. LRP1 expression in colon cancer predicts clinical outcome. *Oncotarget* **2018**, *9*, 8849–8869. [CrossRef]
30. Hamilton, S.R.; Bosman, F.T.; Boffetta, P.; Ilyas, M.; Morreau, H.; Nakamura, S.I. Carcinoma of the colon and rectum. In *WHO Classification of Tumours of the Digestive System*; IARC Press: Lyon, France, 2010; 417p.
31. Sobin, L.; Gospodarowicz, M.; Wittekind, C. *TNM Classification of Malignant Tumours*; Wiley-Blackwell: Oxford, UK, 2010; 30p.
32. Lugli, A.; Kirsch, R.; Ajioka, Y.; Bosman, F.; Cathomas, G.; Dawson, H.; El Zimaity, H.; Fléjou, J.-F.; Hansen, T.P.; Hartmann, A.; et al. Recommendations for reporting tumor budding in colorectal cancer based on the International Tumor Budding Consensus Conference (ITBCC) 2016. *Mod. Pathol.* **2017**, *30*, 1299–1311. [CrossRef]
33. Boulagnon, C.; Dudez, O.; Beaudoux, O.; Dalstein, V.; Kianmanesh, R.; Bouché, O.; Diebold, M.-D. BRAF V600E Gene Mutation in Colonic Adenocarcinomas. Immunohistochemical Detection Using Tissue Microarray and Clinicopathologic Characteristics: An 86 Case Series. *Appl. Immunohistochem. Mol. Morphol.* **2016**, *24*, 88–96. [CrossRef]
34. Cherradi, S.; Martineau, P.; Gongora, C.; Del Rio, M. Claudin gene expression profiles and clinical value in colorectal tumors classified according to their molecular subtype. *Cancer Manag. Res.* **2019**, *11*, 1337–1348. [CrossRef]
35. Gao, J.; Aksoy, B.A.; Dogrusoz, U.; Dresdner, G.; Gross, B.; Sumer, S.O.; Sun, Y.; Jacobsen, A.; Sinha, R.; Larsson, E.; et al. Integrative analysis of complex cancer genomics and clinical profiles using the cBioPortal. *Sci Signal.* **2013**, *6*, pl1. [CrossRef]
36. Cerami, E.; Gao, J.; Dogrusoz, U.; Gross, B.E.; Sumer, S.O.; Aksoy, B.A.; Jacobsen, A.; Byrne, C.J.; Heuer, M.L.; Larsson, E.; et al. The cBio cancer genomics portal: An open platform for exploring multidimensional cancer genomics data. *Cancer Discov.* **2012**, *2*, 401–404. [CrossRef] [PubMed]
37. Leitinger, B. Discoidin domain receptor functions in physiological and pathological conditions. *Int. Rev. Cell Mol. Biol.* **2014**, *310*, 39–87. [PubMed]
38. Valiathan, R.R.; Marco, M.; Leitinger, B.; Kleer, C.G.; Fridman, R. Discoidin domain receptor tyrosine kinases: New players in cancer progression. *Cancer Metastasis Rev.* **2012**, *31*, 295–321. [CrossRef] [PubMed]
39. Guinney, J.; Dienstmann, R.; Wang, X.; de Reyniès, A.; Schlicker, A.; Soneson, C.; Marisa, L.; Roepman, P.; Nyamundanda, G.; Angelino, P.; et al. The consensus molecular subtypes of colorectal cancer. *Nat. Med.* **2015**, *21*, 1350–1356. [CrossRef]
40. Valencia, K.; Ormazábal, C.; Zandueta, C.; Luis-Ravelo, D.; Antón, I.; Pajares, M.J.; Agorreta, J.; Montuenga, L.M.; Martinez-Canarias, S.; Leitinger, B.; et al. Inhibition of Collagen Receptor Discoidin Domain Receptor-1 (DDR1) Reduces Cell Survival, Homing, and Colonization in Lung Cancer Bone Metastasis. *Clin. Cancer Res.* **2012**, *18*, 969–980. [CrossRef]
41. Hu, Y.; Liu, J.; Jiang, B.; Chen, J.; Fu, Z.; Bai, F.; Jiang, J.; Tang, Z. MiR-199a-5p loss up-regulated, DDR1 aggravated colorectal cancer by activating epithelial-to-mesenchymal transition related signaling. *Dig. Dis. Sci.* **2014**, *59*, 2163–2172. [CrossRef]
42. Romayor, I.; Badiola, I.; Olaso, E. Inhibition of DDR1 reduces invasive features of human A375 melanoma, HT29 colon carcinoma and K-HEP hepatoma cells. *Cell Adhes. Migr.* **2020**, *14*, 69–81. [CrossRef]

43. Shen, Q.; Cicinnati, V.R.; Zhang, X.; Iacob, S.; Weber, F.; Sotiropoulos, G.C.; Radtke, A.; Lu, M.; Paul, A.; Gerken, G.; et al. Role of microRNA-199a-5p and discoidin domain receptor 1 in human hepatocellular carcinoma invasion. *Mol. Cancer* **2010**, *9*, 227. [CrossRef] [PubMed]
44. Toy, K.A.; Valiathan, R.R.; Núñez, F.; Kidwell, K.M.; Gonzalez, M.E.; Fridman, R.; Kleer, C.G. Tyrosine kinase discoidin domain receptors DDR1 and DDR2 are coordinately deregulated in triple-negative breast cancer. *Breast Cancer Res. Treat.* **2015**, *150*, 9–18. [CrossRef] [PubMed]

Cancers 2022, 14, 4891

Article

Defining A Liquid Biopsy Profile of Circulating Tumor Cells and Oncosomes in Metastatic Colorectal Cancer for Clinical Utility

Sachin Narayan [1], George Courcoubetis [1], Jeremy Mason [1,2,3], Amin Naghdloo [1], Drahomír Kolenčík [4], Scott D. Patterson [5], Peter Kuhn [1,2,3,6,7,8,*] and Stephanie N. Shishido [1]

1 Michelson Center for Convergent Bioscience, Convergent Science Institute in Cancer, Dornsife College of Letters, Arts and Sciences, University of Southern California, Los Angeles, CA 90089, USA
2 Catherine & Joseph Aresty Department of Urology, Institute of Urology, Keck School of Medicine, University of Southern California, Los Angeles, CA 90033, USA
3 Norris Comprehensive Cancer Center, Keck School of Medicine, University of Southern California, Los Angeles, CA 90033, USA
4 Biomedical Center, Faculty of Medicine in Pilsen, Charles University, 32300 Pilsen, Czech Republic
5 Gilead Sciences, Inc., Lakeside Drive, Foster City, CA 94404, USA
6 Department of Biomedical Engineering, Viterbi School of Engineering, University of Southern California, Los Angeles, CA 90089, USA
7 Department of Biological Sciences, Dornsife College of Letters, Arts, and Sciences, University of Southern California, Los Angeles, CA 90089, USA
8 Department of Aerospace and Mechanical Engineering, Viterbi School of Engineering, University of Southern California, Los Angeles, CA 90089, USA
* Correspondence: pkuhn@usc.edu; Tel.: +1-213-821-3980

Citation: Narayan, S.; Courcoubetis, G.; Mason, J.; Naghdloo, A.; Kolenčík, D.; Patterson, S.D.; Kuhn, P.; Shishido, S.N. Defining A Liquid Biopsy Profile of Circulating Tumor Cells and Oncosomes in Metastatic Colorectal Cancer for Clinical Utility. *Cancers* **2022**, *14*, 4891. https://doi.org/10.3390/cancers14194891

Academic Editors: Stephane Dedieu and Luca Roncucci

Received: 10 September 2022
Accepted: 30 September 2022
Published: 6 October 2022

Simple Summary: Metastatic colorectal cancer (mCRC) is typified by its tumor heterogeneity and changing disease states, suggesting that personalized medicine approaches could be vital to improving clinical practice. As a minimally invasive approach, the liquid biopsy has the potential to be a powerful longitudinal prognostic tool. We investigated mCRC patients' peripheral blood samples using an enrichment-free single-cell approach to capture the broader rare-event population beyond the conventionally detected epithelial-derived circulating tumor cell (CTC). Our analysis reveals a heterogenous profile of CTCs and oncosomes not commonly found in normal donor samples. We identified select rare cell types based on their distinct immunofluorescence expression and morphology across multiple assays. Lastly, we highlight correlations between enumerations of the blood-based analytes and progression-free survival. This study clinically validates an unbiased rare-event approach in the liquid biopsy, motivating future studies to further investigate these analytes for their prognostic potential.

Abstract: Metastatic colorectal cancer (mCRC) is characterized by its extensive disease heterogeneity, suggesting that individualized analysis could be vital to improving patient outcomes. As a minimally invasive approach, the liquid biopsy has the potential to longitudinally monitor heterogeneous analytes. Current platforms primarily utilize enrichment-based approaches for epithelial-derived circulating tumor cells (CTC), but this subtype is infrequent in the peripheral blood (PB) of mCRC patients, leading to the liquid biopsy's relative disuse in this cancer type. In this study, we evaluated 18 PB samples from 10 mCRC patients using the unbiased high-definition single-cell assay (HDSCA). We first employed a rare-event (Landscape) immunofluorescence (IF) protocol, which captured a heterogenous CTC and oncosome population, the likes of which was not observed across 50 normal donor (ND) samples. Subsequent analysis was conducted using a colorectal-targeted IF protocol to assess the frequency of CDX2-expressing CTCs and oncosomes. A multi-assay clustering analysis isolated morphologically distinct subtypes across the two IF stains, demonstrating the value of applying an unbiased single-cell approach to multiple assays in tandem. Rare-event enumerations at a single timepoint and the variation of these events over time correlated with progression-free survival. This study supports the clinical utility of an unbiased approach to interrogating the liquid biopsy in mCRC, representing the heterogeneity within the CTC classification and warranting the further molecular characterization of the rare-event analytes with clinical promise.

Keywords: liquid biopsy; rare cell; circulating tumor cells; oncosomes; colorectal cancer; heterogeneity; multi-assay; high-definition single-cell assay

---

1. Introduction

Colorectal cancer (CRC) is the world's third most common cancer and second leading cause of oncology-related deaths [1]. Most notably, CRC solid tumors are marked by their extensive cellular heterogeneity and proliferation owed to the rapid rate of epithelial self-renewal in the intestines [2–4]. The variety of tumor microenvironments, genetic mutations, and disease subtypes suggests that real-time individualized analysis and subsequent clinical decision-making could improve patient outcomes [5–7].

As a minimally invasive procedure, the liquid biopsy has the potential to be that critical element in the longitudinal evaluation of CRC by characterizing the disease's pathophysiology and mechanisms of metastasis [8–10]. Much of the current liquid biopsy analysis focuses on two primary biomarkers: circulating tumor cells (CTCs) and cell-free DNA (cfDNA). Additionally found in the blood, carcinoembryonic antigen (CEA) is one of the foremost prognostic hallmarks of CRC [11,12]. Previous investigations have correlated CTC counts to CEA levels, with the two used in conjunction to accurately predict survival outcomes [13]. Generally, a higher number of CTCs indicates poorer patient outcomes in CRC [14,15], although nuances emerge when considering morphologically defined CTC subtypes and the change in cell populations over time [16,17]. With the variety of detectable biomarkers in circulation, the liquid biopsy could aid in tackling some of the clinical challenges of CRC.

Previously, CellSearch® (Menarini, Raritan, NJ, USA) was the first platform to receive regulatory approval in CRC via a 510(k) clearance for the enumeration of CTCs to monitor metastatic colorectal, breast and prostate cancer [18]. The platform uses an enrichment-based methodology for the detection of a singular type of CTC defined by the expression of epithelial cell adhesion molecule (EPCAM) and cytokeratin (CK) without the expression of CD45 [19], thereby limiting the liquid biopsy field's current understanding of CTC. CellSearch® uses a uniform threshold for CTC positivity at any timepoint during the patient's treatment. In CRC, this threshold is 3 CTCs/7.5 mL of blood [20]. CellSearch® has shown that a higher frequency of CTCs is associated with poorer overall survival (OS) and progression-free survival (PFS) in CRC [21,22]. However, CellSearch® and similar systems are not commonly utilized by clinicians treating CRC [23,24]. This limited utility could be attributed to a lack of timepoint-specific standards [18] and infrequent CTC kinetics analysis, despite its clinical promise [16]. Most early-generation liquid-biopsy platforms employ enrichment-based approaches that detect a limited CTC population in the peripheral blood (PB) of CRC patients [25–27] and overlook other cellular subtypes with nuanced survival implications [16,28]. In a cancer type known for its tumor heterogeneity like CRC, enrichment-based approaches limit the liquid biopsy's potential clinical utility, thereby warranting an unbiased single-cell approach that focuses on all rarity in the bloodstream.

This study utilized the third-generation high-definition single-cell assay (HDSCA3.0), which is a validated "no cell left behind" immunofluorescence (IF) assay that detects and characterizes all rare events from the liquid biopsy [25,28–33]. Commercialized by Epic Sciences, it has demonstrated clinical utility as a predictive marker in prostate cancer [34–36]. Furthermore, it allows for downstream genomic and proteomic analysis [37,38] and adheres to the standards of the Blood Profiling Atlas Commons [39]. A prior investigation into a cohort of metastatic CRC (mCRC) patients with HDSCA2.0 revealed a 35% CTC positivity rate [25], comparable to CellSearch® positivity rates among similar cohorts [30,37,40]. A subsequent study with the same platform highlighted the importance of CTC subtypes, time of sample collection and changes in cellular populations during treatment in understanding the value of the liquid biopsy in mCRC patient care [16]. Beyond CTC subtypes, prior studies have identified other rare cells such as circulating endothelial cells (CECs) [41,42] with prognostic implications in mCRC. In addition to rare cellular analytes, tumor-derived oncosomes and extracellular vesicles have been shown to promote tumorigenesis and chromosomal deletion across cancer types [43–45]. Now, the third generation of HDSCA detects a heterogeneous CTC and oncosome population with various surface biomarkers and an unbiased computational methodology for the detection of epithelial, mesenchymal, endothelial and immune cells [46,47]. Herein, we analyzed 18 PB samples from 10 mCRC patients using two IF protocols to represent a comprehensive CTC and oncosome liquid-biopsy profile, highlighting previously unidentified rare events and correlating analytes to patient outcomes.

2. Materials and Methods

2.1. Study Design

This study includes a total of 18 PB samples collected between May 2016 to March 2017 from 10 patients with mCRC. Patients were found as part of the GS-US-296-0101 phase I clinical trial (#NCT01803282) evaluating the safety and tolerability of a novel therapeutic in combination with standard-of-care chemotherapy in two different mCRC indications. Apart from their diagnosis of CRC and survival data, no other clinical or demographic information was available for this study cohort per IRB protocol at the time of enrollment. Patients 1, 3, 5, 6 and 9 were first-line inoperable mCRC patients receiving the test compound in combination with mFOLFOX6 and bevacizumab. Patients 2, 4, 7, 8 and 10 were second-line inoperable patients receiving the test compound in combination with FOLFIRI and bevacizumab. PB samples were collected on cycle 1, day 1 and cycle 3, day 1 of therapy, referred to as Draw 1 and Draw 2, respectively. All patients progressed, and PFS was provided for 9 of the 10 patients. In addition, PB samples from 50 normal donors (ND) with no known pathology were collected and provided by Epic Sciences (San Diego, CA, USA).

2.2. Blood Processing

PB samples were collected in 10 mL collection tubes (Cell-free DNA, Streck, La Vista, NE, USA) and were processed as previously described [30,47]. As a brief synopsis, after red blood cell lysis, the nucleated cell fraction was plated as a monolayer of ~3 million cells per slide (Marienfeld, Lauda, Germany) before cryobanking at −80 °C.

2.3. IF Staining Protocols

Samples were analyzed with the previously described workflow for high-resolution imaging and the characterization of tumor cells at a single-cell level [30]. Slides were stained by the IntelliPATH FLX™ autostainer (Biocare Medical LLC, Irvine, CA, USA) in batches of 50. Our validated HDSCA protocols utilize a cocktail of pan-cytokeratin (CK), CD45 antibodies and DAPI [30,31]. In further detail, samples were fixed with a 2% neutral buffered formalin solution (VWR) for 20 min followed by permeabilization using 100% cold methanol for 5 min and blocking nonspecific binding sites with 10% goat serum (Millipore) for 20 min. This is followed by an antibody cocktail consisting of mouse IgG1/Ig2a anti-human cytokeratins 1, 4, 5, 6, 8, 10, 13, 18 and 19 (clones: C-11, PCK-26, CY-90, KS-1A3, M20, A53-B/A2, C2562, Sigma, St. Louis, MO, USA), mouse IgG1 anti-human cytokeratin 19 (clone: RCK108, GA61561-2, Dako, Carpinteria, CA, USA) and mouse anti-human CD45:Alexa Fluor® 647 (clone: F10-89-4, MCA87A647, AbD Serotec, Raleigh, NC, USA). To complete the staining, slides were incubated with Alexa Fluor® 555 goat anti-mouse IgG1 antibody (A21127, Invitrogen, Carlsbad, CA, USA) and 4′,6-diamidino-2-phenylindole (DAPI; D1306, ThermoFisher) prior to being mounted with a glycerol-based aqueous mounting media [48].

Two distinct IF protocols were applied: 1) Landscape and 2) CDX2-targeted. The Landscape protocol identifies epithelial, mesenchymal, endothelial and immune cells through the addition of 100 ug/mL of a goat anti-mouse IgG monoclonal Fab fragments (115-007-003, Jackson ImmunoResearch, West Grove, PA, USA), rabbit IgG anti-human vimentin (Vim) (clone: D21H3, 9854BC, Cell Signaling, Danvers, MA, USA) as a fourth color and mouse IgG1 anti-human CD31:Alexa Fluor® 647 mAb (clone: WM59, MCA1738A647, BioRad, Hercules, CA, USA) to the CD45 channel, hereafter referred to as CD45/CD31 [47]. The CDX2-targeted protocol utilizes the colon-specific CDX2 monoclonal antibody EPR2764Y (Abcam, Cambridge, UK) as a fourth color for further characterization [25]. CDX2 is a transcription factor expressed throughout the intestinal epithelium and has been effectively used as a marker for intestinal carcinomas [49]. The ND samples were stained with the Landscape protocol.

2.4. Scanning and Analysis

Slides were imaged at 2304 frames per slide using automated high-throughput fluorescence scanning microscopy at 100× magnification with exposures and gain set to yield the same background intensity level for normalization purposes. The numeration of cell classifications was converted to concentration based on the sample leukocyte concentration measured at processing and the number of DAPI-positive nuclei detected. White blood cell (WBC) counts in the PB sample were determined automatically (Medonic M-series Hematology Analyzer, Clinical Diagnostic Solutions Inc., Fort Lauderdale, FL, USA). The number of WBCs per slide was utilized in the calculation of the exact amount of blood analyzed, leading to rare-event enumerations presented in events/mL. Cells of interest were further imaged at higher magnification (400×). IF signal expression is categorized as filamentous, diffuse or punctate, as previously described [50].

2.5. Rare Event Detection Approach

Cells were identified via a rare-event detection method termed OCULAR [47,48]. This algorithm uses feature extraction, principal component analysis (PCA) and hierarchical clustering on the principal components to achieve four distinct tasks: fluorescent image feature extraction (761 parameters), rare-event detection (distinguishing between common and rare DAPI-positive events and DAPI-negative events), rare-cell classification and report generation. Image analysis was performed as previously reported [47]. In brief, all events were segmented to generate nuclear and/or cytoplasm masks for feature extraction, which was followed by a dimensionality reduction using principal components and hierarchical clustering to separate common cells (mainly WBCs) and rare cells in each image frame. The manual classification of rare events into CTC or oncosome subgroups was conducted based on biomarker expression in the four fluorescence channels for each IF protocol. Classifications were validated by multiple hematopathologist-trained technical analysts. Previously described as large extracellular vesicles in the context of HDSCA [46,48,51], oncosomes were identified as circular, DAPI-negative events with positive CK expression [44,52,53]. Initially presented as adjacent to nucleated common cells or as individual DAPI-negative events, oncosomes were manually classified and confirmed by trained analysts. The nomenclature for channel-type classifications utilizes those positive channels (Landscape example: CK | Vim = DAPI-positive, CK-positive, Vim-positive, CD45/CD31-negative). Oncosome channel-type classifications are preceded by the abbreviation "Onc" (Landscape example: Onc CK | Vim = DAPI-negative, CK-positive, Vim-positive, CD45/CD31-negative). While rare cell types are predominantly referred to by their channel-type classification, two specific CTC populations, epithelial CTCs (Epi.CTC) and mesenchymal CTCs (Mes.CTC), are evaluated in this study. As previously described [48], Epi.CTCs are CK-positive, Vim-negative and CD45/CD31-negative with a clearly defined nucleus (DAPI). Mes.CTCs are CK-positive, Vim-positive and CD45/CD31-negative and have clearly defined nuclei.

2.6. Multi-Assay Analysis

With sample-matched slides stained by both IF protocols (Landscape and CDX2-targeted), OCULAR's uniform examination of the rare events allowed for a multi-assay analysis. OCULAR derives 761 morphological parameters from IF 100× magnification images of detected cellular events. To perform the multi-assay analysis, we selected a subset of 8 morphometric parameters, consisting of the most representative features. With both IF protocols containing DAPI and CK, the median intensity for these two channels were chosen. In addition, we included the eccentricity and area of both the cell and nucleus, the ratio of nuclear to cellular area and the average distance of the cell outline to the center of the nucleus. Using these 8 shared morphometric features, 5661 rare cells across both assays were grouped together using a hierarchical clustering model. An agglomerative clustering algorithm was used, imported from the scikit-learn library version 0.23.2 [54] in Python. We used a Euclidian metric to compute the distance and the ward linkage criterion. In addition, 111 cells and the oncosome population were removed after manual inspection from the clustering due to highly aberrant nuclear and membrane masking. The cohort's rare-cell population with various cluster assortment options are included as part of an interactive webpage that can be used for additional analyses and discovery (https://pivot.usc.edu/pivot/CRC_MultiAssay.html). The optimal number of clusters from 2 to 16 was selected based on quantitative cluster-separation metrics and the separation of noteworthy, rare cell types. Based on the silhouette average method, 8 clusters provided the optimal separation between the cell groups. When increasing the number of clusters from 8 through 12, cluster 1 is the primary group undergoing rearrangement, but subsequent cluster combinations do not perform as well in the silhouette average metric or in their division of rare cell types of interest. Alongside the clustering approach, a Spearman's rank correlation [55] analysis was performed between the rare-event enumerations of the two IF protocols.

2.7. Survival Analysis

To perform survival analysis on the 9 patients with known progression time, the Spearman's rank correlation coefficient [55] was calculated for all liquid-biopsy analytes versus the PFS. Then, only liquid-biopsy analytes with statistically significant entries were used for subsequent visualization. To visualize the statistically significant entries, the Kaplan–Meier (KM) curves [56] for the patients' PFS were plotted. Then, to depict the effect of a statistically significant liquid-biopsy analyte on PFS, the patients were stratified based on their respective counts per milliliter of blood for the given liquid-biopsy analyte. The stratification was done by using the median counts per milliliter of blood as a threshold, separating the patients into two groups. Finally, the two patient subgroups were plotted together with the original KM curve of the population. The statistical analysis was done using scipy [57], and the KM curves were plotted using scikit-survival library [58] in Python.

2.8. Statistical Analysis

The distinction between the rare-event enumerations of the mCRC and ND samples was determined by a Wilcoxon rank sum test [59]. Statistical correlations between rare-event enumerations were performed using the Spearman's rank correlation coefficient [55]. A correlation was significant if the two-tailed p-value ≤ 0.05. The statistical analysis was done using scipy [57] in Python.

3. Results

3.1. Landscape Rare-Event Detection: Rare Cells and Oncosomes

PB samples were stained with the Landscape IF protocol and analyzed by OCULAR to identify the rare events with biomarkers highlighting epithelial, endothelial, mesenchymal and immune cell origin. Rare-event frequencies, enumerations and sample positivity ($\geq$5 events/mL) from the Landscape and CDX2 IF protocols are reported in Table 1. For the 18 mCRC samples analyzed, on average 0.53 (standard error 0.05, median 0.53, range 0.22–0.97) mL of PB was used for 1 test, thus the sensitivity of the analysis is limited by the blood volume characterized. A representative subset of the rare-cell and oncosome populations detected in the Landscape-stained samples is displayed in Figure 1A, with enumerations, frequencies and comparisons to the ND cohort, and select morphometrics provided in Figure 1B, 1C and 1D, respectively. The sample enumeration of the ND cohort is depicted in Supplemental Figure S1.

Table 1. Rare-event frequencies, enumerations and sample positivity from the Landscape and CDX2-targeted immunofluorescence (IF) protocols. The sample positivity threshold of ≥5 events/mL was determined by comparisons to the rare-event enumerations of a randomly selected normal donor cohort. The frequency of each classification is provided as a percentage of the total rare-event profile for each IF protocol.

IF Protocol	Event Classification	Sample Positivity	Mean (Events/mL)	Standard Error (±Events/mL)	% of Total Rare Events	Median (Events/mL)	Range (Events/mL)
Landscape	DAPI only	15/18	13.69	2.9	3.55	10.84	0.00–51.57
	CK (Epi.CTC)	18-Jul	57	36.36	14.77	3.55	0.00–549.63
	Vim	18-Nov	6.64	1.27	1.72	7.3	0.00–20.16
	CD45/CD31	18-Oct	10.29	2.44	2.67	8.05	0.00–33.16
	CK I Vim (Mes.CTC)	18-Jun	7.6	1.4	1.97	1.4	0.00–91.29
	CK I CD45/CD31	18-Apr	4.66	1.86	1.21	2.13	0.00–30.74
	Vim I CD45/CD31	18-Nov	24.52	8.73	6.35	7.79	0.00–121.00
	CK I Vim I CD45/CD31	15/18	100.4	35.02	26.01	12.89	0.00–453.13
	Onc CK	18-Dec	71.59	38.39	18.55	6.64	1.06–657.60
	Onc CK I Vim	18-Oct	32.72	13.64	8.48	7.38	0.00–217.65
	Onc CK I CD45/CD31	0/18	1.47	0.36	0.38	1.12	0.00–4.47
	Onc CK I Vim I CD45/CD31	14/18	55.37	16.22	14.35	34.32	0.00–268.32
CDX2-targeted	DAPI only	18/18	63.72	12.8	8.33	45.29	16.94–226.94
	CK	18-Dec	88.14	43.95	11.53	11.97	0.00–597.32
	CDX2	18-Dec	11.45	3.28	1.5	7.27	1.04–60.22
	CD45	18-Mar	5.44	3.35	0.71	0	0.00–59.22
	CK I CDX2 (CDX2.CTC)	14/18	19.95	6.71	2.61	11.34	0.00–124.08
	CK I CD45	18-Dec	36.66	13.71	4.79	10.02	0.00–203.63
	CDX2 I CD45	18-Dec	29.37	12.46	3.84	9.8	0.00–185.68
	CK I CDX2 I CD45	14/18	143.67	101.35	18.79	21.79	1.44–1843.34
	Onc CK	15/18	123.55	60.73	16.16	26	2.31–1035.65
	Onc CK I CDX2	18/18	222.4	65.04	29.08	114.87	15.55–1151.64
	Onc CK I CD45	0/18	0.06	0.06	0.01	0	0.00–1.06
	Onc CK I CDX2 I CD45	18-Oct	20.27	7.64	2.65	14.5	0.00–1138.28

For all mCRC samples, total rare-event (total cells and oncosomes) detection had a median of 287.46 (mean 387.74 ± 75.33) events/mL. For ND samples, total rare-event detection had a median of 40.05 (mean 49.96 ± 4.18) events/mL. A significant difference was observed between the mCRC patients and ND ($p < 0.0001$; Figure 1D).

Rare cells comprised 58.25% of the total rare-event profile from the Landscape-stained mCRC samples. The rare cells detected in mCRC patient samples were highly heterogenous in their signal expression and morphology (Figure 1E,F). Total rare-cell detection for the mCRC samples had a median of 124.66 (mean 224.80 ± 51.55) cells/mL. The ND samples presented with a median rare-cell detection of 34.46 (mean 43.21 ± 3.94) cells/mL. A significant difference in total rare-cell detection was observed between the mCRC patients and ND samples ($p = 0.0112$).

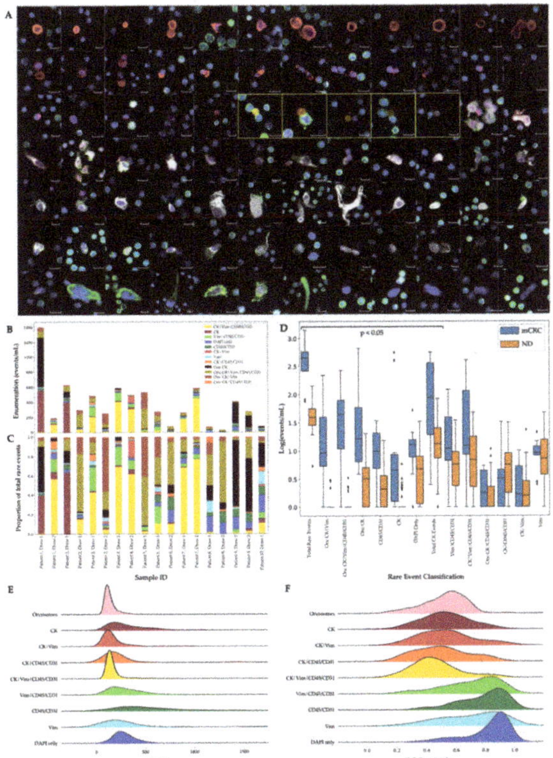

Figure 1. Landscape-stained samples analyzed by OCULAR. (**A**) Representative gallery from the metastatic colorectal cancer (mCRC) cohort showing the morphological heterogeneity of the detected rare events. DAPI: blue, cytokeratin (CK): red, Vim: white, CD45/CD31: green. Events from each of the 10 patients are represented in the gallery. The five oncosomes displayed are bordered by yellow boxes. Events are ordered by decreasing CK signal intensity. Images taken at 400× magnification. Scale bars represent 10 µm. (**B**) Rare-event enumeration (events/mL) and (**C**) frequency (%) per patient and draw. (**D**) Enumeration comparison of Draw 1 mCRC and normal donor samples ordered by statistical significance. Symbols indicate outliers. The first 8 classifications from the left are different between mCRC and normal donors ($p < 0.05$). (**E**) Cellular area and (**F**) cellular eccentricity per rare-event classification detected in the mCRC cohort.

Total CK-positive events were detected with a median of 45.41 (mean 169.66 ± 46.65) events/mL from all mCRC samples. The ND samples had a median of 12.39 (mean 18.96 ± 2.70) CK-positive events/mL. There was a statistically significant difference in total CK-positive event detection between the mCRC patients and ND samples ($p = 0.0070$). The total CK-positive cell population constituted 75.47% of all the rare cells. Epi.CTCs were detected with a median of 3.55 (mean 57.00 ± 36.36) cells/mL from mCRC patient samples, which is a significantly higher incidence compared to ND samples ($p = 0.0023$). Epi.CTCs were only identified in 7 of 18 mCRC samples, with 2 samples containing 92.58% of the total cohort's population. This speaks to the limited frequency of CK-only CTCs in the liquid biopsy of mCRC, here using approximately 0.5 mL of PB. Similarly, Mes.CTCs were only found in 6 of 18 samples and with a median of 1.40 (mean 7.60 ± 1.40) cells/mL in the mCRC cohort. There was not a significant difference in the Mes.CTC count between the mCRC and ND samples. The CK | Vim | CD45/CD31 cells were the most frequent rare cell across the mCRC cohort, but no statistically significance difference was detected between the mCRC and ND samples.

Other detectable rare cells in the mCRC samples included morphologically distinct CD45/CD31-only (median 8.05; mean 10.29 ± 2.44), DAPI-only (median 10.84; mean 13.69 ± 2.90), Vim-only (median 7.30; mean 6.64 ± 1.27) and Vim | CD45/CD31 (median 7.79, mean 24.52 ± 8.73) cells/mL. The DAPI-only cells were detected at a higher prevalence in mCRC samples compared to the ND samples ($p = 0.0047$).

Two morphologically distinct cell types were identified as unique subsets of their broader channel-type classifications. The first was the large, morphologically distinct CD45/CD31 cell population shown in Figure 2A. As a subset of the CD45/CD31 channel-type classification, this cell type was found in 9 of the cohort's 18 samples (Figure 2B). Their distinction highlights the importance of morphological analysis that goes beyond channel-type classifications. The cells of interest possess a punctate CD45/CD31 signal, which is distinct from the diffuse CD45/CD31 signal that typifies the surrounding WBCs. Notable morphometrics quantitatively distinguish this cell group from common WBCs with a larger cell size, larger nuclear size, and higher cellular eccentricity. Even within this category, there is heterogeneity, evidenced by the varying shapes and the nuclear-to-membrane ratios of the cells. These cells are hypothesized to be megakaryocytes.

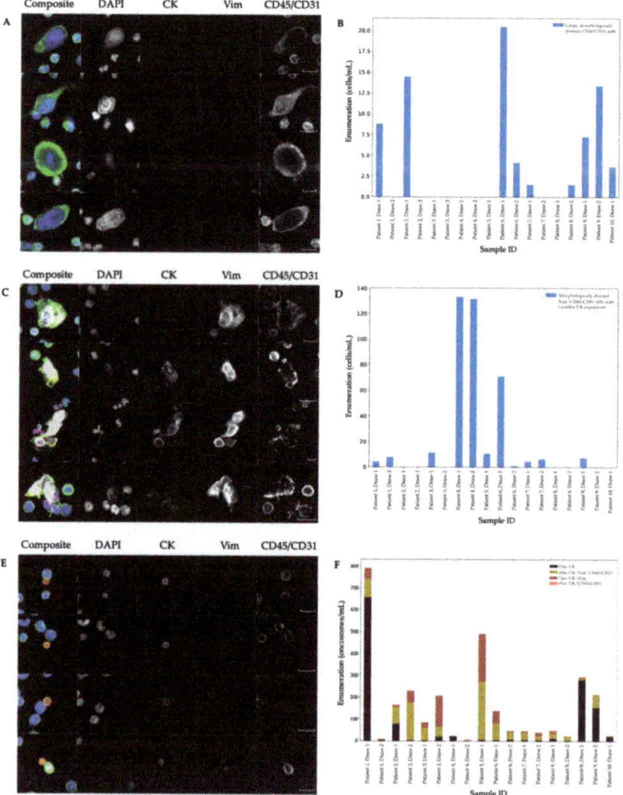

Figure 2. Select rare-event populations in the Landscape-stained metastatic colorectal cancer (mCRC) samples. (**A**) Panel gallery images of the large, morphologically distinct CD45/CD31 cells. (**B**) Large, morphologically distinct CD45/CD31 cell enumeration per patient and draw. (**C**) Panel gallery images of the morphologically distinct Vim | CD45/CD31 cells with variable cytokeratin (CK) expression. (**D**) Morphologically distinct Vim | CD45/CD31 cells with variable CK expression enumeration per patient and draw. (**E**) Panel gallery images of the oncosome population with differential signal expression (**F**) Oncosome enumeration by channel classifications per patient and draw. Images are taken at 400× magnification. Scale bars represent 10 μm.

The second rare-cell group of interest consists of morphologically distinct Vim | CD45/CD31 cells with variable CK expression, as shown in Figure 2C. These cells are found in 11 of 18 samples and across both draws (Figure 2D). The samples from Patient 4 contain the vast majority (81.37%) of the morphologically distinct Vim | CD45/CD31 cells detected in this cohort. The image analysis of these cells revealed a filamentous Vim signal along with a punctate CD45/CD31 signal. The variable CK expression in signal intensity and appearance suggests that there are additional subtypes within this cell type. A total of 45.60% of the total morphologically distinct Vim | CD45/CD31 cell population is CK-positive. Further observation highlights heterogeneous CK expression within the CK-positive subtype, epitomized by the punctate and filamentous signal on the second and third cell of Figure 2C respectively. Beyond the analysis of the IF signal, 35.95% of the morphologically distinct Vim | CD45/CD31 cells found in this cohort are found clustered near one another (bottom of Figure 2C). Additional key morphometrics that distinguish this cell population from surrounding WBCs include a large cell size and eccentric cellular membrane. These cells are hypothesized to be endothelial cells.

In addition to the rare-cell groups, OCULAR identified a sizeable population of oncosomes. Oncosomes accounted for 41.75% of the rare events in the Landscape-stained samples. Morphologically, these vesicles ranged up to the size of neighboring WBCs (~10 μm) and were present in the cellular fraction of blood after centrifugation. As Figure 2E depicts, these vesicles were found in contact with adjacent nucleated cells and in isolation, with 51.09% of this cohort's oncosome population belonging to the latter. Furthermore, their IF signal was diffuse, suggesting an evenly distributed expression across the vesicle. The oncosomes expressing CK were generally the most prevalent (Figure 2F). As the most common subtype, all 18 samples were positive for Onc CK with a median of 6.64 (mean 71.59 ± 38.39) events/mL. The Onc CK | Vim | CD45/CD31 was positive in 16 of 18 samples and had a median of 34.32 (mean 55.37 ± 16.22) events/mL. The Onc CK | Vim was also present in 16 of 18 samples, with a median of 7.38 (mean 32.72 ± 13.64) events/mL. The Onc CK | Vim | CD45/CD31 and Onc CK | Vim counts were found to be highly positively correlated ($p = 0.002$, $\tau = 0.68$).

In comparison to the mCRC and ND samples, 6 specific channel-type rare-event classifications were statistically distinct across the cohorts (Figure 1D). Three of the significantly different channel-type classifications were rare cells that were detected at a higher prevalence in mCRC samples compared to the ND: Epi.CTC ($p = 0.0023$), DAPI only ($p = 0.0494$) and CD45/CD31 ($p = 0.0004$). Three oncosome channel-type classifications were observed at greater numbers in the mCRC patient samples compared to the ND samples: Onc CK ($p = 0.0001$), Onc CK | Vim ($p < 0.0001$) and Onc CK | Vim | CD45/CD31 ($p < 0.0001$).

3.2. Analysis of the CDX2-Targeted Protocol

To complement the vast heterogeneity of rare events across the epithelial, mesenchymal and endothelial cell types presented in the Landscape protocol, the CDX2-targeted protocol was utilized to specifically interrogate and identify circulating rare events of colorectal origin. Slides from the same PB tubes were stained with the CDX2-targeted protocol and analyzed by OCULAR, allowing for a sample-matched study design across two IF assays. All rare-event frequencies, enumerations and sample positivity (≥5 events/mL) from the CDX2-targeted protocol are reported in Table 1. A representative subset of the rare-cell and oncosome populations are displayed in Figure 3A, with enumerations and frequencies of the channel-type classifications in Figure 3B,C. Further analysis of the CDX2-targeted cohort by HDSCA's first generation CK-focused approach is depicted in Supplemental Table S1, Figures S2 and S3.

Rare cells comprised 52.10% of the total rare-event profile from the CDX2-targeted protocol. From the OCULAR analysis, the total CK-positive cell population across all 18 samples had a median of 108.74 (mean 288.43 ± 119.17) cells/mL. The image analysis of the CK-positive cell groups revealed extensive heterogeneity between and within the channel-type classifications. The consistency of the IF signal across the cells of interest varies within categories. Morphological differences were also observed between cells of the same channel type.

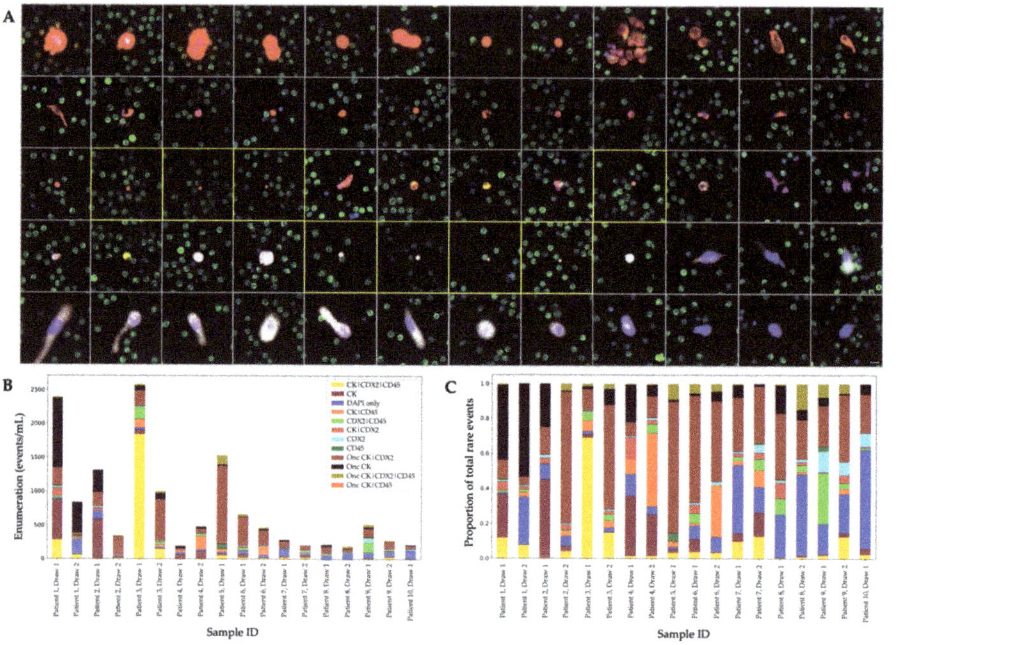

Figure 3. CDX2-targeted samples analyzed by OCULAR. (**A**) Representative rare-event gallery. DAPI: blue, cytokeratin (CK): red, CDX2: white, CD45: green. Events from each of the 10 patients are represented in the gallery. The eight oncosomes displayed are bordered by yellow boxes. Events are ordered by decreasing CK intensity. Images taken at 100× magnification. Scale bar for all images is shown in the bottom right cell, representing 10 μm. (**B**) Enumeration (events/mL) of each channel classification per patient and draw. (**C**) Frequency (%) of each channel classification per patient and draw.

With the added CDX2 marker, we were able to evaluate CTCs for their potential colorectal origin. CDX2-positive CTCs or CDX2.CTCs (also referred to as CK | CDX2 in Table 1) were found in 14 of 18 samples, but not in high frequencies, only constituting 2.52% of the total rare-event profile. The first three images from the top left of Figure 3A are representative of this cell type, and, as the gallery shows, these CDX2.CTCs were also found clustered together.

Beyond the CK-positive populations, additional rare cell types were prominent across the CDX2-stained samples: DAPI-only, CDX2-only, CD45-only and CDX2 | CD45 classifications (Table 1). The CK-negative rare-cell population detected by OCULAR was positive in all 18 samples, with a median of 87.13 (mean 109.98 ± 18.58) cells/mL. Among this group, a subset of large and eccentric cells with punctate CDX2 expression and sizeable nuclei (see the bottom row of Figure 3A) were found in 11 of the 18 samples. Lastly, all 18 of samples were positive for DAPI-only rare cells, with a median of 45.29 (mean 63.72 ± 12.80) cells/mL using the CDX2-targeted protocol.

3.3. Multi-Assay Analysis

A multi-assay comparison was conducted on samples stained both with the CDX2-targeted and Landscape IF protocols. An analysis of the rare-event enumerations from both staining protocols with matched samples revealed various positive and negative correlations between channel-type classification counts (Figure 4A). Specific rare cell types detected by both IF protocols were positively correlated, such as the DAPI-only cells ($p = 0.032$, $\tau = 0.51$) and the CK-only cells ($p = 0.018$, $\tau = 0.55$). Similar positive correlations were found among the aggregate classifications, including the oncosomes ($p = 0.003$, $\tau = 0.65$) and total CK-positive rare-event population ($p = 0.006$, $\tau = 0.62$). A positive association between rare events that differ by one unshared biomarker between the stains potentially indicates a single rare-event type. An example of this is the CK-only cell counts from the Landscape-staining protocol being positively associated with the CK | CDX2 cells from the CDX2 IF protocol ($p = 0.029$, $\tau = 0.51$). A similar pattern was observed within the oncosome population, with the Onc CK | Vim being positively correlated to the Onc CK | CDX2 ($p < 0.001$, $\tau = 0.82$). The significant negative correlations across the assays included the DAPI-only cells from Landscape-staining protocol and the CK | CD45 cells from the CDX2 IF protocol ($p = 0.029$, $\tau = -0.51$). Interestingly, if selected for rarity, the CD45/CD31 population from the Landscape-staining protocol was uncorrelated with the CD45 population from the CDX2-staining protocol ($p = 0.645$, $\tau = 0.12$), indicating that the differential IF expression was likely due to the addition of the CD31 biomarker.

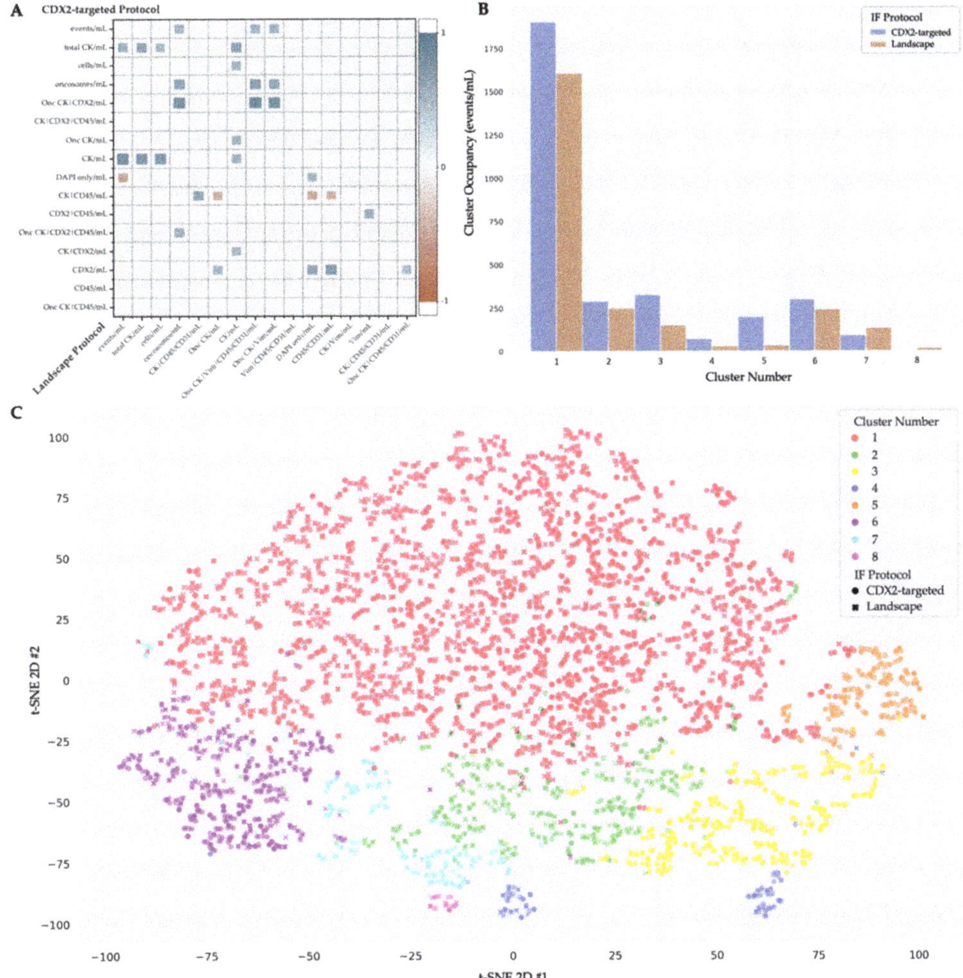

Figure 4. *Cont.*

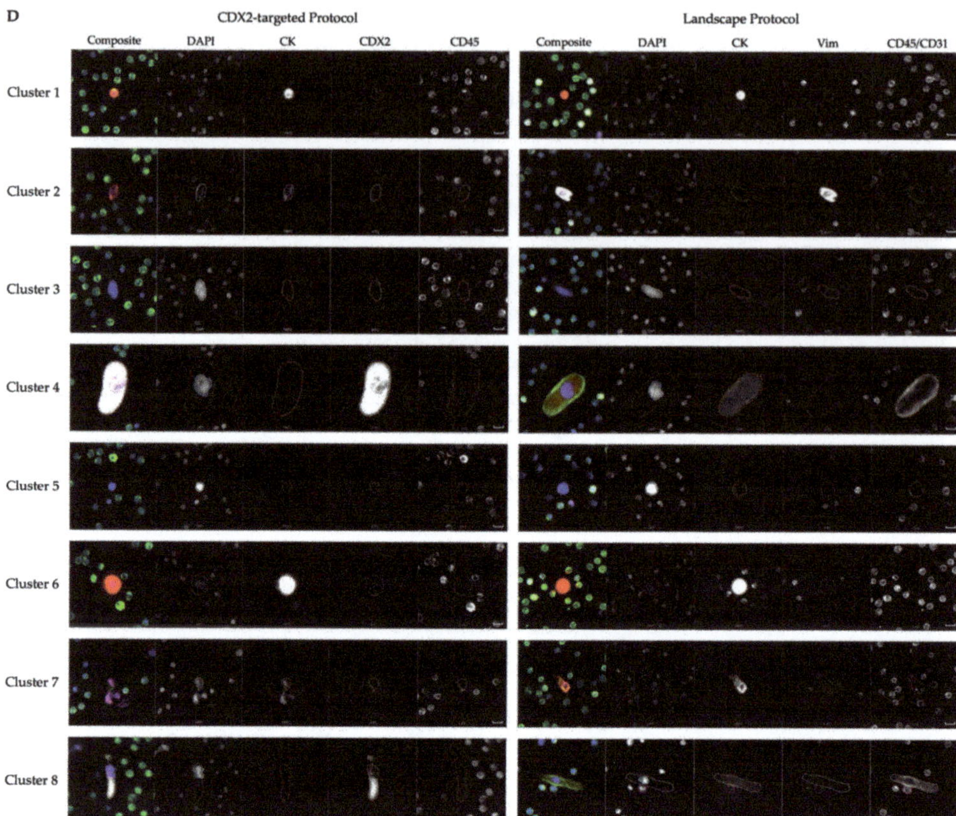

Figure 4. Multi-assay analysis of the Landscape and CDX2-targeted immunofluorescence (IF) assays analyzed by OCULAR. (**A**) Statistically significant ($p \leq 0.05$) rare-event-count correlations across the two IF assays, with the red-to-blue color gradient indicating a negative-to-positive correlation, respectively. (**B**) Cluster occupancy of the rare cells identified by the CDX2-targeted and Landscape stains when using the 8-group hierarchical clustering model. (**C**) t-Stochastic neighbor embedding (t-SNE) plot of the 8 cell clusters comprised of rare cells from both stains, as indicated by the markers' shape. (**D**) Representative gallery of the 8 clusters with cells from both assays. Each row represents a cluster with a cell from the CDX2-targeted protocol on the left and a cell from the Landscape protocol on the right. Images taken at 100× magnification. Scale bars represent 100 µm.

OCULAR presents morphometrics related to the size and shape of the cell and nucleus. A clustering analysis of the morphometric features was conducted to characterize the rare-event types across the two IF staining protocols, improving our understanding of the analytes in liquid biopsy. Hierarchical clustering into eight groups afforded the most discrete separation of the hypothesized megakaryocytes and endothelial cells. It is important to note that these highlighted rare cell types are not entirely separated into their own clusters, indicating the morphological heterogeneity within the cell categories. The distribution of cells from both IF staining protocols into the eight clusters and representative images are provided in Figure 4B–D. Table 2 provides a description of each cluster.

Table 2. Cellular description for the multi-assay cluster analysis. Cluster occupancy of the rare cells identified by the CDX2-targeted and Landscape staining protocols when using the 8-group hierarchical clustering model for multi-assay analysis.

Cluster Number	Cluster Comments	Cells from Landscape	Comments from Landscape	Cells from CDX2-Targeted	Comments from CDX2-Targeted
1	Heterogeneous phenotype with cellular morphology similar to WBCs	1605	Most prominent: CK I Vim I CD45/CD31	1903	Most prominent: CK I CDX2 I CD45
2	Includes endothelial cells	249	Morphologically distinct Vim I CD45/CD31 cells with variable CK expression	291	64 (22%) DAPI-only, 67 (23%) CK and 34 (12%) CK I CDX2 cells
3	Large nuclei, more eccentric than cluster 5	152	101 (66%) DAPI only, 11 (7%) morphologically distinct CD45/CD31 expressing cells	328	279 (85%) DAPI-only
4	Includes megakaryocytes	30	25 (83%) morphologically distinct CD45-/CD31-expressing cells, 5 (17%) small rod-like CD45-/CD31-expressing cells	72	DAPI-only and CDX2-only with similar large morphology
5	Large nuclei, more circular than cluster 3	34	15 (44%) DAPI-only	201	191 (95%) DAPI-only
6	CK only CTCs	242	219 (90%) Epi.CTCs	302	265 (88%) CK and 29 (10%) CK I CDX2 cells
7	Includes endothelial cells	134	84 (63%) morphologically distinct Vim I CD45/CD31 cells with variable CK expression	93	80 (86%) CK only. Morphologically distinct from cluster 6, more elongated.
8	Includes megakaryocytes	18	Morphologically distinct CD45/CD31 cells with variable CK and Vim expression	6	Large, morphologically distinct cells with punctate CDX2 expression

3.4. Clinical Correlation of Liquid-Biopsy Data

Next, we investigated the clinical relevance of the rare events detected in the liquid biopsy. PFS was reported for nine of the ten patients (mean PFS = 6.98 months). A survival analysis of the rare events identified by the Landscape assay in Draw 1 samples revealed a positive correlation between the number of Onc CK I CD45/CD31 and PFS ($p = 0.0372$, $\tau = 0.70$), as patients with $\geq$2.21 oncosomes/mL had an improved survival (Figure 5A). A similar analysis of the populations identified by the CDX2-targeted assay in Draw 2 samples revealed two rare-event types negatively associated with PFS: CK I CD45 cells ($p = 0.0362$, $\tau = -0.77$) and Onc CK ($p = 0.0068$, $\tau = -0.89$). Patients with $\geq$20.61 CK I CD45 cells/mL or $\geq$11.57 Onc CK/mL had poor survival (Figure 5B).

Beyond timepoint-static enumerations, the rare-event kinetics (Figure 5C,D), defined as the change in rare-event subtypes between draws, were significantly associated with PFS. Of the nine patients with survival data, two were missing Draw 2; therefore, only seven patients were included in the kinetics survival analysis. The changes in the CK I CDX2 I CD45 cells ($p = 0.0068$, $\tau = 0.89$) and the Onc CK I Vim I CD45/CD31 ($p = 0.0025$, $\tau = 0.93$) were positively associated with PFS (Figure 5E,F).

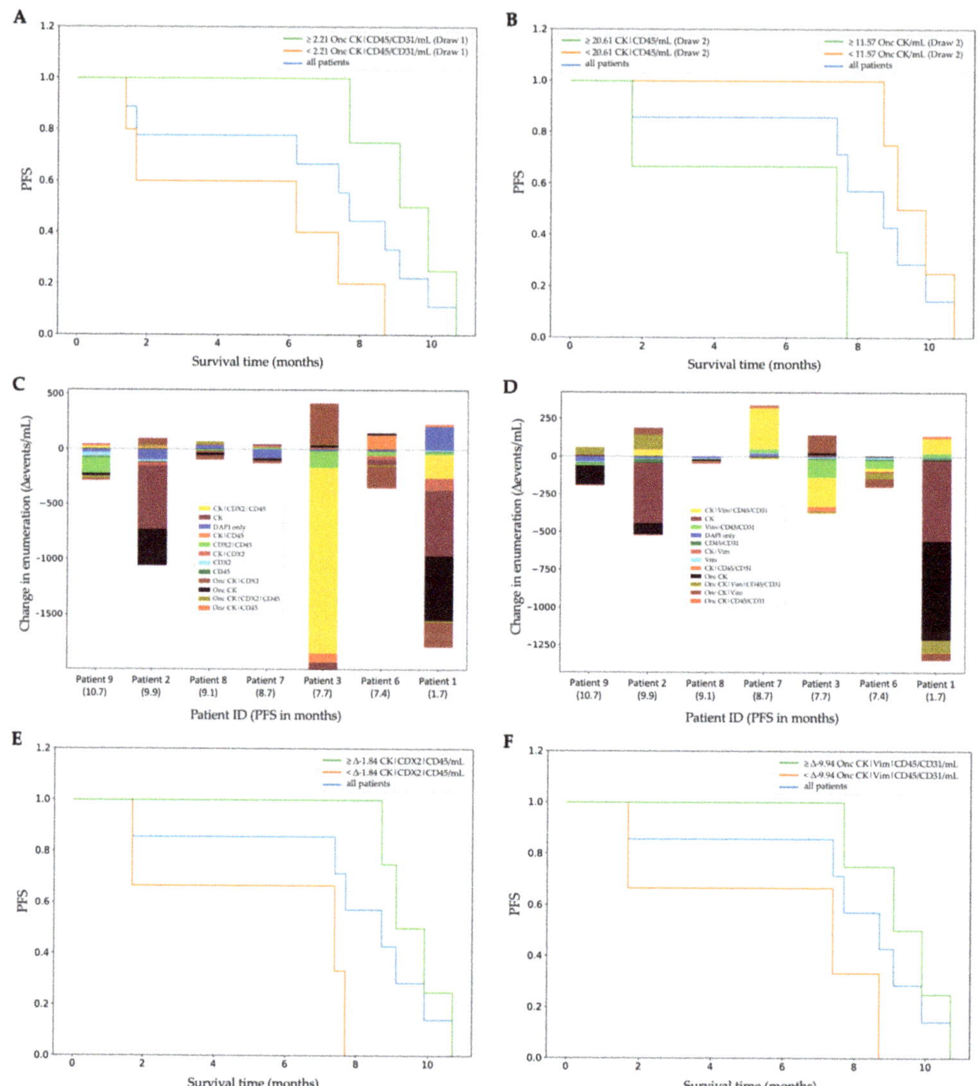

Figure 5. Survival analysis of rare events from the Landscape- and CDX2-targeted immunofluorescence protocols. (**A**) Kaplan-Meier (KM) curve showing that patients with more than the median of 2.21 Onc CK | CD45/CD31/mL found in Draw 1 by the Landscape protocol had longer progression-free survival (PFS). (**B**) KM curve showing two rare events with survival implications found in Draw 2 by the CDX2-targeted protocol. Patients with more than the median of 20.61 CK | CD45 cells/mL or more than the median of 11.57 Onc CK/mL had shorter PFS. (**C**) Rare-event kinetics between Draw 1 and Draw 2 analyzed with the CDX2-targeted protocol, ordered from longest to shortest PFS. (**D**) Rare-event kinetics between Draw 1 and Draw 2 analyzed with the Landscape protocol, ordered from longest to shortest PF€(**E**) KM curve showing that patients with a change of ≥−1.84 CK | CDX2 | CD45 cells/mL from the CDX2-targeted protocol had longer PFS. (**F**) KM curve showing that patients with a change of ≥−9.94 Onc CK | Vim | CD45/CD31/mL from the Landscape protocol had longer PFS.

4. Discussion

As a minimally invasive approach, the liquid biopsy has the potential to significantly advance patient care by addressing current clinical challenges in mCRC. In this study, we show a comprehensive profile of the liquid biopsy that encapsulates the heterogeneous CTC and oncosome populations while providing the initial clinical validation of the liquid biopsy in mCRC patients.

This study corroborates findings of a limited CK-only CTC population in mCRC [25–27]. mCRC is known for its extensive solid tumor heterogeneity, and herein, we show interpatient and intrapatient heterogeneity in the circulating rare-event population. The Landscape protocol aided in the phenotypic characterization of morphologically distinct cell types, allowing for the detection of new rare-cell populations, such as circulating endothelial cells (CECs). Initially identified in the 1970s [60], CEC counts have been shown to be elevated in the liquid biopsy of cancer patients [61]. In mCRC, tumor-derived CECs have been used as a prognostic indicator of clinical response to first-line therapies [62] and patient survival [41,42]. Interestingly, a previous investigation targeting CTCs in CRC instead discovered tumor-derived CEC clusters that starkly differentiated normal donors, treatment-naïve and early-stage patients [63]. The detection of clinically significant CECs while attempting to find CTCs highlights the importance of using an unbiased rare-cell approach to capturing the heterogeneity in the liquid biopsy of mCRC. In this study, the morphologically distinct Vim | CD45/CD31 cells with variable CK expression were found individually and in clusters. Based on their morphology and biomarker expression, we expect that these cells are CECs. While downstream proteomics and genomics are needed to confirm this cell lineage, the filamentous Vim and punctate CD45/CD31 signal captured by the Landscape IF protocol are characteristic of endothelial cells [64,65]. HDSCA has previously identified CECs in the liquid biopsy, the likes of which present a similar morphology and biomarker expression pattern [33].

We additionally identified large, morphologically distinct CD45/CD31 cells in the mCRC samples that appear to be multilobular with sizeable nuclei and punctate cytoplasmic expression. Such cellular features lead us to hypothesize that these events are megakaryocytes. These platelet-producing cells derived from the bone marrow are positive in CD31, have a granular cytoplasm and are large, with up to a 160 μm diameter [66]. Found either in the solid tumor microenvironment or in circulation, megakaryocytes have shown to have prognostic potential in prostate [67] and non-small-cell lung cancer [68]. While platelet indices have served as diagnostic biomarkers in mCRC [69], the novel identification of potential megakaryocytes in this cancer type could have significant clinical implications even in the absence of direct platelet detection. Future studies should characterize these megakaryocyte candidates with downstream proteomics to confirm their lineage and to compare their enumerations to patient-specific clinical factors and survival. This unique cell type further exemplifies the heterogeneity of the circulating cell profile of mCRC and highlights the power of rare-event detection systems to represent it.

This study demonstrates that the utility of the liquid biopsy is not limited to cellular events, emphasizing the importance of detecting vesicles. Vesicles have been described and classified according to the mechanism of cellular release and size, which may be dependent on the method of detection or isolation [44,70]. Oncosomes comprised a significant portion of the rare-event profiles from both the CDX2-targeted and Landscape assays in mCRC patient samples. HDSCA has previously identified oncosomes with similar size and biomarker expression in prostate cancer [46], bladder cancer [48] and upper tract urothelial carcinoma [51], suggesting that these events may be found in a variety of cancer types. Most importantly, oncosome enumerations from both IF protocols correlated with PFS in this cohort. The kinetics of a Landscape-stained oncosome population also correlated significantly with PFS, pointing to the importance of enumerating these analytes over time. The potential for making diagnoses, prognoses and subsequent treatment decisions based off oncosomes, especially in a cancer type like mCRC that does not widely present CK-only CTCs, warrants further studies to molecularly characterize these events.

Herein, we show that oncosomes associated with mCRC tumorigenesis may be useful prognostic biomarkers.

Predicated on this study's sample-matched design, the multi-assay analysis is a novel attempt to overlap IF protocols using the shared features of an unbiased rare-event detection platform. The successful grouping of known cell types into delineated clusters, such as the Epi.CTCs in cluster 6, serves as a proof of concept for applications in a larger cohort. Using only image-based morphometrics, the multi-assay analysis increases the number of evaluable IF biomarkers for a single sample, while maintaining a low monetary cost relative to single-cell proteomics. Cluster 4 best depicts the value of additional biomarkers in understanding the cellular biology. Using the Landscape assay, we detected megakaryocytes with a CD45/CD31-positive signal, while in the CDX2-targeted assay, we detected megakaryocyte-like cells that presented a CDX2-positive but CD45-negative signal. The shared feature analysis allowed for the identification of these cells as similar and the observation of CDX2 antibody binding to these cells. A multi-assay approach is uniquely suited for rare-event analysis, wherein thresholds from multiple IF protocols conducted in tandem could demonstrate the clinical utility of the liquid biopsy.

Phenotypic switches are fundamental to CRC initiation, metastasis and relapse [71–73], thus requiring longitudinal prognostic tools and changing therapies to target the continuously evolving cell types. The minimally invasive liquid biopsy has the potential to address this challenge, and we show initial evidence for clinical utility. The molecular characterization of the rare events detected in this study will elucidate their potential role in mCRC tumorigenesis. The HDSCA3.0 workflow includes the capability for genomic analysis, both SNV (single-nucleotide variation) and CNV (copy-number variation), for both single cells and cell-free DNA (cfDNA) [74–76], as well as targeted multiplexed proteomic analysis [38,75] on samples previously characterized at the morphological and phenotypic level by IF. The data presented here serves to motivate further genomic and/or proteomic analysis of the CTCs and oncosomes detected to validate their neoplastic origin and association with the disease state.

Additional studies are needed with a greater patient sample size, PB draws from multiple timepoints throughout treatment and patient-specific clinical information (KRAS/NRAS status, TNM staging, etc.) to provide the prognostic, diagnostic and predictive utility of the liquid biopsy in the management of mCRC.

5. Conclusions

As one of the world's most prevalent oncological diseases, mCRC poses numerous clinical challenges due to its extensive tumor heterogeneity. This study establishes evidence for the clinical validation of an unbiased rare-event approach to the liquid biopsy. For the first time, we demonstrate the value of a comprehensive CTC and oncosome detection approach in the PB of mCRC. Our results highlight the heterogeneity of the liquid-biopsy profile with the identification of rare-event frequencies unique to mCRC patients. By analyzing both IF expression patterns and morphological parameters, we identify two select rare cell types that warrant future study into their implications in mCRC. Furthermore, we demonstrate the utility of analyzing multiple IF assays in tandem to characterize the heterogeneous populations detected. These findings motivate the further molecular characterization of these analytes and investigation into their predictive power with respect to patient outcomes in mCRC.

Supplementary Materials: The following supporting information can be downloaded at: https://www.mdpi.com/article/10.3390/cancers14194891/s1, Figure S1. Enumeration of the rare events detected in the Landscape-stained ND samples analyzed by OCULAR. Table S1. Rare-event frequencies, enumerations, and sample positivity from the enrichment-based and OCULAR analysis of the samples stained with the CDX2-targeted assay. The sample positivity threshold of $\geq$5 events/mL was determined by comparisons to the rare-event enumerations of an ND cohort. Within each detection approach, the frequency of each classification is provided as a percentage of the total rare-event profile. Figure S2. CK-positive rare events from the CDX2-stained samples analyzed by OCULAR.

(A) Cells positioned on the edge of frames that are intentionally removed by the quality-control process in OCULAR, representative of the 2.04% non-concordance group between the two detection approaches. Frame edges are indicated by the dashed white lines. (B) CK-positive cells, containing representative HD-CTC, CTC-Apoptotic and two CTC-Small candidates as per the CK-focused approach. (C) CK | CDX2 cells, containing a CTC cluster, cells with a dim CK signal, and eccentric rare cells. (D) CK | CD45 cells with varying CD45 expression. (E) CK | CDX2 | CD45 cells. (F) Heterogenous CK positive oncosomes. DAPI: blue, CK: red, CDX2: white, CD45: green. Images taken at 100× magnification. Scale bars represent 10 µm. Figure S3. CDX2-stained samples rare-event enumeration using HDSCA1.0. (A) Enumeration (CTCs/ml) of each CTC subtype using the CTC-focused approach per patient and draw. (B) Frequency (%) of each CTC subtype using the CTC-focused approach per patient and draw.

Author Contributions: Conceptualization, S.N., G.C., J.M. and S.N.S.; methodology, S.N., G.C., J.M. and S.N.S.; validation, S.N., G.C., J.M. and S.N.S.; formal analysis, S.N., G.C., J.M., A.N., D.K. and S.N.S.; patient accrual, S.D.P.; resources, P.K.; data curation, S.N., G.C., J.M. and S.N.S..; writing—original draft preparation, S.N. and S.N.S.; writing—review, all authors; visualization, S.N. and G.C.; supervision, J.M., P.K. and S.N.S.; project administration, J.M., P.K. and S.N.S.; funding acquisition, P.K. All authors have read and agreed to the published version of the manuscript.

Funding: This work received institutional support from USC Michelson Center Convergent Science Institute in Cancer, Gilead Sciences, Inc., Vassiliadis Research Fund, Vicky Joseph Research Fund and Susan Pekarovics.

Institutional Review Board Statement: The study was conducted according to the guidelines of the Declaration of Helsinki and approved by the Institutional Review Board at each clinical site under NTC01803282.

Informed Consent Statement: Informed consent was obtained from all subjects involved in the study.

Data Availability Statement: All data discussed in this manuscript are included in the main manuscript text or Supplementary Materials. Some of the data can be accessed through our website http://pivot.usc.edu/. The imaging data is available through the BloodPAC Data Commons Accession ID "BPDC000124" (https://data.bloodpac.org/discovery/BPDC000124).

Acknowledgments: We thank the patients and their caregivers who consented to this study. We also thank the clinical research staff who contributed to the study. We are grateful to past and current technical staff at CSI-Cancer for the processing of the samples.

Conflicts of Interest: The HDSCA technology described herein is licensed to Epic Sciences. P.K. has ownership in Epic Sciences. All other authors declare no conflict of interest.

References

1. Siegel, R.L.; Miller, K.D.; Goding Sauer, A.; Fedewa, S.A.; Butterly, L.F.; Anderson, J.C.; Cercek, A.; Smith, R.A.; Jemal, A. Colorectal cancer statistics, 2020. *CA A Cancer J. Clin.* **2020**, *70*, 145–164. [CrossRef] [PubMed]
2. Barker, N.; Bartfeld, S.; Clevers, H. Tissue-resident adult stem cell populations of rapidly self-renewing organs. *Cell Stem Cell* **2010**, *7*, 656–670. [CrossRef] [PubMed]
3. Fontana, E.; Nyamundanda, G.; Cunningham, D.; Tu, D.; Cheang, M.C.; Jonker, D.J.; Siu, L.L.; Sclafani, F.; Eason, K.; Ragulan, C. Intratumoral transcriptome heterogeneity is associated with patient prognosis and sidedness in patients with colorectal cancer treated with anti-EGFR therapy from the CO. 20 trial. *JCO Precis. Oncol.* **2020**, *4*, 1152–1162. [CrossRef]
4. Suzuki, Y.; Ng, S.B.; Chua, C.; Leow, W.Q.; Chng, J.; Liu, S.Y.; Ramnarayanan, K.; Gan, A.; Ho, D.L.; Ten, R. Multiregion ultra-deep sequencing reveals early intermixing and variable levels of intratumoral heterogeneity in colorectal cancer. *Mol. Oncol.* **2017**, *11*, 124–139. [CrossRef]
5. Moorcraft, S.Y.; Smyth, E.C.; Cunningham, D. The role of personalized medicine in metastatic colorectal cancer: An evolving landscape. *Ther. Adv. Gastroenterol.* **2013**, *6*, 381–395. [CrossRef] [PubMed]
6. Molinari, C.; Marisi, G.; Passardi, A.; Matteucci, L.; De Maio, G.; Ulivi, P. Heterogeneity in colorectal cancer: A challenge for personalized medicine? *Int. J. Mol. Sci.* **2018**, *19*, 3733. [CrossRef]
7. Kyrochristos, I.D.; Roukos, D.H. Comprehensive intra-individual genomic and transcriptional heterogeneity: Evidence-based Colorectal Cancer Precision Medicine. *Cancer Treat. Rev.* **2019**, *80*, 101894. [CrossRef] [PubMed]
8. Chen, D.; Xu, T.; Wang, S.; Chang, H.; Yu, T.; Zhu, Y.; Chen, J. Liquid Biopsy Applications in the Clinic. *Mol. Diagn. Ther.* **2020**, *24*, 125–132.

9. Esposito, A.; Criscitiello, C.; Locatelli, M.; Milano, M.; Curigliano, G. Liquid biopsies for solid tumors: Understanding tumor heterogeneity and real time monitoring of early resistance to targeted therapies. *Pharmacol. Ther.* **2016**, *157*, 120–124. [CrossRef]
10. Russano, M.; Napolitano, A.; Ribelli, G.; Iuliani, M.; Simonetti, S.; Citarella, F.; Pantano, F.; Dell'Aquila, E.; Anesi, C.; Silvestris, N. Liquid biopsy and tumor heterogeneity in metastatic solid tumors: The potentiality of blood samples. *J. Exp. Clin. Cancer Res.* **2020**, *39*, 95. [CrossRef]
11. Van Cutsem, E.; Cervantes, A.; Nordlinger, B.; Arnold, D. Metastatic colorectal cancer: ESMO Clinical Practice Guidelines for diagnosis, treatment and follow-up. *Ann. Oncol.* **2014**, *25*, iii1–iii9. [CrossRef] [PubMed]
12. Benson, A.B.; Venook, A.P.; Al-Hawary, M.M.; Arain, M.A.; Chen, Y.-J.; Ciombor, K.K.; Cohen, S.; Cooper, H.S.; Deming, D.; Farkas, L.; et al. Colon Cancer, Version 2.2021, NCCN Clinical Practice Guidelines in Oncology. *J. Natl. Compr. Cancer Netw.* **2021**, *19*, 329–359. [CrossRef] [PubMed]
13. Aggarwal, C.; Meropol, N.; Punt, C.; Iannotti, N.; Saidman, B.; Sabbath, K.; Gabrail, N.; Picus, J.; Morse, M.; Mitchell, E. Relationship among circulating tumor cells, CEA and overall survival in patients with metastatic colorectal cancer. *Ann. Oncol.* **2013**, *24*, 420–428. [CrossRef] [PubMed]
14. Cohen, S.J.; Punt, C.J.; Iannotti, N.; Saidman, B.H.; Sabbath, K.D.; Gabrail, N.Y.; Picus, J.; Morse, M.; Mitchell, E.; Miller, M.C.; et al. Relationship of circulating tumor cells to tumor response, progression-free survival, and overall survival in patients with metastatic colorectal cancer. *J. Clin. Oncol.* **2008**, *26*, 3213–3221. [CrossRef] [PubMed]
15. Hendricks, A.; Brandt, B.; Geisen, R.; Dall, K.; Röder, C.; Schafmayer, C.; Becker, T.; Hinz, S.; Sebens, S. Isolation and enumeration of CTC in colorectal cancer patients: Introduction of a novel cell imaging approach and comparison to cellular and molecular detection techniques. *Cancers* **2020**, *12*, 2643. [CrossRef] [PubMed]
16. Kolenčík, D.; Narayan, S.; Thiele, J.-A.; McKinley, D.; Gerdtsson, A.S.; Welter, L.; Hošek, P.; Ostašov, P.; Vyčítal, O.; Brůha, J.; et al. Circulating tumor cell kinetics and morphology from the liquid biopsy predict disease progression in patients with metastatic colorectal cancer following resection. *Cancers* **2022**, *14*, 642. [CrossRef] [PubMed]
17. Kolencik, D.; Shishido, S.N.; Pitule, P.; Mason, J.; Hicks, J.; Kuhn, P. Liquid biopsy in colorectal carcinoma: Clinical applications and challenges. *Cancers* **2020**, *12*, 1376. [CrossRef]
18. Veridex LLC. CellSearch Circulating Tumor Cell Kit Premarket Notification-Expanded Indications for Use—Metastatic Prostate Cancer. 2008. Available online: https://www.accessdata.fda.gov/cdrh_docs/pdf7/k073338.pdf (accessed on 16 December 2008).
19. Allard, W.J.; Matera, J.; Miller, M.C.; Repollet, M.; Connelly, M.C.; Rao, C.; Tibbe, A.G.; Uhr, J.W.; Terstappen, L.W. Tumor cells circulate in the peripheral blood of all major carcinomas but not in healthy subjects or patients with nonmalignant diseases. *Clin. Cancer Res.* **2004**, *10*, 6897–6904. [CrossRef]
20. Miller, M.C.; Doyle, G.V.; Terstappen, L.W. Significance of circulating tumor cells detected by the CellSearch system in patients with metastatic breast colorectal and prostate cancer. *J. Oncol.* **2010**, *2010*, 617421. [CrossRef]
21. Seeberg, L.T.; Waage, A.; Brunborg, C.; Hugenschmidt, H.; Renolen, A.; Stav, I.; Bjornbeth, B.A.; Brudvik, K.W.; Borgen, E.F.; Naume, B.; et al. Circulating tumor cells in patients with colorectal liver metastasis predict impaired survival. *Ann. Surg.* **2015**, *261*, 164–171. [CrossRef]
22. Huang, X.; Gao, P.; Song, Y.; Sun, J.; Chen, X.; Zhao, J.; Xu, H.; Wang, Z. Meta-analysis of the prognostic value of circulating tumor cells detected with the CellSearch System in colorectal cancer. *BMC Cancer* **2015**, *15*, 202. [CrossRef] [PubMed]
23. Mazouji, O.; Ouhajjou, A.; Incitti, R.; Mansour, H. Updates on clinical use of liquid biopsy in colorectal cancer screening, diagnosis, follow-up, and treatment guidance. *Front. Cell Dev. Biol.* **2021**, *9*, 962. [CrossRef] [PubMed]
24. Arneth, B. Update on the types and usage of liquid biopsies in the clinical setting: A systematic review. *BMC Cancer* **2018**, *18*, 527. [CrossRef]
25. Gerdtsson, A.S.; Thiele, J.A.; Shishido, S.N.; Zheng, S.; Schaffer, R.; Bethel, K.; Curley, S.; Lenz, H.J.; Hanna, D.L.; Nieva, J.; et al. Single cell correlation analysis of liquid and solid biopsies in metastatic colorectal cancer. *Oncotarget* **2019**, *10*, 7016–7030. [CrossRef] [PubMed]
26. Sotelo, M.; Sastre, J.; Maestro, M.; Veganzones, S.; Vieitez, J.; Alonso, V.; Gravalos, C.; Escudero, P.; Vera, R.; Aranda, E. Role of circulating tumor cells as prognostic marker in resected stage III colorectal cancer. *Ann. Oncol.* **2015**, *26*, 535–541. [CrossRef]
27. Wang, L.; Balasubramanian, P.; Chen, A.P.; Kummar, S.; Evrard, Y.A.; Kinders, R.J. Promise and limits of the CellSearch platform for evaluating pharmacodynamics in circulating tumor cells. *Semin. Oncol.* **2016**, *43*, 464–475. [CrossRef] [PubMed]
28. Shishido, S.N.; Carlsson, A.; Nieva, J.; Bethel, K.; Hicks, J.B.; Bazhenova, L.; Kuhn, P. Circulating tumor cells as a response monitor in stage IV non-small cell lung cancer. *J. Transl. Med.* **2019**, *17*, 294. [CrossRef]
29. Keomanee-Dizon, K.; Shishido, S.N.; Kuhn, P. Circulating tumor cells: High-throughput imaging of CTCs and bioinformatic analysis. *Recent Results Cancer Res.* **2020**, *215*, 89–104. [CrossRef]
30. Marrinucci, D.; Bethel, K.; Kolatkar, A.; Luttgen, M.S.; Malchiodi, M.; Baehring, F.; Voigt, K.; Lazar, D.; Nieva, J.; Bazhenova, L. Fluid biopsy in patients with metastatic prostate, pancreatic and breast cancers. *Phys. Biol.* **2012**, *9*, 016003. [CrossRef]
31. Rodríguez-Lee, M.; Kolatkar, A.; McCormick, M.; Dago, A.D.; Kendall, J.; Carlsson, N.A.; Bethel, K.; Greenspan, E.J.; Hwang, S.E.; Waitman, K.R. Effect of blood collection tube type and time to processing on the enumeration and high-content characterization of circulating tumor cells using the high-definition single-cell assay. *Arch. Pathol. Lab. Med.* **2018**, *142*, 198–207. [CrossRef]
32. Shishido, S.N.; Welter, L.; Rodriguez-Lee, M.; Kolatkar, A.; Xu, L.; Ruiz, C.; Gerdtsson, A.S.; Restrepo-Vassalli, S.; Carlsson, A.; Larsen, J. Preanalytical variables for the genomic assessment of the cellular and acellular fractions of the liquid biopsy in a cohort of breast cancer patients. *J. Mol. Diagn.* **2020**, *22*, 319–337. [CrossRef] [PubMed]

33. Bethel, K.; Luttgen, M.S.; Damani, S.; Kolatkar, A.; Lamy, R.; Sabouri-Ghomi, M.; Topol, S.; Topol, E.J.; Kuhn, P. Fluid phase biopsy for detection and characterization of circulating endothelial cells in myocardial infarction. *Phys. Biol.* **2014**, *11*, 016002. [CrossRef] [PubMed]
34. Armstrong, A.J.; Halabi, S.; Luo, J.; Nanus, D.M.; Giannakakou, P.; Szmulewitz, R.Z.; Danila, D.C.; Healy, P.; Anand, M.; Rothwell, C.J. Prospective multicenter validation of androgen receptor splice variant 7 and hormone therapy resistance in high-risk castration-resistant prostate cancer: The prophecy study. *J. Clin. Oncol.* **2019**, *37*, 1120. [CrossRef] [PubMed]
35. Scher, H.I.; Graf, R.P.; Schreiber, N.A.; Jayaram, A.; Winquist, E.; McLaughlin, B.; Lu, D.; Fleisher, M.; Orr, S.; Lowes, L. Assessment of the validity of nuclear-localized androgen receptor splice variant 7 in circulating tumor cells as a predictive biomarker for castration-resistant prostate cancer. *JAMA Oncol.* **2018**, *4*, 1179–1186. [CrossRef] [PubMed]
36. Scher, H.I.; Lu, D.; Schreiber, N.A.; Louw, J.; Graf, R.P.; Vargas, H.A.; Johnson, A.; Jendrisak, A.; Bambury, R.; Danila, D. Association of AR-V7 on circulating tumor cells as a treatment-specific biomarker with outcomes and survival in castration-resistant prostate cancer. *JAMA Oncol.* **2016**, *2*, 1441–1449. [CrossRef]
37. Welter, L.; Xu, L.; McKinley, D.; Dago, A.E.; Prabakar, R.K.; Restrepo-Vassalli, S.; Xu, K.; Rodriguez-Lee, M.; Kolatkar, A.; Nevarez, R. Treatment response and tumor evolution: Lessons from an extended series of multianalyte liquid biopsies in a metastatic breast cancer patient. *Mol. Case Stud.* **2020**, *6*, a005819. [CrossRef]
38. Gerdtsson, E.; Pore, M.; Thiele, J.-A.; Gerdtsson, A.S.; Malihi, P.D.; Nevarez, R.; Kolatkar, A.; Velasco, C.R.; Wix, S.; Singh, M. Multiplex protein detection on circulating tumor cells from liquid biopsies using imaging mass cytometry. *Converg. Sci. Phys. Oncol.* **2018**, *4*, 015002. [CrossRef]
39. Febbo, P.G.; Martin, A.M.; Scher, H.I.; Barrett, J.C.; Beaver, J.A.; Beresford, P.J.; Blumenthal, G.M.; Bramlett, K.; Compton, C.; Dittamore, R. Minimum technical data elements for liquid biopsy data submitted to public databases. *Clin. Pharmacol. Ther.* **2020**, *107*, 730–734. [CrossRef]
40. Ghatalia, P.; Smith, C.H.; Winer, A.; Gou, J.; Kiedrowski, L.A.; Slifker, M.; Saltzberg, P.D.; Bubes, N.; Anari, F.M.; Kasireddy, V.; et al. Clinical Utilization Pattern of Liquid Biopsies (LB) to Detect Actionable Driver Mutations, Guide Treatment Decisions and Monitor Disease Burden During Treatment of 33 Metastatic Colorectal Cancer (mCRC) Patients (pts) at a Fox Chase Cancer Center GI Oncology Subspecialty Clinic. *Front. Oncol.* **2018**, *8*, 652. [CrossRef]
41. Matsusaka, S.; Suenaga, M.; Mishima, Y.; Takagi, K.; Terui, Y.; Mizunuma, N.; Hatake, K. Circulating endothelial cells predict for response to bevacizumab-based chemotherapy in metastatic colorectal cancer. *Cancer Chemother. Pharmacol.* **2011**, *68*, 763–768. [CrossRef]
42. Simkens, L.; Tol, J.; Terstappen, L.; Teerenstra, S.; Punt, C.; Nagtegaal, I. The predictive and prognostic value of circulating endothelial cells in advanced colorectal cancer patients receiving first-line chemotherapy and bevacizumab. *Ann. Oncol.* **2010**, *21*, 2447–2448. [CrossRef] [PubMed]
43. Okusha, Y.; Eguchi, T.; Tran, M.T.; Sogawa, C.; Yoshida, K.; Itagaki, M.; Taha, E.A.; Ono, K.; Aoyama, E.; Okamura, H. Extracellular vesicles enriched with moonlighting metalloproteinase are highly transmissive, pro-tumorigenic, and trans-activates cellular communication network factor (CCN2/CTGF): CRISPR against cancer. *Cancers* **2020**, *12*, 881. [CrossRef] [PubMed]
44. Di Vizio, D.; Kim, J.; Hager, M.H.; Morello, M.; Yang, W.; Lafargue, C.J.; True, L.D.; Rubin, M.A.; Adam, R.M.; Beroukhim, R. Oncosome formation in prostate cancer: Association with a region of frequent chromosomal deletion in metastatic disease. *Cancer Res.* **2009**, *69*, 5601–5609. [CrossRef] [PubMed]
45. Ciardiello, C.; Leone, A.; Lanuti, P.; Roca, M.S.; Moccia, T.; Minciacchi, V.R.; Minopoli, M.; Gigantino, V.; De Cecio, R.; Rippa, M. Large oncosomes overexpressing integrin alpha-V promote prostate cancer adhesion and invasion via AKT activation. *J. Exp. Clin. Cancer Res.* **2019**, *38*, 317. [CrossRef]
46. Gerdtsson, A.S.; Setayesh, S.M.; Malihi, P.D.; Ruiz, C.; Carlsson, A.; Nevarez, R.; Matsumoto, N.; Gerdtsson, E.; Zurita, A.; Logothetis, C.; et al. Large extracellular vesicle characterization and association with circulating tumor cells in metastatic castrate resistant prostate cancer. *Cancers* **2021**, *13*, 1056. [CrossRef]
47. Chai, S.; Matsumoto, N.; Storgard, R.; Peng, C.C.; Aparicio, A.; Ormseth, B.; Rappard, K.; Cunningham, K.; Kolatkar, A.; Nevarez, R.; et al. Platelet-coated circulating tumor cells are a predictive biomarker in patients with metastatic castrate-resistant prostate cancer. *Mol. Cancer Res.* **2021**, *19*, 2036–2045. [CrossRef]
48. Shishido, S.N.; Sayeed, S.; Courcoubetis, G.; Djaladat, H.; Miranda, G.; Pienta, K.J.; Nieva, J.; Hansel, D.E.; Desai, M.; Gill, I.S.; et al. Characterization of cellular and acellular analytes from pre-cystectomy liquid biopsies in patients newly diagnosed with primary bladder cancer. *Cancers* **2022**, *14*, 758. [CrossRef]
49. Werling, R.W.; Yaziji, H.; Bacchi, C.E.; Gown, A.M. CDX2, a highly sensitive and specific marker of adenocarcinomas of intestinal origin: An immunohistochemical survey of 476 primary and metastatic carcinomas. *Am. J. Surg. Pathol.* **2003**, *27*, 303–310. [CrossRef]
50. Adams, D.L.; Stefansson, S.; Haudenschild, C.; Martin, S.S.; Charpentier, M.; Chumsri, S.; Cristofanilli, M.; Tang, C.M.; Alpaugh, R.K. Cytometric characterization of circulating tumor cells captured by microfiltration and their correlation to the cellsearch® CTC test. *Cytom. Part A* **2015**, *87*, 137–144. [CrossRef]
51. Shishido, S.N.; Ghoreifi, A.; Sayeed, S.; Courcoubetis, G.; Huang, A.; Ye, B.; Mrutyunjaya, S.; Gill, I.S.; Kuhn, P.; Mason, J.; et al. Liquid biopsy landscape in patients with primary upper tract Urothelial Carcinoma. *Cancers* **2022**, *14*, 3007. [CrossRef]
52. Minciacchi, V.R.; Freeman, M.R.; Di Vizio, D. Extracellular vesicles in cancer: Exosomes, microvesicles and the emerging role of large oncosomes. *Semin. Cell Dev. Biol.* **2015**, *40*, 41–51. [CrossRef] [PubMed]

53. Di Vizio, D.; Morello, M.; Dudley, A.C.; Schow, P.W.; Adam, R.M.; Morley, S.; Mulholland, D.; Rotinen, M.; Hager, M.H.; Insabato, L. Large oncosomes in human prostate cancer tissues and in the circulation of mice with metastatic disease. *Am. J. Pathol.* **2012**, *181*, 1573–1584. [CrossRef] [PubMed]
54. Pedregosa, F.; Varoquaux, G.; Gramfort, A.; Michel, V.; Thirion, B.; Grisel, O.; Blondel, M.; Prettenhofer, P.; Weiss, R.; Dubourg, V. Scikit-learn: Machine learning in Python. *J. Mach. Learn. Res.* **2011**, *12*, 2825–2830.
55. Spearman, C. The proof and measurement of association between two things. *Am. J. Psychol.* **1987**, *100*, 441–471. [CrossRef]
56. Kaplan, E.L.; Meier, P. Nonparametric estimation from incomplete observations. *J. Am. Stat. Assoc.* **1958**, *53*, 457–481. [CrossRef]
57. Virtanen, P.; Gommers, R.; Oliphant, T.E.; Haberland, M.; Reddy, T.; Cournapeau, D.; Burovski, E.; Peterson, P.; Weckesser, W.; Bright, J. SciPy 1.0: Fundamental algorithms for scientific computing in Python. *Nat. Methods* **2020**, *17*, 261–272. [CrossRef]
58. Pölsterl, S. Scikit-survival: A library for time-to-event analysis built on top of scikit-learn. *J. Mach. Learn. Res.* **2020**, *21*, 8747–8752.
59. Sharpe, A.; McIntosh, M.; Lawrie, S.M. Research methods, statistics and evidence-based practice. In *Companion to Psychiatric Studies*, 8th ed.; Churchill Livingstone: London, UK, 2010; p. 157.
60. Hladovec, J.; Rossmann, P. Circulating endothelial cells isolated together with platelets and the experimental modification of their counts in rats. *Thromb. Res.* **1973**, *3*, 665–674. [CrossRef]
61. Mancuso, P.; Burlini, A.; Pruneri, G.; Goldhirsch, A.; Martinelli, G.; Bertolini, F. Resting and activated endothelial cells are increased in the peripheral blood of cancer patients. *Blood J. Am. Soc. Hematol.* **2001**, *97*, 3658–3661. [CrossRef]
62. Ronzoni, M.; Manzoni, M.; Mariucci, S.; Loupakis, F.; Brugnatelli, S.; Bencardino, K.; Rovati, B.; Tinelli, C.; Falcone, A.; Villa, E. Circulating endothelial cells and endothelial progenitors as predictive markers of clinical response to bevacizumab-based first-line treatment in advanced colorectal cancer patients. *Ann. Oncol.* **2010**, *21*, 2382–2389. [CrossRef]
63. Cima, I.; Kong, S.L.; Sengupta, D.; Tan, I.B.; Phyo, W.M.; Lee, D.; Hu, M.; Iliescu, C.; Alexander, I.; Goh, W.L. Tumor-derived circulating endothelial cell clusters in colorectal cancer. *Sci. Transl. Med.* **2016**, *8*, 345ra389. [CrossRef] [PubMed]
64. Dave, J.M.; Bayless, K.J. Vimentin as an integral regulator of cell adhesion and endothelial sprouting. *Microcirculation* **2014**, *21*, 333–344. [CrossRef] [PubMed]
65. DeLisser, H.; Newman, P.; Albelda, S. Platelet endothelial cell adhesion molecule (CD31). *Curr. Top. Microbiol. Immunol.* **1993**, *184*, 37–45. [PubMed]
66. Metcalf, D.; MacDonald, H.; Odartchenko, N.; Sordat, B. Growth of mouse megakaryocyte colonies in vitro. *Proc. Natl. Acad. Sci. USA* **1975**, *72*, 1744–1748. [CrossRef]
67. Xu, L.; Mao, X.; Guo, T.; Chan, P.Y.; Shaw, G.; Hines, J.; Stankiewicz, E.; Wang, Y.; Oliver, R.T.D.; Ahmad, A.S. The novel association of circulating tumor cells and circulating megakaryocytes with prostate cancer prognosis. *Clin. Cancer Res.* **2017**, *23*, 5112–5122. [CrossRef] [PubMed]
68. Huang, W.; Zhao, S.; Xu, W.; Zhang, Z.; Ding, X.; He, J.; Liang, W. Presence of intra-tumoral CD61+ megakaryocytes predicts poor prognosis in non-small cell lung cancer. *Transl. Lung Cancer Res.* **2019**, *8*, 323. [CrossRef]
69. Zhu, X.; Cao, Y.; Lu, P.; Kang, Y.; Lin, Z.; Hao, T.; Song, Y. Evaluation of platelet indices as diagnostic biomarkers for colorectal cancer. *Sci. Rep.* **2018**, *8*, 11814. [CrossRef] [PubMed]
70. Crescitelli, R.; Lässer, C.; Lötvall, J. Isolation and characterization of extracellular vesicle subpopulations from tissues. *Nature Protocols* **2021**, *16*, 1548–1580. [CrossRef]
71. Schwitalla, S.; Fingerle, A.A.; Cammareri, P.; Nebelsiek, T.; Göktuna, S.I.; Ziegler, P.K.; Canli, O.; Heijmans, J.; Huels, D.J.; Moreaux, G. Intestinal tumorigenesis initiated by dedifferentiation and acquisition of stem-cell-like properties. *Cell* **2013**, *152*, 25–38. [CrossRef]
72. Merlos-Suárez, A.; Barriga, F.M.; Jung, P.; Iglesias, M.; Céspedes, M.V.; Rossell, D.; Sevillano, M.; Hernando-Momblona, X.; da Silva-Diz, V.; Muñoz, P. The intestinal stem cell signature identifies colorectal cancer stem cells and predicts disease relapse. *Cell Stem Cell* **2011**, *8*, 511–524. [CrossRef]
73. Fumagalli, A.; Oost, K.C.; Kester, L.; Morgner, J.; Bornes, L.; Bruens, L.; Spaargaren, L.; Azkanaz, M.; Schelfhorst, T.; Beerling, E. Plasticity of Lgr5-negative cancer cells drives metastasis in colorectal cancer. *Cell Stem Cell* **2020**, *26*, 569–578.e7. [CrossRef] [PubMed]
74. Dago, A.E.; Stepansky, A.; Carlsson, A.; Luttgen, M.; Kendall, J.; Baslan, T.; Kolatkar, A.; Wigler, M.; Bethel, K.; Gross, M.E. Rapid phenotypic and genomic change in response to therapeutic pressure in prostate cancer inferred by high content analysis of single circulating tumor cells. *PLoS ONE* **2014**, *9*, e101777. [CrossRef] [PubMed]
75. Malihi, P.D.; Morikado, M.; Welter, L.; Liu, S.T.; Miller, E.T.; Cadaneanu, R.M.; Knudsen, B.S.; Lewis, M.S.; Carlsson, A.; Velasco, C.R. Clonal diversity revealed by morphoproteomic and copy number profiles of single prostate cancer cells at diagnosis. *Converg. Sci. Phys. Oncol.* **2018**, *4*, 015003. [CrossRef] [PubMed]
76. Ruiz, C.; Li, J.; Luttgen, M.S.; Kolatkar, A.; Kendall, J.T.; Flores, E.; Topp, Z.; Samlowski, W.E.; McClay, E.; Bethel, K. Limited genomic heterogeneity of circulating melanoma cells in advanced stage patients. *Phys. Biol.* **2015**, *12*, 016008. [CrossRef]

 *cancers*

Article

Snail Overexpression Alters the microRNA Content of Extracellular Vesicles Released from HT29 Colorectal Cancer Cells and Activates Pro-Inflammatory State In Vivo

Izabela Papiewska-Pająk [1,*], Patrycja Przygodzka [1], Damian Krzyżanowski [1], Kamila Soboska [1,2], Izabela Szulc-Kiełbik [1], Olga Stasikowska-Kanicka [3], Joanna Boncela [1], Małgorzata Wągrowska-Danilewicz [3] and M. Anna Kowalska [1,4,*]

1 Institute of Medical Biology, Polish Academy of Sciences, 93-232 Lodz, Poland; pprzygodzka@cbm.pan.pl (P.P.); dkrzyzanowski@cbm.pan.pl (D.K.); ksoboska@cbm.pan.pl (K.S.); iszulc@cbm.pan.pl (I.S.-K.); jboncela@cbm.pan.pl (J.B.)
2 Faculty of Biology and Environmental Protection, University of Lodz, 90-236 Lodz, Poland
3 Department of Diagnostic Techniques in Pathomorphology, Medical University of Lodz, 90-419 Lodz, Poland; olga.stasikowska@umed.lodz.pl (O.S.-K.); malgorzata.wagrowska-danilewicz@umed.lodz.pl (M.W.-D.)
4 Department of Pediatrics, The Children's Hospital of Philadelphia, Philadelphia, PA 19104, USA
* Correspondence: ipapiewska-pajak@cbm.pan.pl (I.P.-P.); mkowalska@cbm.pan.pl (M.A.K.)

Simple Summary: Knowledge of the factors that help migration of carcinoma cells is important for prevention of metastasis. Cancer cells release small particles, extracellular vesicles (EVs) that contain such factors. The aim of this study was to assess if the content of EVs changes through different stages of colorectal cancer (CRC) and evaluate how this process affects cancer progression in vivo in mouse CRC model. We found that EVs released from cells that have migratory properties contain different factors then EVs released from original tumor cells. We also show here that EVs can be incorporated into other cells that facilitate metastasis and change their properties depending on the EVs content. The content of cell-released EVs may also serve as a biomarker that denotes the stage of CRC and may be a target to prevent cancer progression.

Abstract: During metastasis, cancer cells undergo phenotype changes in the epithelial-mesenchymal transition (EMT) process. Extracellular vesicles (EVs) released by cancer cells are the mediators of intercellular communication and play a role in metastatic process. Knowledge of factors that influence the modifications of the pre-metastatic niche for the migrating carcinoma cells is important for prevention of metastasis. We focus here on how cancer progression is affected by EVs released from either epithelial-like HT29-cells or from cells that are in early EMT stage triggered by Snail transcription factor (HT29-Snail). We found that EVs released from HT29-Snail, as compared to HT29-pcDNA cells, have a different microRNA profile. We observed the presence of interstitial pneumonias in the lungs of mice injected with HT29-Snail cells and the percent of mice with lung inflammation was higher after injection of HT29-Snail-EVs. Incorporation of EVs released from HT29-pcDNA, but not released from HT29-Snail, leads to the increased secretion of IL-8 from macrophages. We conclude that Snail modifications of CRC cells towards more invasive phenotype also alter the microRNA cargo of released EVs. The content of cell-released EVs may serve as a biomarker that denotes the stage of CRC and EVs-specific microRNAs may be a target to prevent cancer progression.

Keywords: Snail transcription factor; extracellular vesicles; colon cancer; pre-metastatic niche

Citation: Papiewska-Pająk, I.; Przygodzka, P.; Krzyżanowski, D.; Soboska, K.; Szulc-Kiełbik, I.; Stasikowska-Kanicka, O.; Boncela, J.; Wągrowska-Danilewicz, M.; Kowalska, M.A. Snail Overexpression Alters the microRNA Content of Extracellular Vesicles Released from HT29 Colorectal Cancer Cells and Activates Pro-Inflammatory State In Vivo. *Cancers* **2021**, *13*, 172. https://doi.org/10.3390/cancers13020172

Received: 9 October 2020
Accepted: 29 December 2020
Published: 6 January 2021

1. Introduction

Colorectal cancer (CRC) is the third most common cancer in men and second in women worldwide. It is predicted that the number of deaths will rise to 1.1 million per year worldwide by 2030 [1]. Formation of distant metastases are the primary cause of CRC treatment failure and patient death [2]. In the metastasis process, to enter the circulation, cancer cells lose their epithelial features, exhibit decreased polarity and intercellular adhesion, undergo cytoskeletal reorganization and become more motile. This phenotypic transformation of epithelial carcinoma into mesenchymal-like cells (EMT) is triggered by several factors including Snail transcription factor that plays crucial role at the beginning of the EMT process [3].

Crucial for cancer progression is also the intercellular cross-talk and subsequent regulation of both local and distant microenvironments. Kaplan et al. [4] introduced the concept termed "pre-metastatic niche" that is defined as the microenvironment that facilitates the formation of metastasis and consists of immune and stromal cells and also the components of the primary tumour. A number of niche-promoting molecules has been identified in various metastasis mouse models [5]. Monocytes and macrophages associated with tumour environment have been widely recognized as immunological effectors [5]. Macrophages display functional plasticity in response to local microenvironment stimuli and participate in cancer-related inflammation, matrix remodelling, immune escape and ultimately in cancer metastasis. Additionally, cancer cell-derived extracellular vesicles (EVs) are recognized as significant contributors to different aspects of modifications of pre-metastatic niches [6].

EVs are defined as a heterogeneous group of a phospholipid bilayer particles that are released to the extracellular environment by most cells in the organism and are present in all body fluids [7]. EVs can be divided into main three subpopulations including exosomes (exo), microvesicles/ectosomes (MVs) and apoptotic bodies (APOs) that differ in size, formation process and content [8]. EVs contain a variety of bioactive molecules, including proteins, lipids and multiple nucleic acid species, including the most extensively studied class of microRNA (miRNA/miR) [6,9]. The cargo of EVs may either reflect the cell of origin or can be actively sorted [10,11]. Recent reviews summarize the importance of EVs as communicators between cells that accelerate cancer progression and metastasis [12,13]. EVs can stimulate angiogenesis, matrix remodelling and modulation of immune response [9,14–16]. Factors, including cytokines, chemokines and EVs that are released from cancer or other cells that are linked to inflammation, influence the environment of pre-metastatic niches in distant organs [17].

We have previously shown that Snail, as an early regulator of EMT, affects the transcriptome and miRNA profile of human HT29 CRC cells and changes HT29 cells phenotype to pro-migratory [18,19]. Here we show that HT29 cells overexpressing Snail, in comparison to epithelial-like HT29 cells, release EVs with different miRNA content and postulate that HT29-Snail-EVs modify the activity of cells constituting the pre-metastatic niche, thus facilitating cancer progression.

2. Results

2.1. Characterization of Extracellular Vesicles (EVs) Released from Control HT29 and HT29-Snail Cells

2.1.1. Size and Purity

The preparation of EVs free of cellular debris, FBS-derived vesicles or any non-EVs RNAs and proteins is critical for any analyses. We applied a commonly used procedure for isolation of EVs that we used in a previous study [18,19]. Cells release vesicles that according to current nomenclature can be classified as exosomes (exo, 30–120 nm), microvesicles (MVs, 0.1–1.0 μm) and apoptotic bodies (0.8–5.0 μm) [20]. We purified EVs that are the mixture of exo and MVs. EVs sizes were similar for all HT29 clones with sizes of 80 to 240 nm (Figure 1A).

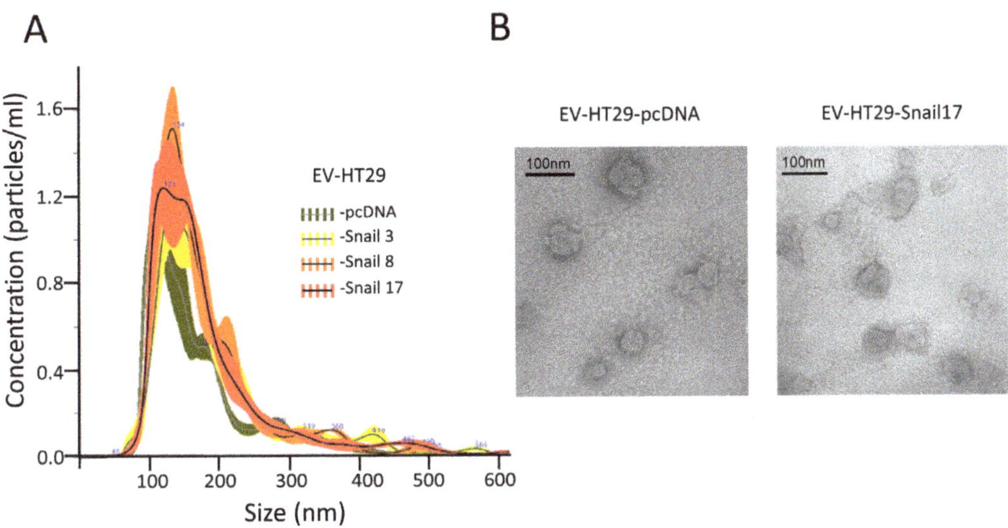

Figure 1. Characterization of extracellular vesicles (EVs) released by HT29 clones. EVs were isolated from conditioned media after 24 h of incubation in FBS-free culture media of HT29 clones. (**A**) NTA showed similar numbers and sizes of EVs isolated from HT29 clones. (**B**) Representative Electron microscopic image of EVs derived from HT29-cDNA and HT29-Snail 17. Scale bar, 100 nm.

Similar vesicles number (~5.2×10^{13}) for each preparation of EVs were isolated from the same number of cells (~6.8×10^{8}). The spheroid morphology of EVs as well as their size were confirmed by TEM (Figure 1B). EVs were enriched by markers such as CD63, CD9 and flotillin-1 as compared to lysates of cells of origin (Figure S1). They also contained 70 kDa heat-shock protein (HSP70) and were Annexin V positive; the purity of EVs was confirmed by the absence of cytochrome c and GM130 (Figure S1).

2.1.2. miRNA Content

The miRNA profiling was performed by Next Generation Sequencing (NGS) analysis of the total mRNA isolated from EVs. There were no significant differences in the number of detectable miRNAs between EVs of control and either of HT29-Snail clones (Figure S2). Further, the two-way hierarchical clustering of miRNA was performed (Figure 2A). As presented in Figure 2B,C, we identified 23 miRNAs differentially expressed in EVs released from HT29-Snail 3, 30 miRNAs from HT29-Snail 8, and 48 miRNAs from HT29-Snail 17.

Three miRNAs: let-7i, miR-205, and miR-130b were up-regulated while five miRNAs: miR-1246, miR-3131, miR-375, miR-552-3p and miR-552-5p were down-regulated in HT29-Snail-derived EVs (Figure 2C). Additionally, as it is shown in Table S1, among other miRNAs altered more than two times in EVs released from HT29-Snail clones, miR-483-5p (HT29-Snail 17-EVs) was upregulated and miR-142-5p (HT29-Snail 8, -17-EVs) as well as miR-203a and miR-203b-3p (HT29-Snail 17-EVs) were downregulated. Moreover, as shown in Table S2, there were also miRNAs upregulated in EVs released from HT29-Snail cells with a lower fold change. Those differentially expressed miRNAs (marked in red in Tables S1 and S2) were reported to be important in cancer progression as pointed out in the Discussion section.

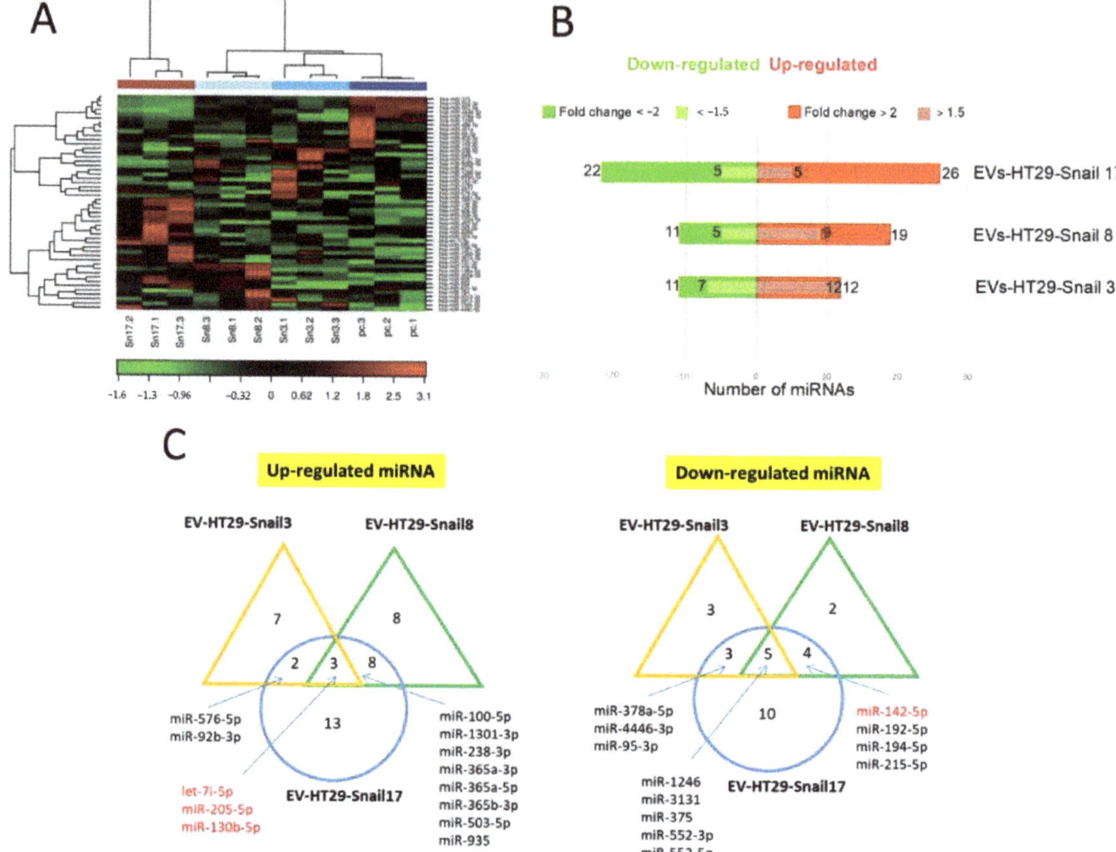

Figure 2. Changes in the expression of miRNAs in EVs released by HT29-Snail cells as compared to EVs from HT29-pcDNA cells. (**A**) Heatmap and unsupervised hierarchical clustering by sample and miRNA. (**B**) Number of differentially expressed miRNAs detected by NGS analysis, that were either significantly upregulated (red) or downregulated (green) in EVs derived from HT29-Snail-3, -8 and -17 versus EVs from control cells. (**C**) Venn diagrams show differentially expressed miRNAs. The most regulated miRNAs in all three or two clone-EVs are marked. (corrected FDR $p < 0.05$ and fold change > 2). miRs marked in red are discussed in the text.

2.1.3. Gene Ontology (GO) Enrichment Analysis

To identify the biological processes that might be triggered in the response to elevated Snail levels in the HT29 clones that are shedding EVs, Gene Ontology (GO) enrichment analysis was performed to identify GO terms that are significantly associated with differentially expressed microRNAs in EVs released by the clones. Positive regulation of thyrosine phosphorylation of STAT-3 and -1 was at the top of Biological Process in the GO analysis (Figure S3).

2.2. *In Vivo Studies*

To look for the in vivo effect of Snail on the cancer progression we injected athymic mice subcutaneously (s.c.) with the same number (1.5×10^6/mouse) of control (HT29-pcDNA) and HT29-Snail 17 cells. Clone 17 contains the most elevated Snail expression ([18] and Figure S5), and their EVs have the highest number of differentially expressed miRNA (Figure 2 and Tables S1 and S2). Further, animals of both groups were injected intravenously (i.v. day 8, 12, 15 and 18) with the same amount (10 μg/mouse) of HT29-pcDNA-EVs, and observed for 28 days. Administration of HT29-pcDNA-EVs did not have statistical impact on neither tumour growth nor plasma MCP-1/CCL2 levels (Figure 3A,C, Table S3) within the groups of mice injected with either HT29-pcDNA cells or HT29-Snail cells. However, there was a ~20% difference in the growth rate of tumour (day 12 to 28) between the control and HT29-Snail mice, regardless of the injection of EVs (Figure 3A). Growth of tumours in mice administered s.c. with HT29-Snail cells was slower in comparison to administration of control cells, which is not surprising since HT29-Snail cells are already in the early EMT stage [18,19]. We have also observed an increase in the concentration of monocyte chemotactic protein-1 (MCP-1/CCL2) at day -7 (Figure 3B) and -28 (Figure 3C) in the plasma of mice that bore a tumour induced by injection of HT29-Snail17 cells as compared to control mice injected with HT29-pcDNA cells. We did not observe any metastatic sites of HT29 cells in the lungs, livers or intestines after 28 days of tumour growth. However, early interstitial pneumonias were observed after 28 days in animals injected with HT29-Snail17 cells as compared to animals injected with HT29-pcDNA (Table 1, group IV and II, respectively). Injection of EVs released from control HT29-pcDNA cells slightly increased the number of mice with lung infection in control group of mice that bore a tumor induced by injection of HT29-pcDNA cells (Table 1, group II and III). Based on the fact that HT29-Snail cells injected s.c. into mice may secrete additional HT29-Snail-EVs, we have additionally injected such EVs into the group of mice bearing tumors induced by HT-29-Snail cells. We did not observe any changes in either tumor growth or MCP-1/CCL2 levels (Table S3). However, in this group (Table 1, group VI) we observed the highest number of animals presenting lung inflammation (80%) As presented on Figure 3D, inflammatory infiltration from lymphocytes and thickened interalveolar septum were observed in lungs of those mice.

Table 1. Incidence of lung infection in mice with tumours inflicted by injection of control HT29-pcDNA or HT29-Snail cells followed by injections of various EVs.

Group	Cell Injection Day 0	EV Injection Day 8, 12, 15, 18	Lung Infection Day 28
I	None	None	0/5 (0%)
II	HT29-pcDNA	None	0/5 (0%)
III	HT29-pcDNA	HT29-pcDNA	1/5 (20%)
IV	HT29-Snail17	None	2/5 (40%)
V	HT29-Snail17	HT29-pcDNA	2/5 (40%)
VI	HT29-Snail17	HT29-Snail17	4/5 (80%)

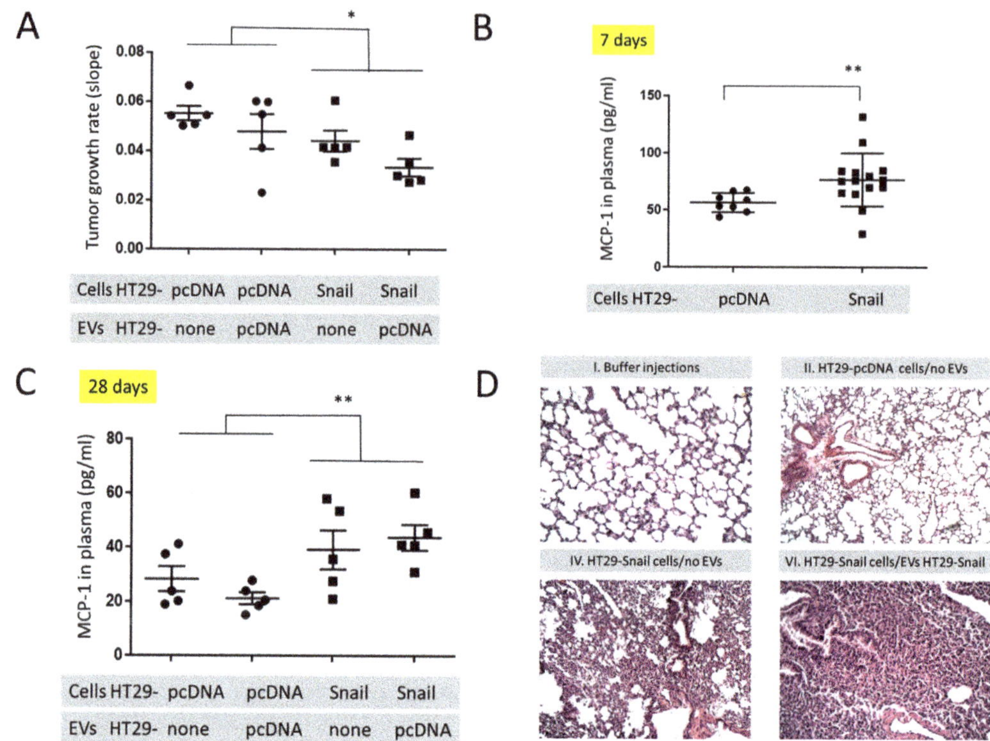

Figure 3. In vivo mouse studies in animals injected s.c. with HT29-pcDNA- (small dot) or HT29-Snail17- (small cube) cells. (**A**) The growth rate of tumours (day 12–28) in animals with HT29-pcDNA or HT29-Snail administered s.c. and additionally injected i.v. with EVs released by HT29-pcDNA, mean value ± SEM: 0.052 ± 0.04 and 0.039 ± 0.03, respectively (**B**) Levels of monocyte chemoattractant protein-1 (MCP-1/CCL2) measured in the plasma of mice at day 7 (before first injection of EVs). (**C**) Levels of monocyte chemoattractant protein-1 (MCP-1/CCL2) measured in the plasma of mice at day 28 (end of experiment). Mean value ± SEM for animals injected s.c. with HT29-pcDNA and HT29-Snail were 24.8 ± 2.7 and 41.5 ± 4.1, respectively. (**D**) Representative H&E staining of murine lungs. 100× magnification. The images of groups IV and VI (see Table 1) show interstitial pneumonia pictures. n = 10–15, * $p < 0.02$; ** $p < 0.01$.

2.3. In Vitro Studies

2.3.1. Uptake of EVs by THP-1 Derived Macrophages (TDM)

The uptake of EVs by TDM was visualized by confocal imaging (Figure 4A,B). EVs obtained from either control HT29-pcDNA or HT29-Snail clones -3, -8 and -17 were incorporated into TDM in FBS-free media for 4 h. The viability of TDM cells was not changed by incorporation of any EVs (Figure S4).

2.3.2. Effect of EVs Released from HT29 Clones on Macrophage Activity

We examined whether the incorporation of EVs released by HT29-Snail cells that represent the intermediate state of EMT, affects the activity of macrophages differently than incorporation of EVs released from epithelial-like HT29 cells (Figure 5).

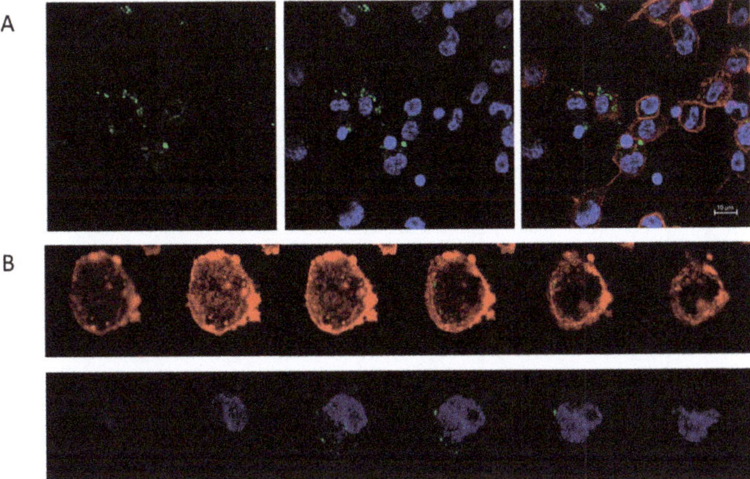

Figure 4. Uptake of EVs derived from Snail-HT29 cells (clone 17) by TDM. HT29-Snail EVs were labeled with PKH67 dye (green). Nuclei were stained with Hoechst (blue), actin was stained with Phalloidin Texas Red (red). (**A**) Representative confocal laser scanning microscope. (**B**) Z-stack analysis of cells and incorporated EVs. Note: Viability of TDM cells was not affected by the incorporation of EVs (Figure S4).

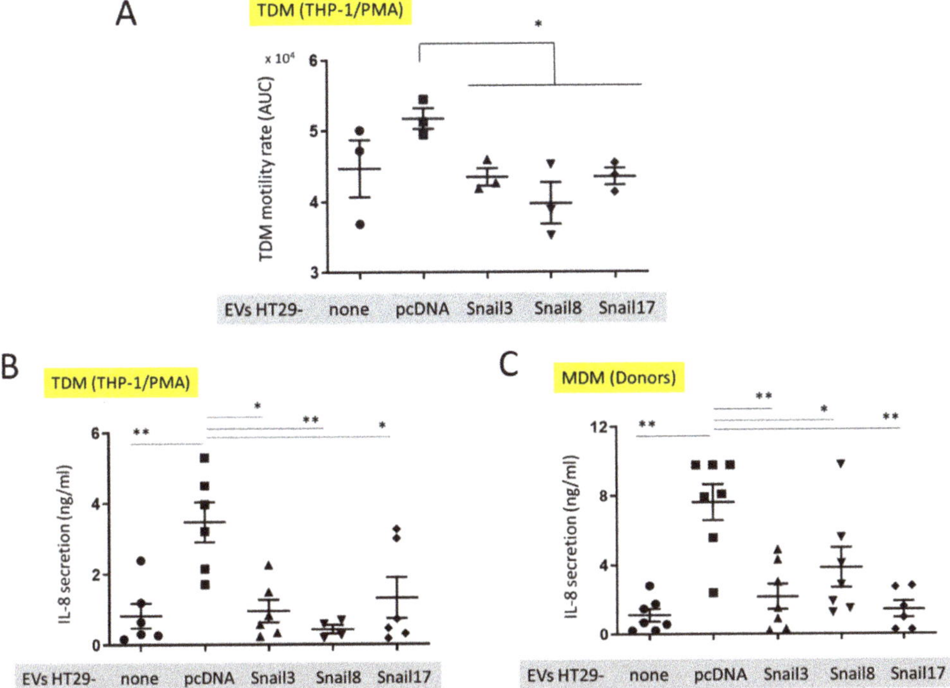

Figure 5. In vitro effect of EVs on macrophages. (**A**) Motility of TDM cells before (small dot) after the uptake of EVs isolated from control HT29-pcDNA (small cube) or HT29-Snail cells (clone 3- small triangle, clone 8- small upside triangle and clone 17- diamond). (**B**) IL-8 release from TDM after treatment with EVs indicated. (**C**) IL-8 release from macrophages isolated from healthy donors blood (MDM) and incubated with EVs. $n = 3$–7, * $p < 0.05$; ** $p < 0.01$.

The motility of TDM was evaluated by measuring the cell speed changes in time. There was a slight, but not significant, increase in the motility of TDM that had incorporated EVs released from HT29-pcDNA as compared to control TDM cells (Figure 5A). However, the motility of TDM cells with incorporated EVs released from HT29-Snail cells was inhibited in comparison to cells containing HT29-pcDNA-EVs. (Figure 5A). We have also quantified the level of cytokines secreted by TDM. We observed a significantly increased level of IL-8 released by TDM (Figure 5B) treated by EVs released from HT29-pcDNA cells, as compared to non-treated TDMs (no EVs). However, incorporation of HT29-Snail-EVs resulted in decreased IL-8 levels (Figure 5B). Next, we performed the analysis of cytokines released by macrophages derived from monocytes isolated from healthy donors and differentiated with GM-CSF (MDM). Consistent with the previous results, MDM cells incubated with EVs released from HT29-Snail clones secreted less IL-8 as compared to cells incubated with control HT29-pcDNA-EVs (Figure 5C).

3. Discussion

Our group has recently reported that overexpression of Snail drives HT29 colon cancer cell to a partial-EMT and modulates the expression of specific protein transcripts and miRNAs [18,19,21]. We reasoned that cancer cells at various EMT stages will release extracellular vesicles of different content and thus differently affect recipient cells. Changes in the protein levels in EVs released from cells stimulated by Snail were already observed [22]. As a rule, the cargo loaded into EVs reflects the status of cancer cells [23], but potential preferential packaging into EVs has also been suggested [11]. The most extensively studied class of factors transported between cells by EVs are miRNAs that regulate the translation of target mRNAs in recipient cells [10,24]. We show here that overexpression of Snail in HT29 cells significantly triggers changes on individual miRNAs levels.

We are aware of the fact that our HT29-Snail clones differ in the amount of Snail overexpressed in their cells. In this study, we used several HT29-Snail clones with clone 3 representing the lower and clone 17 the highest Snail level [18]. Thus, each of the clones may reflect a slightly different EMT stage.

In EVs released from all three HT29-Snail clones let-7i, miR-205 and miR-130b-5p miRNAs were highly upregulated. There are conflicting reports concerning the role of miR-205 and let-7i in cancer [25–27]. Whether particular miRNA is considered as tumor suppressor or onco-miRNA appears to be dependent on the specific cancer and tumor-environment [28,29]. Increased migratory properties of HT29-Snail cells with elevated miR-205 and let-7i expression was shown previously by us, pointing on their role in CRC progression [19]. The mRNA targets for miR-205-5p and let7i-5p were also shown in our previous study. Potentiated miR-205 expression was correlated with Dynamin 3 (DNM3) mRNA decrease. DNM3 is considered as a cancer suppressor. Thus, from the tumour point of view redundant during progression of the disease and decreased in HT29-Snail cells [19].

We observed miR-130b-5p enrichment in EVs released from all our HT29-Snail clones. In contrast to miR-205 and let-7i up-regulation in EVs, which mirrored intracellular changes in miRNAs expression, miR-130b-5p was not up-regulated in HT29-Snail cells [19]. This suggests that miR-130b-5p could be actively sorted into vesicles released from cells that overexpress Snail and thus undergo EMT. MiR-130b was identified in the microvesicles of leukemia K562 cells, but, unlike in CRC cells in this study, at an equal expression level as in cells of origin [30]. In various human tumor types altered miR-130b expression has been implicated as either promoting or suppressing tumorigenesis; miR-130b is significantly downregulated in pituitary adenomas and endometrial cancer [31,32] whereas it is upregulated in bladder cancer, melanoma, metastatic renal carcinoma [33–35]. The contribution of miR-130b to CRC progression is also the subject of the debate. In cell lines SW-480 and SW-620 over-expression of miR-130b downregulates integrin β1, leading to the impaired migration and invasion of CRC cells [36] whereas another analysis of a series of CRC cell lines showed that miR-130b acts as an efficient inducer of EMT in vivo and in vitro, likely through up-regulation of Snail and ZEB1 transcription factors. The effects of miR-130b on promoting cell migration and invasion of CRC cells with poor prognosis for colorectal cancer was also indicated [37]. Whether the miR-130b that is packed into EVs released from HT29-Snail clones affects positively colorectal cancer progression is not definitely comprehensible from our studies. However, our in vivo studies (Table 1) suggest that the presence of EVs released from CRC cells that undergo EMT and contain packed miR-130b may lead to increased lung inflammation that facilitates cancer progression. This is in agreement with the previously identified effects [37] and suggests that miR-130b may be a target to attempt to slow the CRC progression. Additionally, the presence of miR-130b on EVs released from CRC cells can serve as a biomarker of an advanced stage of CRC.

We found increased amount of miR-483-5p in EVs released from HT29-Snail clone 17 that express the highest amount of Snail and was used in our in vivo studies. Increased content of miR-483-5p was observed in exosomes isolated from plasma of a CRC-patient in various stages and was found in EVs from SW480 CRC line suggesting its diagnostic potential [38]. We have also observed that miR-221, miR-222, and miR-125a were overexpressed in HT29-Snail-EVs, although with lower fold change. MiR-221 is among commonly upregulated miRNAs in CRC in tumors [39] and in patient's serum [40] while miR-222-3p promotes macrophage polarization and differentiation to M2 phenotype in vitro and in vivo, which enhance the progression of epithelial ovarian cancer [41]. Additionally, the decrease in miRNA-34a and miRNA-203b in EVs released from HT29-Snail confirms the previously described reports about repression of miRNA-34 and miRNA-203 by Snail as a part of the EMT program in cells [42,43] that leads to cancer progression. Thus, our analysis of the differences in individual miRNA expression shows that EVs released from cells in early EMT stage can carry miRNAs that promote cell migration and invasion of CRC cells and are associated with poor prognosis for colorectal cancer. Further, the positive regulation of STAT3 is the most significant GO term for Biological Process (Figure S3). Cytokine-driven JAK/STAT3 pathway plays an important role in the processes of signal transduction and is associated with the hyperproliferative and invasive phenotype of CRC cells [44].

Tumors induced by injection of HT29-Snail, as compared to control HT20-pcDNA cells, tend to have a lower rate of growth (Figure 3A). This observation is in an agreement with the inhibition of cell proliferation within the growing tumor due to the cell EMT progression caused by Snail [18]. At the same time we have observed the increased amount of MCP-1/CCL2 in the plasma of mice bearing the tumours induced by HT29-Snail. These data are in agreement with the earlier findings suggesting that Snail, as EMT inducer, can induce MCP-1/CCL2 production [45]. MCP-1/CCL2 is a potent chemoattractant for circulating blood monocytes via binding to its receptor CCR2 but is not an effective chemoattractant for differentiated monocyte-derived macrophages [46,47]. In parallel with the results of inhibited tumour growth and increased production of MCP-1/CCL2 we have observed the increased number of early interstitial pneumonias in mice injected with HT29-Snail cells, as compared to HT29-pcDNA cells (Table 1, group IV and II). Additionally, almost all mice bearing tumour induced by HT29-Snail cells that were also injected with EVs released from HT29-Snail cells, presented the appearance of early interstitial pneumonias (Table 1, group VI), while injection of HT29-pcDNA-EV had no effect (Table 1, group V and IV, respectively). As HT29-Snail-EVs were injected only into the group that was bearing tumours induced by injection of HT29-Snail cells, we cannot exclude the possibility that HT29-Snail-EVs would have an effect on control (HT29-pcDNA) mice and that remains to be elucidated.

Our in vitro findings (Figure 5) suggest that EVs released from HT29-Snail cells that are in an early EMT stage affect macrophages differently than the EVs released from epithelial-like HT29 cells. We observed the inhibition of random motility of macrophages that were treated with HT29-Snail-EVs cells as compared to control EVs released from HT29-pcDNA cells. Thus, the macrophages that are affected by HT29-Snail-EVs may become retained in pre-metastatic niche. We have also found the differences in the secretion of IL-8 by macrophages (Figure 5B,C) that may be attributed to the variability in EVs miRNA profiles. We found for example that miR-142-5p is downregulated in EVs released from HT29-Snail 8 and -17, so relatively there is higher amount of miR-142-5p in control EVs. High concentrations of miR-142-5p were observed in ulcerative colitis (UC), the inflammatory disease that frequently leads to development of colorectal cancer. MiR-142-5p levels were negatively correlated with the expression of suppressor of cytokine signalling 1 (SOCS1) in UC patients [48]. Further, miR-142-5p increased the secretion of IL-6 and IL-8 in TNFα-treated-HT29 [48]. Thus, we postulate that downregulation of IL-8 secretion macrophages treated with HT29-Snail-EVs might be associated to some extend with lower levels of miR-142-5p.

4. Materials and Methods

4.1. Cell Culture and Differentiation

The HT29 cell line (cells: colon, disease: colorectal adenocarcinoma) was obtained from American Type Culture Collection (Manassas, VA, USA) and cultured in McCoy's 5A medium (LifeTechnologies, Waltham, MA, USA), supplemented with 10% FBS (LifeTechnologies) and antibiotics—streptomycin and penicillin (P/S) (Sigma-Aldrich, St. Louis, MO, USA), primocin (Invivogen, San Diego, CA, USA). The cells were periodically tested for mycoplasma using the PlasmoTest (Invivogen). For isolation of EVs released by HT29, serum-free medium with P/S was used to rule out the effect of exosomes of foetal bovine serum.

THP-1 monocyte/macrophage-like cell line from American Type Culture Collection (Manassas, VA, USA) was cultured in RPMI-1640 culture medium supplemented with 1 mM sodium pyruvate, 10% FBS, 0.05 mM 2-ME, P/S and primocin. For differentiation of THP-1 monocytes into THP-1-derived macrophages (TDM), cells were cultured with 20 ng/mL of phorbol ester (PMA) in culture medium for 48 h. Isolations of monocytes from healthy donors were performed as described previously [49]. To generate monocyte-derived macrophages (MDM) human monocytes were cultured 7 days with 10 ng/mL of GM-CSF (Thermo Scientific, Waltham, MA, USA) in RPMI-1640 supplemented with 10% human serum type AB from Sigma-Aldrich (St. Louis, MO, USA). All cell cultures were performed in a 90–95% humidified atmosphere of 5% CO_2.

4.2. HT29 Stable Clone Generation and Isolation of EVs from Culture Supernatants

The pcDNA3.1 vector (Invitrogen, Carlsbad, CA, USA) and pcDNA3.1 vector expressing Snail was obtained from Prof. Muh-Hwa Yang (Institute of Clinical Medicine, National Yang-Ming University, Taipei, Taiwan). HT29 nucleofection and clone generation was performed as previously described [18]. Western blot showing the levels of Snail in clones -3, -8, and -17 is shown in Supplemental Figure S5. The HT29 cell line and HT29-pcDNA control clone were authenticated by ATCC using Short Tandem Repeat (STR) analysis. Extracellular vesicles were isolated by differential centrifugations and subsequent ultra-centrifugations as described previously [19]. HT29-Snail 3, 8, 17 and HT29-pcDNA clones were grown to 70–80% confluence on 15-cm dishes, washed three times with empty medium to remove vesicles present in FBS of culture medium and then cultured for 24 h to obtain conditioned medium. Next the medium was collected and centrifuged ($350\times g$ for 10 min) to remove floating cells. The supernatants were then collected and centrifuged at $2000\times g$ for 20 min to remove APOs. Finally, EVs pellets were obtained after ultra-centrifugation for 1.5 h at $100{,}000\times g$ using OPTIMA L-80 Ultracentrifuge and Type 45 Ti Rotor, Fixed Angle (Beckman Coulter, Inc., Brea, CA, USA). The pellets were next washed by diluting in PBS and centrifuging for 1.5 h at $100{,}000\times g$. All centrifugations were performed at 4 °C. Finally, EVs pellets were resuspend in PBS for Western blots or in appropriate media for functional experiments.

4.3. Nanoparticle Tracking Analysis (NTA)

EV size distribution and quantification of vesicles were analyzed by NTA using a NanoSight NS300 System (Malvern Panalytical Ltd., Malvern, UK) by a courtesy of the representative of company (A.P. Instruments, Warsaw, Poland).

4.4. Transmission Electron Microscopy (TEM)

TEM assay was used to evaluate the shape and size of EVs. Ten microliters of the sample were placed on 200-mesh copper grids with a carbon surface. The samples were negatively stained with 2% uranyl acetate for 1 min. and dried at room temperature. The transmission electron microscopy images were obtained using JEOL-1010 (Akishima, Japan).

4.5. miRNA Isolation from Extracellular Vesicles and miRNA Content Analysis

EVs pellets were treated with RNAze A (20 µg/mL in PBS). Total RNA was isolated and quality control of RNA was performed as described earlier [19]. Next-generation sequencing (NGS) analysis of the miRNAs was performed by Exiqon (https://www.exiqon.com/small-rna-ngs). For the comparisons between control EVs released from HT29-pcDNA and EVs released from clones HT29-Snail 3, -8, -17, the Benjamini-Hochberg FDR corrected p-values were calculated.

4.6. Animal Studies

The animal experiments were performed in the Center for Experimental Medicine Medical University of Bialystok (PL), in compliance with the Local Ethical Committee for Experiments on Animals in Olsztyn. Female CByJ.Cg- Foxn1<nu>/ccmdb mice (~20 g, 6–8 weeks old) were obtained from the same Center. Animals in group I were left as controls. The primary tumours were established by inoculating HT29-pcDNA or HT29-Snail 17 cells (1.5×10^6/100 µL PBS) subcutaneously into the flank of the mice (group II and IV respectively). Tumour volumes were calculated: (length x width^2)/2. Animals from group II and IV were additionally injected intravenously (tail vein) with 10 µg/mouse of indicated EVs (groups III, V and VI). Blood was collected from the retroorbital sinus (~50µL at day 7) and during the section of animals at the end of the procedure (day 28). Collected organs (liver, lungs, large intestine) were divided and fixed and stained with haematoxylin and eosin (H & E). Tumour growth was estimated by fitting each animal's tumour growth to an exponential model. To fit data to this model, at first low volumes were truncated (up to 12 day) and $\log_{10}$ tumour volume versus time for each animal was plotted. The slopes and R^2 values for the fits were calculated using linear regression [50].

4.7. PKH67 Labelling of Extracellular Vesicles and Their Uptake into TDM Cells

Extracellular vesicles were labelled using PKH67 Fluorescent Cell Linker kit (Sigma-Aldrich) according to the manufacturer's instructions, with minor modifications [51]. After 4 h of PKH67-labelled EV incorporation into TDM, cells were fixed, treated with 0.1% Triton X-100 and incubated in sequence with Texas Red®-X phalloidin (F-actin marker) and Hoechst 33,342 (cell-permeant nuclear dye). The uptake was visualized using a confocal microscope (Nikon D-Eclipse C1) and analysed with EZ-C1 software.

4.8. Cell Motility Measurements

Measurements of cell motility were performed between the 7th and 25th hour after beginning of incubation of TDM with EVs using HoloMonitor M4 (Phase Holographic Imaging PHI AB, Lund, Sweden—Courtesy of the representative of company). Cell speed was calculated using the App Suite software of the same company. A plot of cell speed versus time was generated and the area under curve (AUC) was computed using GraphPad Prism 7.05 software.

4.9. Measurements of Cytokine Release

Macrophages were incubated with EVs in FBS-free RPMI-1640 culture medium for 24 h, the medium collected and centrifuged for 20 min at 18,000× *g*. IL-8 levels in the cells supernatants were analysed by flow cytometry according to the manufacturer's procedure using the BD Cytometric Bead Array (CBA) Human Inflammatory Cytokines Kit. The BD Cytometric Bead Array (CBA) Mouse Inflammation Kit was used to measure MCP-1/CCL2 in the plasma of mice. Data were analysed using FCAP Array^{TM} Software Version 3.0 (BD Life Sciences—Biosciences, San Jose, CA, USA).

4.10. Statistical Analyses

Data are presented as mean ± SEM. All experiments were performed at least in triplicate. The Shapiro-Wilk test was used to confirm the Gaussian distributions of raw data. Analysis of variance (ANOVA) was used for multiple comparisons. The Kruskal-Wallis analysis was performed to test the differences between groups of data with non-normal distributions. For analyses of two groups, the appropriate Student's *t* test (or the Welch's test for unequal variances) was performed to test the differences between groups for normally distributed data. p value less than 0.05 was considered statistically significant. All statistical analysis was performed using GraphPad Prism 7.05 software.

5. Conclusions

We conclude that Snail that modifies CRC cells towards a more invasive phenotype, can also alter microRNA cargo of cell-released EVs. Thus, the content of cell-released EVs may serve as a biomarker that defines the stage of CRC and either Snail, or the different microRNAs that is carried by EVs to the destination sites, which serves as a pre-metastatic niche, may be a target to prevent cancer progression. We also point to the macrophages that may reside in the tumour pre-metastatic niche, as one of the possible EVs recipient cells that is modified during growing CRC invasiveness.

Supplementary Materials: The following are available online at https://www.mdpi.com/2072-6694/13/2/172/s1, Figure S1: Western blot of EVs markers, Figure S2: Global efficiency of miRNA processing, Figure S3: Gene Ontology Enrichment Analysis, Figure S4: Cell viability assay, Figure S5: Snail transcription factor expression in HT29 clones, Table S1: Clone-specific miRNA evaluated during analysis of differentially expressed extracellular vesicles miRNA (fold change ≤ -2.0 for down-regulation and ≥ 2.0 for up-regulation) between each clone overexpressing Snail and control clone, Table S2: Differentially expressed miRNA evaluated during analysis of extracellular vesicles miRNA between each clone overexpressing Snail and control clone, Table S3: The growth rate of tumours and MCP-1/CCL2 levels in plasma of mice injected s.c. with of control HT29-pcDNA or HT29-Snail cells followed by i.v. injections of various EVs.

Author Contributions: Conceptualization: I.P.-P. and M.A.K.; methodology, I.P.-P., P.P., O.S.-K., J.B. and M.W.-D.; software: D.K.; validation: I.P.-P., P.P., M.W.-D. and M.A.K.; formal analysis: I.P.-P., P.P. and D.K.; investigation: I.P.-P., D.K., K.S., I.S.-K., J.B.; data curation: I.P.-P., P.P. and M.A.K.; writing—original draft preparation: I.P.-P., P.P., M.A.K.; writing—review and editing: M.A.K., J.B. and M.W.-D.; visualization: I.P.-P., D.K., O.S.-K. and M.W.-D.; supervision: M.A.K.; project administration: I.P.-P. and M.A.K.; funding acquisition: M.A.K., P.P. and J.B. All authors have read and agreed to the published version of the manuscript reported.

Funding: This research was funded by the National Science Centre, Cracow, Poland, grant number 2011/02/A/NZ3/00068 and 2018/29/B/NZ5/01756 (M.A.K), 2016/22/E/NZ3/00341 (P.P) and IBM PAS Statutory Funds (J.B.). The APC was funded by IBM PAS.

Institutional Review Board Statement: The study was conducted according to the guidelines of the Declaration of Helsinki, and approved by the Local Ethical Committee for Experiments on Animals in Olsztyn (approval number 55/2017, date: 2017/07/25).

Informed Consent Statement: Not applicable.

Data Availability Statement: Publicly available datasets were analyzed in this study. This data can be found here: https://www.ncbi.nlm.nih.gov/sra/PRJNA674779.

Acknowledgments: The Authors gratefully acknowledge Łucja Balcerzak and Sylwia Michlewska from Laboratory of Microscopic Imaging and Specialized Biological Techniques, Faculty of Biology and Environmental Protection, University of Lodz, Poland, for transmission electron microscopy (TEM) imaging of the EVs samples.

Conflicts of Interest: The authors declare no conflict of interest. The funders had no role in the design of the study; in the collection, analyses, or interpretation of data; in the writing of the manuscript, or in the decision to publish the results.

References

1. Arnold, M.; Sierra, M.S.; Laversanne, M.; Soerjomataram, I.; Jemal, A.; Bray, F. Global patterns and trends in colorectal cancer incidence and mortality. *Gut* **2017**, *66*, 683–691. [CrossRef]
2. Brenner, H.; Kloor, M.; Pox, C.P. Colorectal cancer. *Lancet* **2014**, *383*, 1490–1502. [CrossRef]
3. Papiewska-Pajak, I.; Kowalska, M.A.; Boncela, J. Expression and activity of SNAIL transcription factor during Epithelial to Mesenchymal Transition (EMT) in cancer progression. *Postepy Hig. Med. Dosw.* **2016**, *70*, 968–980. [CrossRef] [PubMed]
4. Kaplan, R.N.; Rafii, S.; Lyden, D. Preparing the "soil": The premetastatic niche. *Cancer Res.* **2006**, *66*, 11089–11093. [CrossRef] [PubMed]
5. Liu, Y.; Cao, X. Characteristics and Significance of the Pre-metastatic Niche. *Cancer Cell* **2016**, *30*, 668–681. [CrossRef]
6. Zhao, H.; Achreja, A.; Iessi, E.; Logozzi, M.; Mizzoni, D.; Di Raimo, R.; Nagrath, D.; Fais, S. The key role of extracellular vesicles in the metastatic process. *Biochim. Biophys. Acta Rev. Cancer* **2018**, *1869*, 64–77. [CrossRef]
7. Admyre, C.; Johansson, S.M.; Qazi, K.R.; Filén, J.J.; Lahesmaa, R.; Norman, M.; Neve, E.P.; Scheynius, A.; Gabrielsson, S. Exosomes with immune modulatory features are present in human breast milk. *J. Immunol.* **2007**, *179*, 1969–1978. [CrossRef]
8. Doyle, L.M.; Wang, M.Z. Overview of Extracellular Vesicles, Their Origin, Composition, Purpose, and Methods for Exosome Isolation and Analysis. *Cells* **2019**, *8*, 727. [CrossRef]
9. Manning, S.; Danielson, K.M. The immunomodulatory role of tumor-derived extracellular vesicles in colorectal cancer. *Immunol. Cell Biol.* **2018**. [CrossRef]
10. Mittelbrunn, M.; Gutierrez-Vazquez, C.; Villarroya-Beltri, C.; Gonzalez, S.; Sanchez-Cabo, F.; Gonzalez, M.A.; Bernad, A.; Sanchez-Madrid, F. Unidirectional transfer of microRNA-loaded exosomes from T cells to antigen-presenting cells. *Nat. Commun.* **2011**, *2*, 282. [CrossRef]
11. Cha, D.J.; Franklin, J.L.; Dou, Y.; Liu, Q.; Higginbotham, J.N.; Demory Beckler, M.; Weaver, A.M.; Vickers, K.; Prasad, N.; Levy, S.; et al. KRAS-dependent sorting of miRNA to exosomes. *eLife* **2015**, *4*, e07197. [CrossRef] [PubMed]
12. Jabalee, J.; Towle, R.; Garnis, C. The Role of Extracellular Vesicles in Cancer: Cargo, Function, and Therapeutic Implications. *Cells* **2018**, *7*, 93. [CrossRef]
13. Becker, A.; Thakur, B.K.; Weiss, J.M.; Kim, H.S.; Peinado, H.; Lyden, D. Extracellular Vesicles in Cancer: Cell-to-Cell Mediators of Metastasis. *Cancer Cell* **2016**, *30*, 836–848. [CrossRef] [PubMed]
14. Rackov, G.; Garcia-Romero, N.; Esteban-Rubio, S.; Carrion-Navarro, J.; Belda-Iniesta, C.; Ayuso-Sacido, A. Vesicle-Mediated Control of Cell Function: The Role of Extracellular Matrix and Microenvironment. *Front. Physiol.* **2018**, *9*, 651. [CrossRef] [PubMed]
15. Song, W.; Yan, D.; Wei, T.; Liu, Q.; Zhou, X.; Liu, J. Tumor-derived extracellular vesicles in angiogenesis. *Biomed. Pharm.* **2018**, *102*, 1203–1208. [CrossRef]
16. Xiao, M.; Zhang, J.; Chen, W.; Chen, W. M1-like tumor-associated macrophages activated by exosome-transferred THBS1 promote malignant migration in oral squamous cell carcinoma. *J. Exp. Clin. Cancer Res.* **2018**, *37*, 143. [CrossRef]
17. Atretkhany, K.N.; Drutskaya, M.S.; Nedospasov, S.A.; Grivennikov, S.I.; Kuprash, D.V. Chemokines, cytokines and exosomes help tumors to shape inflammatory microenvironment. *Pharm. Ther.* **2016**, *168*, 98–112. [CrossRef]
18. Przygodzka, P.; Papiewska-Pajak, I.; Bogusz, H.; Kryczka, J.; Sobierajska, K.; Kowalska, M.A.; Boncela, J. Neuromedin U is upregulated by Snail at early stages of EMT in HT29 colon cancer cells. *Biochim. Biophys. Acta* **2016**, *1860*, 2445–2453. [CrossRef]
19. Przygodzka, P.; Papiewska-Pajak, I.; Bogusz-Koziarska, H.; Sochacka, E.; Boncela, J.; Kowalska, M.A. Regulation of miRNAs by Snail during epithelial-to-mesenchymal transition in HT29 colon cancer cells. *Sci. Rep.* **2019**, *9*, 2165. [CrossRef]
20. Willms, E.; Cabanas, C.; Mager, I.; Wood, M.J.A.; Vader, P. Extracellular Vesicle Heterogeneity: Subpopulations, Isolation Techniques, and Diverse Functions in Cancer Progression. *Front. Immunol.* **2018**, *9*, 738. [CrossRef]
21. Kryczka, J.; Papiewska-Pajak, I.; Kowalska, M.A.; Boncela, J. Cathepsin B Is Upregulated and Mediates ECM Degradation in Colon Adenocarcinoma HT29 Cells Overexpressing Snail. *Cells* **2019**, *8*, 203. [CrossRef] [PubMed]
22. Papiewska-Pająk, I.; Krzyżanowski, D.; Katela, M.; Rivet, R.; Michlewska, S.; Przygodzka, P.; Kowalska, M.A.; Brézillon, S. Glypican-1 Level Is Elevated in Extracellular Vesicles Released from MC38 Colon Adenocarcinoma Cells Overexpressing Snail. *Cells* **2020**, *9*, 1585. [CrossRef] [PubMed]

23. Zaborowski, M.P.; Balaj, L.; Breakefield, X.O.; Lai, C.P. Extracellular Vesicles: Composition, Biological Relevance, and Methods of Study. *Bioscience* **2015**, *65*, 783–797. [CrossRef] [PubMed]
24. Redzic, J.S.; Balaj, L.; van der Vos, K.E.; Breakefield, X.O. Extracellular RNA mediates and marks cancer progression. *Semin. Cancer Biol.* **2014**, *28*, 14–23. [CrossRef]
25. Orang, A.V.; Safaralizadeh, R.; Hosseinpour Feizi, M.A.; Somi, M.H. Diagnostic and prognostic value of miR-205 in colorectal cancer. *Asian Pac. J. Cancer Prev.* **2014**, *15*, 4033–4037. [CrossRef]
26. Eyking, A.; Reis, H.; Frank, M.; Gerken, G.; Schmid, K.W.; Cario, E. MiR-205 and MiR-373 Are Associated with Aggressive Human Mucinous Colorectal Cancer. *PLoS ONE* **2016**, *11*, e0156871. [CrossRef]
27. Boyerinas, B.; Park, S.M.; Hau, A.; Murmann, A.E.; Peter, M.E. The role of let-7 in cell differentiation and cancer. *Endocr. Relat. Cancer* **2010**, *17*, F19–F36. [CrossRef]
28. Svoronos, A.A.; Engelman, D.M.; Slack, F.J. OncomiR or Tumor Suppressor? The Duplicity of MicroRNAs in Cancer. *Cancer Res.* **2016**, *76*, 3666–3670. [CrossRef]
29. Garzon, R.; Calin, G.A.; Croce, C.M. MicroRNAs in Cancer. *Annu. Rev. Med.* **2009**, *60*, 167–179. [CrossRef]
30. Chen, X.; Xiong, W.; Li, H. Comparison of microRNA expression profiles in K562-cells-derived microvesicles and parental cells, and analysis of their roles in leukemia. *Oncol. Lett.* **2016**, *12*, 4937–4948. [CrossRef]
31. Leone, V.; Langella, C.; D'Angelo, D.; Mussnich, P.; Wierinckx, A.; Terracciano, L.; Raverot, G.; Lachuer, J.; Rotondi, S.; Jaffrain-Rea, M.L.; et al. Mir-23b and miR-130b expression is downregulated in pituitary adenomas. *Mol. Cell. Endocrinol.* **2014**, *390*, 1–7. [CrossRef] [PubMed]
32. Dong, P.; Karaayvaz, M.; Jia, N.; Kaneuchi, M.; Hamada, J.; Watari, H.; Sudo, S.; Ju, J.; Sakuragi, N. Mutant p53 gain-of-function induces epithelial-mesenchymal transition through modulation of the miR-130b-ZEB1 axis. *Oncogene* **2013**, *32*, 3286–3295. [CrossRef] [PubMed]
33. Wu, X.; Weng, L.; Li, X.; Guo, C.; Pal, S.K.; Jin, J.M.; Li, Y.; Nelson, R.A.; Mu, B.; Onami, S.H.; et al. Identification of a 4-microRNA signature for clear cell renal cell carcinoma metastasis and prognosis. *PLoS ONE* **2012**, *7*, e35661. [CrossRef] [PubMed]
34. Scheffer, A.R.; Holdenrieder, S.; Kristiansen, G.; von Ruecker, A.; Muller, S.C.; Ellinger, J. Circulating microRNAs in serum: Novel biomarkers for patients with bladder cancer? *World J. Urol.* **2014**, *32*, 353–358. [CrossRef]
35. Sand, M.; Skrygan, M.; Sand, D.; Georgas, D.; Gambichler, T.; Hahn, S.A.; Altmeyer, P.; Bechara, F.G. Comparative microarray analysis of microRNA expression profiles in primary cutaneous malignant melanoma, cutaneous malignant melanoma metastases, and benign melanocytic nevi. *Cell Tissue Res.* **2013**, *351*, 85–98. [CrossRef]
36. Zhao, Y.; Miao, G.; Li, Y.; Isaji, T.; Gu, J.; Li, J.; Qi, R. MicroRNA- 130b suppresses migration and invasion of colorectal cancer cells through downregulation of integrin beta1 [corrected]. *PLoS ONE* **2014**, *9*, e87938. [CrossRef]
37. Colangelo, T.; Fucci, A.; Votino, C.; Sabatino, L.; Pancione, M.; Laudanna, C.; Binaschi, M.; Bigioni, M.; Maggi, C.A.; Parente, D.; et al. MicroRNA-130b promotes tumor development and is associated with poor prognosis in colorectal cancer. *Neoplasia* **2013**, *15*, 1086–1099. [CrossRef]
38. Chen, M.; Xu, R.; Rai, A.; Suwakulsiri, W.; Izumikawa, K.; Ishikawa, H.; Greening, D.W.; Takahashi, N.; Simpson, R.J. Distinct shed microvesicle and exosome microRNA signatures reveal diagnostic markers for colorectal cancer. *PLoS ONE* **2019**, *14*, e0210003. [CrossRef]
39. Volinia, S.; Calin, G.A.; Liu, C.G.; Ambs, S.; Cimmino, A.; Petrocca, F.; Visone, R.; Iorio, M.; Roldo, C.; Ferracin, M.; et al. A microRNA expression signature of human solid tumors defines cancer gene targets. *Proc. Natl. Acad. Sci. USA* **2006**, *103*, 2257–2261. [CrossRef]
40. Chen, X.; Ba, Y.; Ma, L.; Cai, X.; Yin, Y.; Wang, K.; Guo, J.; Zhang, Y.; Chen, J.; Guo, X.; et al. Characterization of microRNAs in serum: A novel class of biomarkers for diagnosis of cancer and other diseases. *Cell Res.* **2008**, *18*, 997–1006. [CrossRef]
41. Ying, X.; Wu, Q.; Wu, X.; Zhu, Q.; Wang, X.; Jiang, L.; Chen, X.; Wang, X. Epithelial ovarian cancer-secreted exosomal miR-222-3p induces polarization of tumor-associated macrophages. *Oncotarget* **2016**, *7*, 43076–43087. [CrossRef] [PubMed]
42. Siemens, H.; Jackstadt, R.; Hunten, S.; Kaller, M.; Menssen, A.; Gotz, U.; Hermeking, H. miR-34 and SNAIL form a double-negative feedback loop to regulate epithelial-mesenchymal transitions. *Cell Cycle* **2011**, *10*, 4256–4271. [CrossRef] [PubMed]
43. Moes, M.; Le Bechec, A.; Crespo, I.; Laurini, C.; Halavatyi, A.; Vetter, G.; Del Sol, A.; Friederich, E. A novel network integrating a miRNA-203/SNAI1 feedback loop which regulates epithelial to mesenchymal transition. *PLoS ONE* **2012**, *7*, e35440. [CrossRef] [PubMed]
44. Gordziel, C.; Bratsch, J.; Moriggl, R.; Knosel, T.; Friedrich, K. Both STAT1 and STAT3 are favourable prognostic determinants in colorectal carcinoma. *Br. J. Cancer* **2013**, *109*, 138–146. [CrossRef]
45. Hsu, D.S.; Wang, H.J.; Tai, S.K.; Chou, C.H.; Hsieh, C.H.; Chiu, P.H.; Chen, N.J.; Yang, M.H. Acetylation of snail modulates the cytokinome of cancer cells to enhance the recruitment of macrophages. *Cancer Cell* **2014**, *26*, 534–548. [CrossRef]
46. Yoshimura, T. cDNA cloning of guinea pig monocyte chemoattractant protein-1 and expression of the recombinant protein. *J. Immunol.* **1993**, *150*, 5025–5032.
47. Phillips, R.J.; Lutz, M.; Premack, B. Differential signaling mechanisms regulate expression of CC chemokine receptor-2 during monocyte maturation. *J. Inflamm.* **2005**, *2*, 14. [CrossRef]

48. Han, J.; Li, Y.; Zhang, H.; Guo, J.; Wang, X.; Khang, Y.; Luo, Y.; Wu, M.; Zhang, X. MicroRNA-142-5p facilitates the pathogenesis of ulcerative colitis by regulating SOCS1. *Int. J. Clin. Exp. Pathol.* **2018**, *11*, 5735–5744.
49. Szulc-Kielbik, I.; Brzezinska, M.; Kielbik, M.; Brzostek, A.; Dziadek, J.; Kania, K.; Sulowska, Z.; Krupa, A.; Klink, M. Mycobacterium tuberculosis RecA is indispensable for inhibition of the mitogen-activated protein kinase-dependent bactericidal activity of THP-1-derived macrophages in vitro. *FEBS J.* **2015**, *282*, 1289–1306. [CrossRef]
50. Hather, G.; Liu, R.; Bandi, S.; Mettetal, J.; Manfredi, M.; Shyu, W.C.; Donelan, J.; Chakravarty, A. Growth rate analysis and efficient experimental design for tumor xenograft studies. *Cancer Inf.* **2014**, *13*, 65–72. [CrossRef]
51. Chiba, M.; Watanabe, N.; Watanabe, M.; Sakamoto, M.; Sato, A.; Fujisaki, M.; Kubota, S.; Monzen, S.; Maruyama, A.; Nanashima, N.; et al. Exosomes derived from SW480 colorectal cancer cells promote cell migration in HepG2 hepatocellular cancer cells via the mitogen-activated protein kinase pathway. *Int. J. Oncol.* **2016**, *48*, 305–312. [CrossRef] [PubMed]

MDPI

Article

Performance of the Use of Genetic Information to Assess the Risk of Colorectal Cancer in the Basque Population

Koldo Garcia-Etxebarria [1,2,*], Ane Etxart [3], Maialen Barrero [3], Beatriz Nafria [3], Nerea Miren Segues Merino [3], Irati Romero-Garmendia [4], Andre Franke [5], Mauro D'Amato [6,7,8] and Luis Bujanda [2,3]

1 Biodonostia, Gastrointestinal Genetics Group, 20014 San Sebastián, Spain
2 Centro de Investigación Biomédica en Red de Enfermedades Hepáticas y Digestivas (CIBERehd), 08036 Barcelona, Spain
3 Biodonostia, Gastrointestinal Disease Group, Universidad del País Vasco (UPV/EHU), 20014 San Sebastián, Spain
4 Department of Genetics, Physical Anthropology and Animal Physiology, University of the Basque Country (Universidad del País Vasco/Euskal Herriko Unibertsitatea), 48940 Leioa, Spain
5 Institute of Clinical Molecular Biology, Christian-Albrechts-University of Kiel, 24105 Kiel, Germany
6 Gastrointestinal Genetics Lab, CIC bioGUNE, Basque Research and Technology Alliance, 48160 Derio, Spain
7 IKERBASQUE, Basque Foundation for Sciences, 48009 Bilbao, Spain
8 Department of Medicine and Surgery, LUM University, 70010 Casamassima, Italy
* Correspondence: koldo.garcia@biodonostia.org

Simple Summary: The risk of developing colorectal cancer (CRC) is partially associated with genetics. Different studies have provided valuable genetic information to understand the biology behind CRC and to build models of genetic risk. However, the study of the applicability of such genetic information within the Basque population is limited. Thus, our objectives were to find out if the genetic variants associated with CRC in other populations are the same in the Basque population and to assess the performance of the use of genetic information to calculate the risk of developing CRC. We found that the available genetic information can be applied to the Basque population, although local genetic variation can affect its use. Our findings will help to refine the use of CRC genetic risk calculation in the Basque population, and we expect that our findings could be useful for other populations.

Abstract: Although the genetic contribution to colorectal cancer (CRC) has been studied in various populations, studies on the applicability of available genetic information in the Basque population are scarce. In total, 835 CRC cases and 940 controls from the Basque population were genotyped and genome-wide association studies were carried out. Mendelian Randomization analyses were used to discover the effect of modifiable risk factors and microbiota on CRC. In total, 25 polygenic risk score models were evaluated to assess their performance in CRC risk calculation. Moreover, 492 inflammatory bowel disease cases were used to assess whether that genetic information would not confuse both conditions. Five suggestive ($p < 5 \times 10^{-6}$) *loci* were associated with CRC risk, where genes previously associated with CRC were located (e.g., *ABCA12*, *ATIC* or *ERBB4*). Moreover, the analyses of CRC locations detected additional genes consistent with the biology of CRC. The possible contribution of cholesterol, BMI, Firmicutes and Cyanobacteria to CRC risk was detected by Mendelian Randomization. Finally, although polygenic risk score models showed variable performance, the best model performed correctly regardless of the location and did not misclassify inflammatory bowel disease cases. Our results are consistent with CRC biology and genetic risk models and could be applied to assess CRC risk in the Basque population.

Keywords: colorectal cancer; genome-wide association study; Mendelian randomization; polygenic risk scores

Citation: Garcia-Etxebarria, K.; Etxart, A.; Barrero, M.; Nafria, B.; Segues Merino, N.M.; Romero-Garmendia, I.; Franke, A.; D'Amato, M.; Bujanda, L. Performance of the Use of Genetic Information to Assess the Risk of Colorectal Cancer in the Basque Population. *Cancers* **2022**, *14*, 4193. https://doi.org/10.3390/cancers14174193

Academic Editor: Stephane Dedieu

Received: 16 June 2022
Accepted: 26 August 2022
Published: 29 August 2022

Publisher's Note: MDPI stays neutral with regard to jurisdictional claims in published maps and institutional affiliations.

1. Introduction

In total, 10% of the cancers diagnosed in the world are colorectal cancers (CRC) and, in addition, CRC is the second cause of cancer death in developed countries [1,2]. The development of CRC can be sporadic or due to inflammatory processes [3]; the risk of CRC is influenced by the environment, genetics, and microbial composition [4,5]. Since CRC is a major public health issue, different strategies for its early detection and prognosis have been proposed and developed [6].

As mentioned, genetic factors are involved in CRC risk, or they can be associated with other risk factors related to CRC. As a consequence, their utility as biomarkers has been explored: their role in CRC risk has been studied by analyzing specific genetic variants [7–9], as well as, genome-wide association studies (GWAS) [10]. Moreover, the effect of genetic information on modifiable risk factors (e.g., lipids level) on CRC has been analyzed using Mendelian Randomization analyses [11,12], a method to estimate causal effects if specific assumptions are fulfilled. In addition, it has been detected that some genetic variants involved in the abundance of some microbial groups are related to CRC risk [13]. Finally, it has been proposed that polygenic risk scores (PRS) derived from different genetic studies are useful to predict the risk of CRC of one individual based on the carriership of risk genetic variants, among other factors [14,15].

Previously, 48 SNPs associated with CRC were analyzed in 230 CRC cases and 230 controls from the Basque population [16]. From those analyzed SNPs, only rs6687758 SNP was associated with CRC risk, and the application of those 48 SNPs as a model to predict PRS risk was successful [16]. Indeed, the Basque population has a particular genetic history compared to the rest of the European population, since the migrations associated with the Steppe pastoralism had less effect on that population, therefore, genetic variants from populations that lived in Europe in the Neolithic [17] or Iron Age [18] could be higher. Previously, a genetic study of this cohort showed that it was useful to study the effect of local genetic variants on the risk and ability to predict the risk of complex diseases [19]. In addition, according to the data available from the Basque Statistic Institute (https://en.eustat.eus, accessed: 1 August 2022), between 2016 and 2019, in the Basque Autonomous Community (Northern Spain) CRC caused 8356 hospitalizations (on average, 95.58 hospitalizations per 100.000 habitants per year), while in the rest of Spain there were 101.12 hospitalizations per 100.000 habitants per year (between 2016–2019, according with Instituto Nacional de Estadística, https://www.ine.es, accessed: 1 August 2022), and in Europe, there were 123.45 hospitalizations per 100.000 habitants per year (between 2016–2019, according to Eurostat, https://ec.europa.eu/eurostat, accessed: 1 August 2022).

In the present study, we analyze a larger Basque cohort (835 cases and 940 controls) to detect the risk factors for CRC that can be explained or inferred from the genetic component of CRC using genome-wide association studies and Mendelian Randomization to assess the applicability of existing CRC PRS models on this population.

2. Materials and Methods

2.1. Recruitment

CRC cases were diagnosed using standard criteria and the samples used in this study were obtained in the standard clinical practice, after informed consent, in Hospital Universitario Donostia (San Sebastian, Spain). The samples of non-CRC controls were obtained through the Basque Biobank; the samples were sourced from healthy blood donors (the age range to be eligible to be a blood donor is 18–65). The information of those blood donors is anonymized and only information about sex and age is made available. In total, 869 cases were recruited, and 987 controls were used.

The present study was approved by the Local Ethics Committee (Comité de Ética de la Investigación con medicamentos de Euskadi, code: PI+CES-BIOEF 2017-10).

2.2. Genotyping and Imputation

Illumina Global Screening Array was used to genotype the DNA samples of the individuals analyzed in this work. For this, Illumina iScan high-throughput screening system was used in the Institute of Clinical Molecular Biology (Kiel, Germany). Raw intensities were transformed to alleles using the GenCall algorithm available in Illumina GenomeStudio software.

Then, the called genotypes and samples were filtered using the following criteria: samples with $\geq$5% missing rates; markers with non-called alleles; markers with missing call rates > 0.05; related samples (PI-HAT > 0.1875); samples whose genotyped sex could not be determined; samples with high heterozygosity rate (more than 3 times SD from the mean) were excluded. In addition, only autosomal SNPs were kept; markers with Hardy–Weinberg equilibrium $p < 1 \times 10^{-5}$; markers whose P of difference in missingness between cases and control was $<1 \times 10^{-5}$; samples that were outliers, identified using principal component analysis (deviation of more than 6 times interquartile range), using FlashPCA (v2.0) [20], were removed.

Additional SNPs were imputed using the Sanger Imputation service. Release 1.1 of the Haplotype Reference Consortium was used as a reference panel, and the EAGLE2+PBWT pipeline was used to carry out the imputation [21–23]. Once imputed, markers with INFO score < 0.80, MAF < 0.01 and non-biallelic markers were removed.

After genotyping, quality control and imputation, 5,399,981 SNPs from 1775 individuals (835 cases and 940 controls) were kept.

2.3. Genetic Analyses

2.3.1. Admixture Analysis

Genotyped SNPs were pruned using Plink (v1.90) [24] and SNPs from regions with high linkage disequilibrium were removed. Admixture (v1.3) [25] was used to analyze the admixture of the samples of our cohort, with settings K between 1 and 10, and using the results with the lowest cross-validation value.

2.3.2. Genome-Wide Association Study

GWAS analyses of CRC cases and non-CRC controls were performed using logistic regression implemented in Plink [24], adjusting by sex, age and the first 4 principal components. In addition, GWAS of right colon cancer, left colon cancer, and rectal cancer vs non-CRC controls, as well as right colon cancer vs left colon cancer, and colon cancer vs rectal cancer were carried out using logistic regression implemented in Plink, and adjusting by sex, age and first 4 principal components.

To compare our results with SNPs previously associated with CRC, SNPs associated with the "Colorectal cancer" term (EFO_0005842) and studied in populations of European origin were retrieved from GWAS Catalog [26]. In total, 209 SNP from 34 studies were retrieved.

Moreover, CRC patients were compared to 492 inflammatory bowel disease patients without CRC [19] to find genetic differences in our cohort. To perform that analysis, a logistic regression implemented in Plink, adjusting by sex, age and first 4 principal components, was used. In addition, a comparison of CRC patients against the mentioned inflammatory bowel disease patients plus controls was carried out.

2.3.3. Mendelian Randomization Analyses

For carrying out Mendelian Randomization (MR) analyses TwoSampleMR (v0.5.6) [27] and gsmr (v1.0.9) [28] packages from R language (v4.0.5) were used [29], as we have used previously to study the effect of modifiable risk factors in CRC risk [13].

First, we selected the modifiable risk factors based on a previous work [12] which analyzed modifiable risk factors using Mendelian Randomization that affects CRC (BMI, cholesterol, triglycerides, selenium, iron, vitamin B12, metabolism, body fat percentage, waist circumference, IL6 receptor and height). Then, we retrieved the instruments available in MRC-IEU (https://gwas.mrcieu.ac.uk, accessed: 14 February 2022) of those traits through TwoSampleMR [27]. In addition, to analyze the effect of the microbiota in CRC cancer, we retrieved instruments of bacterial phyla which are available from MiBioGen consortium data [30].

Then, the analysis was carried out if 10 or more instruments were available, and HEIDI outlier analysis was used to discard heterogenous instruments. The strength of the instruments was measured by the F-statistic: $F = R^2(N - K - 1)/K(1 - R^2)$, where R^2 is the variance explained by genetic variance, N is the sample size, and K is the number of instruments [31]. In addition, I^2 was calculated using TwoSampleMR R Package.

The MR analyses were carried out using Inverse Variance Weighted, Weighted Median and MR Egger methods. In addition, the heterogeneity Q test and pleiotropy test available in TwoSampleMR R Package were used as sensitivity tests. The analysis was applied to all CRC cases, as well as, right colon cancer, left colon cancer and rectal cancer analyses.

2.3.4. Polygenic Risk Scores

Polygenic risk scores (PRS) were retrieved from PGS Catalog [32]. 29 scores available in the "Colorectal cancer" term (EFO_0005842) derived using cohorts with >90% samples of European ancestry and whose assembly version was known were used for the PRS analysis [33–41]. From those 29 panels, our cohort had available SNPs to apply in 25 of them. In addition, the PRS used previously in the Basque population was tested [16]. The weights of the SNPs present in our data were applied in our cohort using Plink [24]. The performance of the PRS was measured by comparing the PRS score distribution of CRC cases and non-CRC controls using a T-test using R language [29]; the effect size of the T-test was calculated using Cohen's d through the package rstatix (https://CRAN.R-project.org/package=rstatix, accessed: 28 April 2022) of R language, the area under de curve, sensitivity and specificity was calculated using pROC package of R language. The 95% of confidence interval of the area under the curve was calculated using that package and the DeLong method.

In addition, CRC PRS were applied in 492 patients with inflammatory bowel disease without CRC [19] to measure the ability to distinguish both conditions.

Additional statistical analyses and graphics were done using R language [29].

3. Results

In this study, we have analyzed 835 CRC cases and 940 population-based controls (Table 1). In the cases and the controls, around two-thirds of the individuals were males (63.47% and 67.13%, respectively), and cases were older (average age, 73.54) than the controls (average age, 41.53). The majority of the CRC patients were in stages II and III (37.61% and 26.71%, respectively), with located tumors in the rectum (28.14%) and left colon (26.23%) (Table 1).

The individuals with modern European ancestry overlapped with the Iberian population of 1000 Genomes data, while the ancient European ancestry was distanced from European populations (Supplementary Figure S1A). In addition, the PC1 of the principal component analysis of the samples was determined by the ancestry component of our cohort (Supplementary Figure S1B).

Table 1. Demographics of the participants.

	Cases	Controls
N	835	940
Male (%)	530 (63.47%)	631 (67.13%)
Female (%)	305 (36.53%)	309 (32.87%)
Age (SE)	73.54 (11.38)	41.53 (11.79)
Stage		
0	37 (4.43%)	
I	130 (15.57%)	
II	314 (37.61%)	
III	223 (26.71%)	
IV	105 (12.57%)	
Undetermined	26 (3.11%)	
Location		
Right	170 (20.36%)	
Left	219 (26.23%)	
Rectal	235 (28.14%)	
Unspecific	211 (25.27%)	

3.1. Genome-Wide Association Studies

The genome-wide association study of all CRC cases showed five suggestive ($p < 5 \times 10^{-6}$) signals (Table 2). The most significant SNP was rs77317240, located in chromosome 2 and upstream of *ABCA12* and *ATIC* genes ($p = 5.8 \times 10^{-7}$; OR = 6.4; CI 95%, 3.1–13.2). Other suggestive SNPs were located in *ERBB4* and *MAGI2* genes, and downstream of the *IL15* gene (Table 2).

Table 2. Suggestive signals ($p < 5 \times 10^{-6}$) detected in colorectal cancer and the locations. Gene, gene where is located the SNP or nearest gene 100kb upstream or downstream from the SNP. OR, odds ratio. CI 95%, confidence interval of 95% of the odds ratio. Freq, frequency of A1 in Basque cohort. Freq EUR, frequency of A1 in European populations of 1 KG.

Lead SNP	Position	Gene	A1	A2	OR (CI 95%)	*p*-Value	Freq	Freq EUR
Colorectal cancer vs. controls								
rs79374732	2:212815957	*ERBB4*	T	C	8.5 (3.4–21.0)	4.5×10^{-6}	0.032	0.022
rs77317240	2:216091445	Upstream of *ABCA12* and *ATIC*	T	C	6.4 (3.1–13.2)	5.8×10^{-7}	0.039	0.024
rs116443146	4:142699393	Downstream of *IL15*	G	A	16.3 (5.0–53.8)	4.4×10^{-6}	0.013	0.02
rs34931968	7:79055118	*MAGI2*	T	G	29.7 (7.1–124.3)	3.4×10^{-6}	0.011	0.01
rs1693967	16:86289580	*LINC01081*	G	A	11.4 (4.1–32.1)	3.9×10^{-6}	0.017	0.024
Right colon cancer vs. controls								
rs3004681	1:69054715	Downstream of *DEPDC1*	T	G	11.8 (4.3–32.7)	2.0×10^{-6}	0.062	0.073
rs77445470	1:226800066	Downstream of *STUM* and *ITPKB*	G	C	18.5 (5.3–64.5)	4.8×10^{-6}	0.044	0.055
rs76653793	4:47962934	*CNGA1, LOC101927157*	G	T	21.7 (6.4–73.8)	7.9×10^{-7}	0.028	0.036
rs142444738	4:106095747	*TET2, TET2-AS1*	A	G	51.1 (9.6–270.9)	3.8×10^{-6}	0.011	0.005
rs4696337	4:153602674	*TMEM154, LOC105377495*	A	C	35.8 (8.2–156.2)	2.0×10^{-6}	0.023	0.023
rs139432545	4:174624195		G	A	48.4 (9.6–244.9)	2.7×10^{-6}	0.012	0.022
rs13211079	6:36977349	*FGD2*	G	C	43.9 (9.2–210.2)	2.2×10^{-6}	0.019	0.012
rs190591066	7:89988294	*GTPBP10*	A	G	40.6 (8.8–186.4)	1.9×10^{-6}	0.017	0.011
rs75772232	8:83689525		T	C	15.8 (4.9–51.2)	4.3×10^{-6}	0.039	0.045
rs118025264	9:119407781	*ASTN2, LOC105376240*	T	C	25.7 (6.4–102.7)	4.3×10^{-6}	0.026	0.022
rs16933489	12:5572210	*NTF3*	T	C	34.9 (9.1–133.3)	2.0×10^{-7}	0.02	0.044
rs78263620	18:72995680	*TSHZ1*	T	C	43.6 (9.2–207.9)	2.2×10^{-6}	0.011	0.019
rs148452202	19:2527577	*GNG7*	A	G	34.6 (8.3–144.8)	1.2×10^{-6}	0.022	0.01
rs35914129	19:48115566	*BICRA*	T	G	56.2 (11.2–283.0)	1.0×10^{-6}	0.013	0.009
rs28495197	22:36050632	*APOL6*	T	C	39.9 (9.1–174.2)	9.4×10^{-7}	0.023	0.017
rs117820381	22:40738486	Downstream of *TNRC6B*, upstream of *ADSL*	A	G	37.0 (8.4–163.1)	1.8×10^{-6}	0.013	0.028

Table 2. *Cont.*

Lead SNP	Position	Gene	A1	A2	OR (CI 95%)	*p*-Value	Freq	Freq EUR
				Left colon cancer vs. controls				
rs112033525	2:23176856		T	G	39.4 (8.2–189.6)	4.5×10^{-6}	0.017	0.015
rs139367040	2:173950614	*MAP3K20*	T	C	33.0 (7.7–142.5)	2.8×10^{-6}	0.019	0.014
rs72774468	9:137697318	*COL5A1*	C	T	15.1 (5.0–45.3)	1.3×10^{-6}	0.035	0.051
rs114144417	16:48116976	*ABCC12*	T	C	149.8 (20.2–1112.0)	9.7×10^{-7}	0.01	0.008
rs17721600	17:27268513	*PHF12, LOC101927018*	A	G	25.9 (6.9–97.7)	1.6×10^{-6}	0.037	0.053
rs140107269	18:1828990		T	C	26.8 (6.6–109.2)	4.4×10^{-6}	0.023	0.027
rs62093285	18:49252189		A	G	12.8 (4.3–38.4)	4.9×10^{-6}	0.044	0.035
				Rectal cancer vs. controls				
rs78144988	1:102199388	*LINC01709*	C	T	54.9 (11.2–268.4)	7.6×10^{-7}	0.013	0.018
rs13403794	2:9785060	Upstream of *YWHAQ* and *ADAM17*	C	T	65.5 (12.0–355.9)	1.3×10^{-6}	0.012	0.021
rs354856	2:142433670	*LRP1B, LOC107985779*	C	T	17.4 (5.5–55.0)	1.1×10^{-6}	0.027	0.062
rs116443146	4:142699393	Downstream of *IL15*	G	A	40.3 (9.2–176.9)	9.7×10^{-7}	0.013	0.02
rs72909399	6:86581045		T	G	74.7 (13.5–414.7)	8.1×10^{-7}	0.014	0.03
rs71516114	8:784674	*DLGAP2*	C	T	5.2 (2.6–10.4)	2.7×10^{-6}	0.111	0.112
rs61848097	10:50134508	*WDFY4, LRRC18*	G	A	8.6 (3.5–21.0)	2.9×10^{-6}	0.073	0.089
rs77470802	14:27547598	*LOC105370420*	G	T	12.4 (4.2–36.5)	4.6×10^{-6}	0.027	0.033
rs76799782	14:91624544	*DGLUCY*	A	G	18.9 (5.4–65.4)	3.8×10^{-6}	0.029	0.039
rs141553824	16:50380386	*BRD7*	C	T	45.8 (10.4–202.4)	4.5×10^{-7}	0.017	0.05
				Left colon cancer vs. right colon cancer				
rs4655303	1:213834643	*LOC105372912*	T	A	2.2 (1.6–3.0)	3.6×10^{-6}	0.43	0.377
rs62005704	14:53465150	Downstream of *DDHD1*, upstream of *FERMT2*	A	G	0.4 (0.3–0.6)	9.8×10^{-7}	0.464	0.503
				Rectal cancer vs. colon cancer				
rs73171906	7:147986529	*CNTNAP2*	T	C	2.2 (1.6–2.9)	6.4×10^{-7}	0.23	0.154
rs9773025	8:6674458	*XKR5*	G	A	0.5 (0.3–0.6)	1.5×10^{-6}	0.414	0.468
rs79619562	21:38742422	*DYRK1A*	C	T	2.7 (1.8–4.1)	1.8×10^{-6}	0.1	0.093

When cancer locations were analyzed separately different signals were detected (Table 2): 16 in right colon cancer (the most significant signal was located in the *NTF3* gene), 7 in left colon cancer (the most significant signal was located in the *ABCC12* gene), and 10 in rectal cancer (the most significant signal was located in *BRD7* gene). When locations were compared (Table 2), 2 signals were detected when comparing left and right colon cancers (the most significant genetic variant was located in the *FERMT2* gene) and 3 when comparing rectal vs colon cancers (the most significant genetic variant was located in *CNTNAP2* gene).

Among the SNPs previously associated with CRC (Supplementary Table S1), 16 SNPs (7.65% of SNPs previously associated) showed nominal association in our cohort. When those SNPs were analyzed by the location of cancer, 9 (4.31%) were nominally significant in right colon cancer, 12 (5.74%) in left colon cancer (including rs6687758, an SNP previously associated with CRC in the Basque population) and 12 (5.74%) in rectal cancer. Among the 31 SNPs previously associated with CRC in more than one study (Supplementary Table S1), 5 SNPs (16.13%) showed nominal association in CRC; 3 (9.68%) in right colon cancer; 3 (9.68%) in left colon cancer and 1 (3.23%) in rectal cancer.

Regarding the comparison with inflammatory bowel disease (Table 3), 11 genomic regions had suggestive different frequencies. Among them, the signal located upstream of the *ATP8B4* gene (rs541295) reached a genome-wide significant *p*-value ($p = 1.8 \times 10^{-8}$). When colorectal cancer was compared with the pool of controls and inflammatory bowel disease (Table 3), the most significant signal in CRC vs controls (upstream of the *ABCA12* and *ATIC* genes) was detected. In addition, 4 of the signals detected when CRC was compared with inflammatory bowel disease patients were suggestive: in the HLA region, in the *DLGAP2* gene, downstream of the *PTCHD3* gene and upstream of the *ATP8B4* gene.

Table 3. Suggestive signals ($p < 5 \times 10^{-6}$) detected in the comparison of colorectal cancer and inflammatory bowel disease. Gene, gene where is located the SNP or nearest gene 100kb upstream or downstream from the SNP. OR, odds ratio. CI 95%, confidence interval of 95% of the odds ratio. Freq, frequency of A1 in Basque cohort. Freq EUR, frequency of A1 in European populations of 1 KG.

Lead SNP	Position	Gene	A1	A2	OR (CI 95%)	*p*-Value	Freq	Freq EUR
		Colorectal cancer vs inflammatory bowel disease						
rs35493687	1:41285292	*KCNQ4*	A	C	0.4 (0.3–0.6)	4.2×10^{-6}	0.122	0.147
rs76845271	2:73665817	*ALMS1*	T	G	0.3 (0.2–0.5)	2.9×10^{-6}	0.043	0.048
rs6738805	2:231083171	*SP110*	C	T	0.4 (0.3–0.6)	4.6×10^{-7}	0.135	0.128
rs10007784	4:81977690	*BMP3*	C	T	0.5 (0.4–0.7)	1.8×10^{-6}	0.228	0.222
rs181206673	5:25834969		C	G	0.3 (0.1–0.5)	4.1×10^{-6}	0.039	0.0467
rs72840740	6:18745458		C	T	0.1 (0.0–0.2)	1.1×10^{-6}	0.014	0.03
rs9271365	6:32586794	Downstream of *HLA-DRB1* and upstream of *HLA-DQA1*	G	T	1.8 (1.4–2.3)	2.2×10^{-6}	0.353	0.388
rs951197	6:103210765		C	A	0.5 (0.4–0.7)	5.6×10^{-7}	0.476	0.446
rs1875664	8:827824	*DLGAP2*	G	A	2.3 (1.6–3.3)	2.8×10^{-6}	0.128	0.161
rs988874	10:27684660	Downstream of *PTCHD3*	A	T	0.5 (0.3–0.6)	1.6×10^{-6}	0.174	0.157
rs541295	15:50056050	Upstream of *ATP8B4*	G	A	0.2 (0.1–0.4)	1.8×10^{-8}	0.055	0.022
		Colorectal cancer vs. controls + inflammatory bowel disease						
rs7550486	1:14777040	*KAZN*	C	T	0.6 (0.5–0.7)	1.3×10^{-6}	0.498	0.475
rs115681984	2:216032071	Upstream of *ABCA12* and *ATIC*	T	C	4.2 (2.4–7.1)	2.6×10^{-7}	0.034	0.026
rs72840741	6:18747455		G	A	0.1 (0.0–0.2)	1.8×10^{-6}	0.014	0.03
rs5002178	6:32611590	*HLA-DQA1*	G	A	0.6 (0.5–0.7)	6.8×10^{-7}	0.33	0.374
rs951197	6:103210765		C	A	0.6 (0.5–0.7)	2.4×10^{-7}	0.484	0.446
rs1875664	8:827824	*DLGAP2*	G	A	2.2 (1.6–3.0)	3.24×10^{-7}	0.124	0.161
rs988874	10:27684660	Downstream of *PTCHD3*	A	T	0.5 (0.3–0.6)	2.0×10^{-6}	0.171	0.157
rs150840049	14:59165709	Downstream of *DACT1*	C	T	0.1 (0.1–0.3)	2.6×10^{-6}	0.025	0.052
rs541295	15:50056050	Upstream of *ATP8B4*	G	A	0.2 (0.1–0.4)	5.3×10^{-8}	0.045	0.022

3.2. Mendelian Randomization

Mendelian Randomization analyses were carried out to analyze the effect of modifiable risk factors and the abundance of bacterial phyla on CRC risk. The instruments used seemed appropriate (Supplementary Table S2), although the modifiable risk factors were stronger than bacterial phyla (F-statistic between 55.82–211.35 in the former, 18.73–20.28 in the latter).

When analyzing the effect of modifiable risk factors on CRC, there were no significant results (Figure 1A, Supplementary Table S3). However, when the locations of CRC were separately analyzed, the MR Egger method showed the effect of total cholesterol (beta = 2.4 ± 1.1; $p = 0.0395$) on left-sided colon cancer risk, and the effect of BMI (beta = 8.7 ± 3.3; $p = 0.0094$) in rectal cancer risk. In the latter, pleiotropic effects were detected ($p = 0.0112$, Supplementary Table S3). In addition, Inverse Variance Weighted method showed the effect of LDL cholesterol (beta = 1.56 ± 0.64; $p = 0.0148$) on left-sided colon cancer risk.

In the case of bacterial phyla (Figure 1B, Supplementary Table S4), according to MR Egger method, Firmicutes phylum showed a significant effect on CRC and left colon cancer (beta=3.6 ± 1.7; $p = 0.0364$; beta = 6.4 ± 2.8; $p = 0.0282$, respectively), although pleiotropy was detected in both cases ($p = 0.0347$; $p = 0.0456$, respectively, Supplementary Table S4), as well as, heterogeneity in the used instruments (Q-test $p = 0.0336$ and $p = 0.0107$, respectively, Supplementary Table S4). In the case of Inverse Variance Weighted, there was an inverse effect of Cyanobacteria abundance on CRC risk and left colon cancer risk (beta = -0.86 ± 0.39; $p = 0.0299$; beta = -1.66 ± 0.68; $p = 0.014$, respectively).

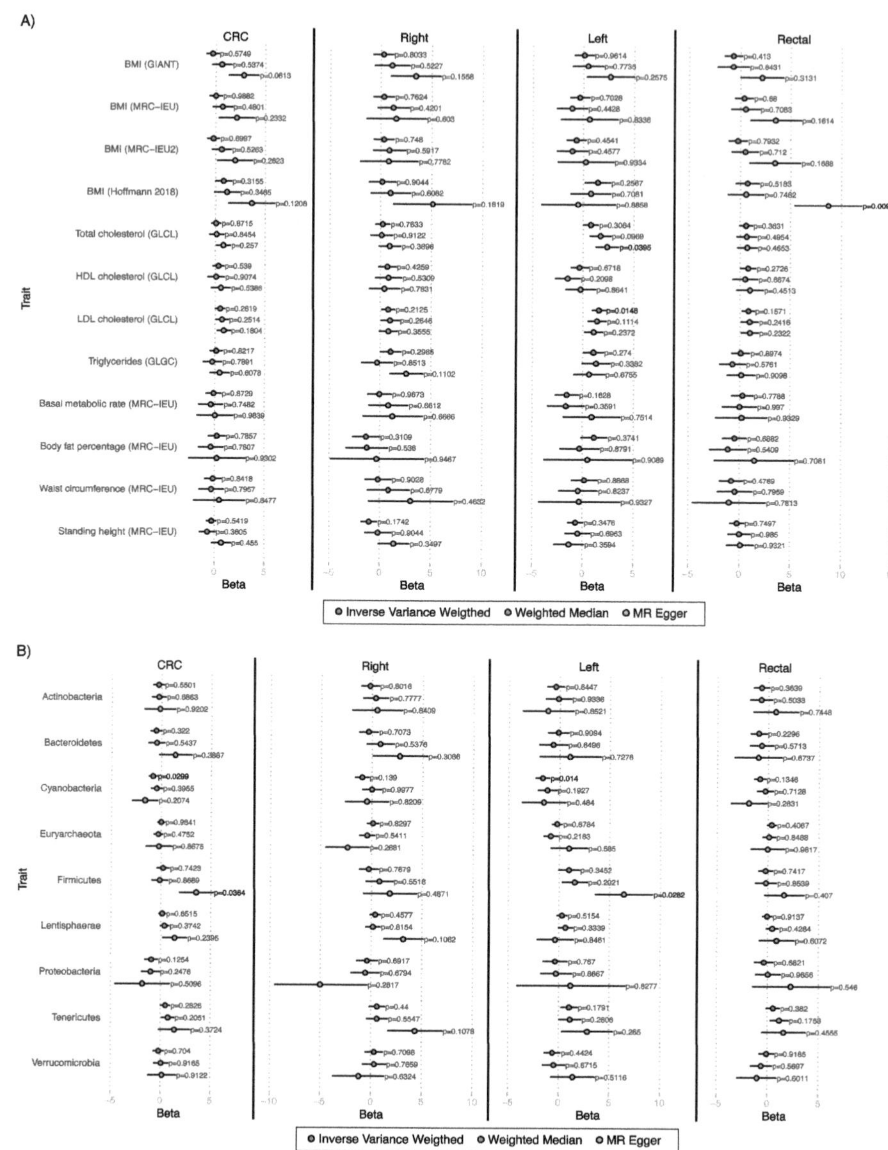

Figure 1. Mendelian Randomization results. The beta value and standard error are depicted for each method used. (**A**) Modifiable risk factors. (**B**) Bacterial phyla.

3.3. Polygenic Risk Scores

Polygenic risk scores for our cohort were built using 25 different models available in PGS Catalog for CRC. From all of them (Figures 2 and 3A), PGS000785 showed the best discrimination between the PRS values for cases and controls (T-test $p = 2.12 \times 10^{-14}$; small effect according to Cohen's d), as well as, the best AUC value (0.6, CI 95% 0.58–0.62); followed by PGS000734 and PGS000765 (both $p = 2.64 \times 10^{-13}$; small effect according to Cohen's d; AUC of 0.6, CI 95% 0.57–0.61). In addition, the PRS used previously in a Basque cohort showed lower significance ($p = 0.0003$; negligible effect according to Cohen's d) and AUC value (0.55, CI 95% 0.52–0.56).

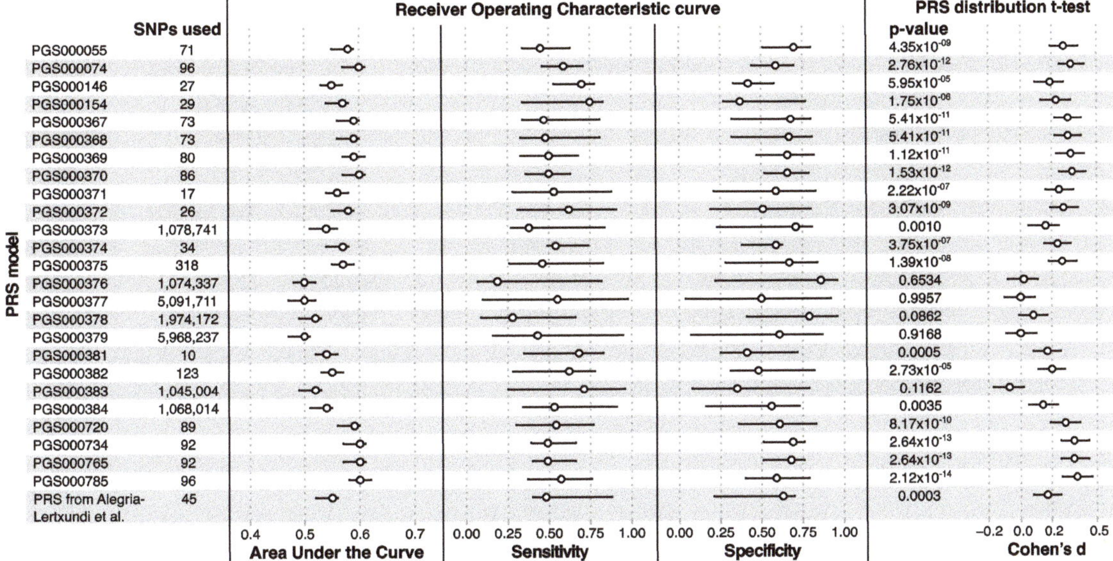

Figure 2. Performance of different Polygenic Risk Score sets. For each parameter, the value and 95% of confidence interval are depicted. Cohen's d, the effect size of the T-test and 95% of the confidence interval, <0.2 negligible effects, 0.2–0.5 small effect, 0.5–0.8 moderate effect, >0.8 large effects.

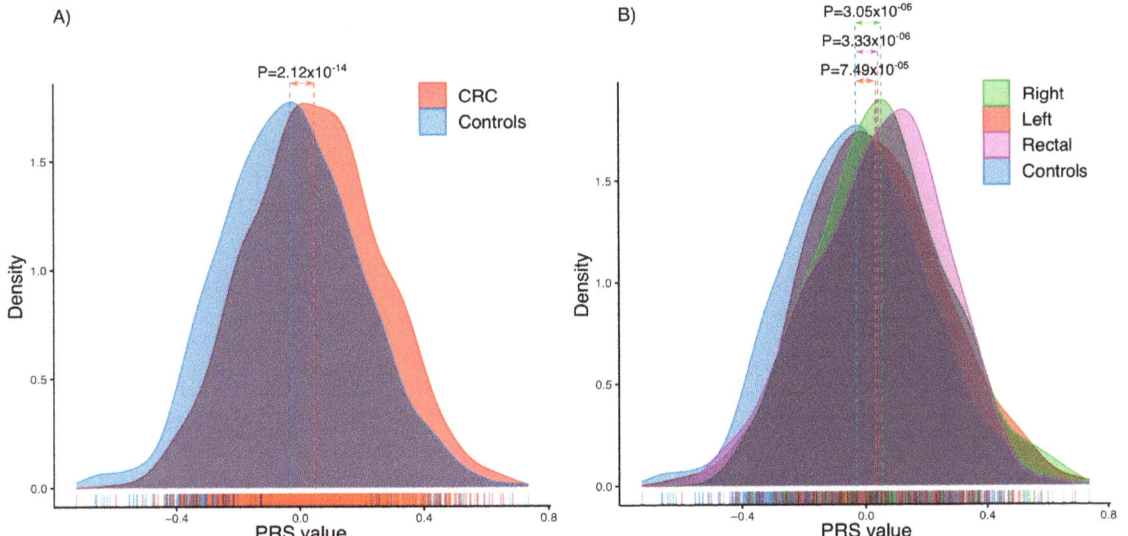

Figure 3. *Cont.*

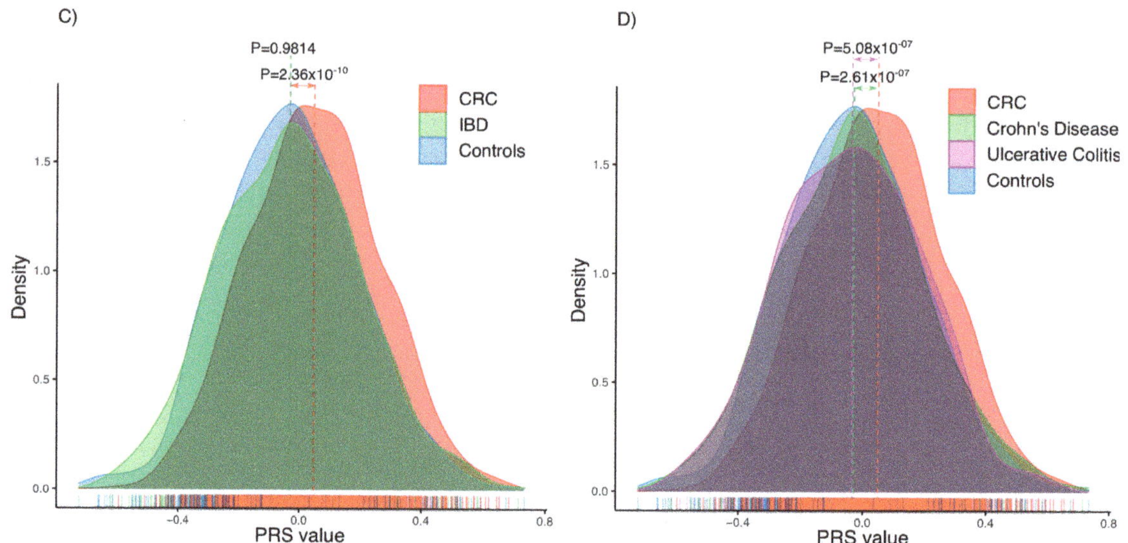

Figure 3. Distribution of polygenic risk score based on PGS000785 set. P, the *p*-value of the T-test. (**A**) Colorectal cancer. (**B**) According to the location. (**C**) Comparison with patients with irritable bowel disease. (**D**) Comparison with patients of main types of irritable bowel disease.

The PGS000785 PRS model had a good performance regardless of the location of CRC (Figure 3B): the distribution of the PRS score was significantly higher in right colon cancer (p = 3.05 $\times$ 10^{-6}), left colon cancer (p = 7.49 $\times$ 10^{-6}) and rectal cancer (p = 3.33 $\times$ 10^{-6}) compared to controls, while there were no significant differences comparing locations. In addition, that model was able to differentiate inflammatory bowel disease patients from colorectal cancer patients (p = 2.36 $\times$ 10^{-10}, Figure 3C), regardless of the type of inflammatory bowel disease (Crohn's Disease, p = 2.61 $\times$ 10^{-7}; Ulcerative colitis, p = 5.08 $\times$ 10^{-7}; Figure 3D).

4. Discussion

The development of colorectal cancer (CRC) is influenced by environmental factors [4], microbiome composition [5] and genetic factors. In this work, we have analyzed the contribution of the genetic component to CRC risk in the Basque population, a population with a particular genetic history. That particular genetic history was reflected in the principal component analysis and, as it was done before [19], adjusting for PCs is enough to avoid artifacts due to the presence of two ancestries in the population.

Previously, selected SNPs were analyzed in CRC in the Basque population [16] and, in this study, we have used a GWAS approach and increased the sample size. In that previous work, the SNP rs6687758 was nominally significant [16] and we have been able to detect the nominal significance of that SNP in left colon cancer, as well as more genetic variants. We are aware that the sample size affected the results we obtained, and, for example, few previously associated SNPs with CRC were detected in our study. However, we were able to find nominally significant results for the SNPs detected in more than one study. In addition, the majority of SNPs detected in previous studies were not detected in other studies. Thus, the genetic risk of CRC could be partially due to local variation, therefore, it seems appropriate for the genetic analysis of CRC in new populations.

The most significant signal in CRC, although it was not genome-wide significant, was located between *ABCA12* and *ATIC* genes. It has been reported that the expression of *ABCA12* is upregulated in CRC [42,43], its expression is higher in the colon than in the rectum [43], and its expression is higher in colorectal adenoma than in hyperplastic polyp [44]. In the case of the *ATIC* gene, it has been proposed that its expression could be a prognostic marker for colon adenocarcinoma [45]; its presence in small extracellular vesicles in serum is useful to differentiate early colorectal neoplasia from advanced colorectal neoplasia [46].

Another suggestive signal was located on the *ERBB4* gene. In cell culture and mice, it has been observed that *ERBB4* expression and signaling can prevent apoptosis of the cells in an inflammatory environment [47], therefore, its chronic overexpression could contribute to the appearance of tumors, since apoptosis of colonic cells is inhibited [48]. In humans, it has been reported the overexpression of *ERBB4* in CRC and that tumors with high levels of this receptor could have enhanced cell survival [49]. In addition, it has been suggested that the expression of *ERBB4* is associated with unfavorable clinical outcomes in CRC [50] and that it could be a marker of a higher risk of recurrence [51]. Additionally, it has been reported that *ERRB4* expression is positively associated with lymph node metastasis [50]; that *ERBB4* could play a relevant role in a gene network associated with progression from colon adenocarcinoma to liver metastases [52], and that *ERBB4* could be part of a pathway that enhances the invasion of CRC cells [53].

Additional suggestive signals were located in the *MAGI2* gene and downstream of the *IL15* gene. The SNP rs34931968 detected in our cohort is located in the *MAGI2* gene, upstream of a lncRNA that is next to *MAGI2* (called *MAGI2-AS3*), a lncRNA that has been involved in CRC [54–56]. In addition, the SNP rs34931968 is in linkage disequilibrium with an SNP (rs7783388) involved in CRC throughout changes in *MAGI2-AS3* expression [56]. In the case of *IL15*, its expression has been associated with the outcome of CRC [57].

When the locations of the tumors were analyzed separately, other possible relevant genes were detected. In right colon cancer, the most significant signal was located in *NTF3*, a gene implicated in unfavorable prognosis in hepatocellular carcinoma [58,59]; in left colon cancer *ABCC12* gene, another ATP-binding cassette as the previously discussed *ABCA12*; in rectal cancer *BRD7* gene, a possible oncogene involved in CRC progression [60]. In addition, in rectal cancer the SNP rs13403794 was detected, an SNP located upstream of *ADAM17*, which is a gene that is part of the signaling pathway involved in colorectal cancer progression and chemoresistance [61]. When locations were compared, additional genes were detected: *FERMT2*, whose overexpression in CRC has been detected and associated with cell growth [62]; *CNTNAP2*, a gene that has not been associated with CRC. It has been observed that the genetic mechanisms behind CRC could be different depending on its location [63] and the differences in the genetic variants detected in our study are consistent with that suggestion.

On the whole, considering the biological role of some of the genes where the suggestive genetic variants were located, those genetic variants could be markers of the progression of CRC, at least in the Basque population, although follow-up analyses are needed to confirm their potential utility as markers.

Various modifiable risk factors have been observed to affect CRC risk [11,12,64], but we were not able to find those effects when all CRC patients were analyzed. However, when each location was analyzed, the effect of genetic risk to higher cholesterol levels (general levels or LDL) on left colon cancer and higher BMI on rectal cancer were detected, as has been suggested previously for CRC [11–13,64,65]. Although we tried to replicate the results obtained using Mendelian Randomization in previous works [12,13] and the traits and instruments used seem appropriate to replicate them, the results we obtained were limited or were detected only by one method. It could be possible that the size and characteristics of our cohort and GWAS analyses complicate the finding of clear causalities, since the traits we used to have strong instruments to avoid the biases of our cohort.

The genetic signature of the abundance of Firmicutes was associated with a higher risk of CRC and left colon cancer in our cohort, although the results should be taken with caution since heterogeneity was detected. In addition, that association had a pleiotropic effect, that is, rather than the presence of Firmicutes affecting the risk of CRC (cause and effect), there is a shared genetic component that affects both (common biologic mechanism).

It has been described the importance of the microbiota in CRC risk and development [66,67], the differences in its composition between left and right colon cancer [68–70] and shared genetic variants in CRC risk and the abundance of Firmicutes [13]. Although the connection we have detected between CRC and Firmicutes is based only on their shared genetic variants, it has been observed that the involvement of Firmicutes in CRC risk was variable [68–71]: some genera of Firmicutes were enriched in CRC while others were depleted. In the case of Cyanobacteria, a higher abundance of that phylum has been observed in colorectal adenomas [72], and in animal models, it has been observed a higher abundance of Cyanobacteria when oxaliplatin is administered [73]. Therefore, follow-up analyses of Firmicutes and Cyanobacteria as a marker of CRC risk in the Basque cohort are needed. Although the involvement of Firmicutes and Cyanobacteria in CRC seems biologically possible, their connection through Mendelian Randomization in our work seems weak, since they have been detected only by one method. In addition, although the study of the effect of host genetics on microbial abundance has been a valuable resource [30], it could be possible that the available instruments are not still appropriate to carry out Mendelian Randomization analyses, at least in our cohort.

Finally, polygenic risk scores (PRS) have been proposed as a tool for risk prediction in colorectal cancer [15]. We applied several publicly available PRS models, and their performance was variable. The best model was built using different sources available in GWAS Catalog and the interplay between genetic risk and modifiable risk factors [37]. In the case of CRC, that work suggested that PRS was the primary determinant of risk stratification in their application of the PRS model in UK Biobank data [37]. Although our cohort has a slightly different genetic background, since there is a higher genetic component of ancient European ancestry, the application of the PRS was able to differentiate CRC cases from controls, regardless of the location of the tumors. Since the AUC was low and the effect small, additional genetic or non-genetic risk factors should be incorporated to build a model for better discrimination. In addition, this PRS did not confuse CRC and inflammatory bowel disease or its main types in our cohort, suggesting that when there are overlapping symptoms, the use of that PRS would not misclassify an IBD patient as a CRC patient. In addition, we found genetic variants that could be used to discriminate between CRC and inflammatory bowel disease in our cohort, although follow-up analyses are needed. Regarding the PRS previously used in Basques [16], the performance in our data was not as good as the best model, but the controls showed lower PRS than CRC cases ($p = 0.003$), similar to the previous analysis of Basques ($p = 0.002$ for the unweighted values, $p = 0.036$ for weighted values) [16]. Therefore, the incorporation of a different set of SNPs for the development of more precise PRS models is still necessary, and the performance of PRS models should be investigated in additional samples of this population.

Considering the results obtained in the different analyses we have carried out since the results are quite consistent with previous results, genetic CRC risk in the Basque population seems to be similar to other European populations. The suggestive signals from the GWAS were consistent with CRC biology, although in some variants the frequency in the Basque population was quite different. Mendelian Randomization analyses did not find clear causal relationships, although the traits used were reported to affect CRC risk in other cohorts, therefore, follow-up studies are needed to assess if our results are due to methodological constraints or differences in the specific mechanisms. Finally, the application of polygenic risk scores based on European populations seemed a feasible approach to capture the CRC risk in the Basque population, although they can be improved. Thus, as happened in inflammatory bowel disease [19], the genetic architecture of CRC risk in the Basque population is similar to other European populations but local genetic variation shapes the risk.

5. Conclusions

In conclusion, we have analyzed the genetic component of the risk of CRC in the Basque population. Although the sample size was limited and there were constraints in the analyses due to the cohort used, we detected genetic factors whose involvement in the risk of CRC is consistent with the biological mechanisms of CRC, and we identified plausible genetic markers and an appropriate polygenic risk score model to assess the genetic contribution to CRC risk in this population. In the future, those genetic factors and the polygenic risk score model should be validated in follow-up studies.

Supplementary Materials: The following supporting information can be downloaded at: https://www.mdpi.com/article/10.3390/cancers14174193/s1, Figure S1: PCA plots of analyzed Basque cohort; Table S1: Results of SNPs previously associated with CRC; Supplementary Table S2: Sensitivity analyses of used instruments in Mendelian Randomization analyses; Supplementary Table S3: Results of Mendelian Randomization analyses using modifiable risk factors as exposures; Supplementary Table S4: Results of Mendelian Randomization analyses using bacterial phyla as exposures.

Author Contributions: Conceptualization, K.G.-E., M.D. and L.B.; methodology, K.G.-E.; formal analysis, K.G.-E., I.R.-G.; resources, A.E., M.B., B.N., N.M.S.M., A.F., M.D., L.B.; data curation, A.E., M.B., B.N., N.M.S.M.; writing—original draft preparation, K.G.-E.; writing—review and editing, K.G.-E., I.R.-G., L.B.; supervision, K.G.-E.; funding acquisition, M.D., L.B. All authors have read and agreed to the published version of the manuscript.

Funding: This work was partially founded by Gipuzkoako Foru Aldundia/Diputación Foral de Gipuzkoa (Code: 111/17).

Institutional Review Board Statement: The study was conducted in accordance with the Declaration of Helsinki, and approved by Comité de Ética de la Investigación con medicamentos de Euskadi (code: PI+CES-BIOEF 2017-10).

Informed Consent Statement: Informed consent was obtained from all subjects involved in the study.

Data Availability Statement: The data presented in this study are available on request from the authors. The data are not publicly available due to ethical reasons (genotype data cannot be shared).

Acknowledgments: Samples and data used in the present work were provided by the Basque Biobank (http://www.biobancovasco.org, accessed: 1 August 2022). We want to thank Miguel Ángel Vesga from the Basque Centre of Transfusion and Human Tissues for providing the access to control samples.

Conflicts of Interest: The authors declare no conflict of interest.

References

1. Stewart, B.; Wild, C. *World Cancer Report 2014*; IARC Publications: Lyon, France, 2014; ISBN 978-92-832-0443-5.
2. Ferlay, J.; Soerjomataram, I.; Dikshit, R.; Eser, S.; Mathers, C.; Rebelo, M.; Parkin, D.M.; Forman, D.; Bray, F. Cancer incidence and mortality worldwide: Sources, methods and major patterns in GLOBOCAN 2012. *Int. J. Cancer* **2015**, *136*, E359–E386. [CrossRef] [PubMed]
3. Lasry, A.; Zinger, A.; Ben-Neriah, Y. Inflammatory networks underlying colorectal cancer. *Nat. Immunol.* **2016**, *17*, 230–240. [CrossRef]
4. Cross, A.J.; Ferrucci, L.M.; Risch, A.; Graubard, B.I.; Ward, M.H.; Park, Y.; Hollenbeck, A.R.; Schatzkin, A.; Sinha, R. A large prospective study of meat consumption and colorectal cancer risk: An investigation of potential mechanisms underlying this association. *Cancer Res.* **2010**, *70*, 2406–2414. [CrossRef] [PubMed]
5. Wong, S.H.; Yu, J. Gut microbiota in colorectal cancer: Mechanisms of action and clinical applications. *Nat. Rev. Gastroenterol. Hepatol.* **2019**, *16*, 690–704. [CrossRef]
6. Levin, B.; Lieberman, D.A.; McFarland, B.; Smith, R.A.; Brooks, D.; Andrews, K.S.; Dash, C.; Giardiello, F.M.; Glick, S.; Levin, T.R.; et al. Screening and surveillance for the early detection of colorectal cancer and adenomatous polyps, 2008: A joint guideline from the American Cancer Society, the US Multi-Society Task Force on Colorectal Cancer, and the American College of Radiology. *CA. Cancer J. Clin.* **2008**, *58*, 130–160. [CrossRef] [PubMed]
7. Abulí, A.; Fernández-Rozadilla, C.; Alonso-Espinaco, V.; Muñoz, J.; Gonzalo, V.; Bessa, X.; González, D.; Clofent, J.; Cubiella, J.; Morillas, J.D.; et al. Case-control study for colorectal cancer genetic susceptibility in EPICOLON: Previously identified variants and mucins. *BMC Cancer* **2011**, *11*, 339. [CrossRef]
8. Abulí, A.; Bessa, X.; González, J.R.; Ruizponte, C.; Cáceres, A.; Muñoz, J.; Gonzalo, V.; Balaguer, F.; Fernándezrozadilla, C.; González, D.; et al. Susceptibility genetic variants associated with colorectal cancer risk correlate with cancer phenotype. *Gastroenterology* **2010**, *139*, 788–796.e6. [CrossRef] [PubMed]

9. Burns, M.B.; Montassier, E.; Abrahante, J.; Priya, S.; Niccum, D.E.; Khoruts, A.; Starr, T.K.; Knights, D.; Blekhman, R. Colorectal cancer mutational profiles correlate with defined microbial communities in the tumor microenvironment. *PLoS Genet.* **2018**, *14*, 1–24. [CrossRef]
10. Law, P.J.; Timofeeva, M.; Fernandez-Rozadilla, C.; Broderick, P.; Studd, J.; Fernandez-Tajes, J.; Farrington, S.; Svinti, V.; Palles, C.; Orlando, G.; et al. Association analyses identify 31 new risk loci for colorectal cancer susceptibility. *Nat. Commun.* **2019**, *10*, 1–15. [CrossRef]
11. Rodriguez-broadbent, H.; Law, P.J.; Sud, A.; Palin, K.; Tuupanen, S.; Kaasinen, E.; Sarin, A.; Ripatti, S.; Eriksson, J.G. Mendelian randomisation implicates hyperlipidaemia as a risk factor for colorectal cancer. *Int. J. Cancer* **2017**, *140*, 2701–2708. [CrossRef]
12. Cornish, A.J.; Law, P.J.; Timofeeva, M.; Palin, K.; Farrington, S.M.; Palles, C.; Jenkins, M.A.; Casey, G.; Brenner, H.; Chang-Claude, J.; et al. Modifiable pathways for colorectal cancer: A mendelian randomisation analysis. *Lancet Gastroenterol. Hepatol.* **2020**, *5*, 55–62. [CrossRef]
13. Garcia-Etxebarria, K.; Clos-Garcia, M.; Telleria, O.; Nafría, B.; Alonso, C.; Iruarrizaga-Lejarreta, M.; Franke, A.; Crespo, A.; Iglesias, A.; Cubiella, J.; et al. Interplay between genome, metabolome and microbiome in colorectal cancer. *Cancers* **2021**, *13*, 6216. [CrossRef] [PubMed]
14. Ibáñez-Sanz, G.; Diéz-Villanueva, A.; Alonso, M.H.; Rodríguez-Moranta, F.; Pérez-Gómez, B.; Bustamante, M.; Martin, V.; Llorca, J.; Amiano, P.; Ardanaz, E.; et al. Risk Model for Colorectal Cancer in Spanish Population Using Environmental and Genetic Factors: Results from the MCC-Spain study. *Sci. Rep.* **2017**, *7*, 43263. [CrossRef]
15. Thomas, M.; Sakoda, L.C.; Hoffmeister, M.; Rosenthal, E.A.; Lee, J.K.; van Duijnhoven, F.J.B.; Platz, E.A.; Wu, A.H.; Dampier, C.H.; de la Chapelle, A.; et al. Genome-wide Modeling of Polygenic Risk Score in Colorectal Cancer Risk. *Am. J. Hum. Genet.* **2020**, *107*, 432–444. [CrossRef] [PubMed]
16. Alegria-Lertxundi, I.; Aguirre, C.; Bujanda, L.; Fernández, F.J.; Polo, F.; Ordovás, J.M.; Carmen Etxezarraga, M.; Zabalza, I.; Larzabal, M.; Portillo, I.; et al. Single nucleotide polymorphisms associated with susceptibility for development of colorectal cancer: Case-control study in a Basque population. *PLoS ONE* **2019**, *14*, e0225779. [CrossRef]
17. Günther, T.; Valdiosera, C.; Malmström, H.; Ureña, I.; Rodriguez-Varela, R.; Sverrisdóttir, Ó.O.; Daskalaki, E.A.; Skoglund, P.; Naidoo, T.; Svensson, E.M.; et al. Ancient genomes link early farmers from Atapuerca in Spain to modern-day Basques. *Proc. Natl. Acad. Sci. USA* **2015**, *112*, 11917–11922. [CrossRef]
18. Olalde, I.; Mallick, S.; Patterson, N.; Rohland, N.; Villalba-Mouco, V.; Silva, M.; Dulias, K.; Edwards, C.J.; Gandini, F.; Pala, M.; et al. The genomic history of the Iberian Peninsula over the past 8000 years. *Science (80-)* **2019**, *363*, 1230–1234. [CrossRef]
19. Garcia-Etxebarria, K.; Merino, O.; Gaite-Reguero, A.; Rodrigues, P.M.; Herrarte, A.; Etxart, A.; Ellinghaus, D.; Alonso-Galan, H.; Franke, A.; Marigorta, U.M.; et al. Local genetic variation of inflammatory bowel disease in Basque population and its effect in risk prediction. *Sci. Rep.* **2022**, *12*, 3386. [CrossRef] [PubMed]
20. Abraham, G.; Qiu, Y.; Inouye, M. FlashPCA2: Principal component analysis of Biobank-scale genotype datasets. *Bioinformatics* **2017**, *33*, 2776–2778. [CrossRef] [PubMed]
21. Loh, P.R.; Danecek, P.; Palamara, P.F.; Fuchsberger, C.; Reshef, Y.A.; Finucane, H.K.; Schoenherr, S.; Forer, L.; McCarthy, S.; Abecasis, G.R.; et al. Reference-based phasing using the Haplotype Reference Consortium panel. *Nat. Genet.* **2016**, *48*, 1443–1448. [CrossRef] [PubMed]
22. Durbin, R. Efficient haplotype matching and storage using the positional Burrows-Wheeler transform (PBWT). *Bioinformatics* **2014**, *30*, 1266–1272. [CrossRef]
23. McCarthy, S.; Das, S.; Kretzschmar, W.; Delaneau, O.; Wood, A.R.; Teumer, A.; Kang, H.M.; Fuchsberger, C.; Danecek, P.; Sharp, K.; et al. A reference panel of 64,976 haplotypes for genotype imputation. *Nat. Genet.* **2016**, *48*, 1279–1283. [CrossRef]
24. Chang, C.C.; Chow, C.C.; Tellier, L.C.A.M.; Vattikuti, S.; Purcell, S.M.; Lee, J.J. Second-generation PLINK: Rising to the challenge of larger and richer datasets. *Gigascience* **2015**, *4*, 7. [CrossRef] [PubMed]
25. Alexander, D.H.; Novembre, J.; Lange, K. Fast model-based estimation of ancestry in unrelated individuals. *Genome Res.* **2009**, *19*, 1655–1664. [CrossRef] [PubMed]
26. Buniello, A.; Macarthur, J.A.L.; Cerezo, M.; Harris, L.W.; Hayhurst, J.; Malangone, C.; McMahon, A.; Morales, J.; Mountjoy, E.; Sollis, E.; et al. The NHGRI-EBI GWAS Catalog of published genome-wide association studies, targeted arrays and summary statistics 2019. *Nucleic Acids Res.* **2019**, *47*, D1005–D1012. [CrossRef]
27. Hemani, G.; Zheng, J.; Elsworth, B.; Wade, K.H.; Haberland, V.; Baird, D.; Laurin, C.; Burgess, S.; Bowden, J.; Langdon, R.; et al. The MR-base platform supports systematic causal inference across the human phenome. *Elife* **2018**, *7*, 1–29. [CrossRef]
28. Zhu, Z.; Zheng, Z.; Zhang, F.; Wu, Y.; Trzaskowski, M.; Maier, R.; Robinson, M.R.; McGrath, J.J.; Visscher, P.M.; Wray, N.R.; et al. Causal associations between risk factors and common diseases inferred from GWAS summary data. *Nat. Commun.* **2018**, *9*, 224. [CrossRef]
29. R Development Core Team. *R: A Language and Eviroment for Statistical Computing*; R Foundation for Statistical Computing: Vienna, Austria, 2008.
30. Kurilshikov, A.; Medina-Gomez, C.; Bacigalupe, R.; Radjabzadeh, D.; Wang, J.; Demirkan, A.; Le Roy, C.I.; Raygoza Garay, J.A.; Finnicum, C.T.; Liu, X.; et al. Large-scale association analyses identify host factors influencing human gut microbiome composition. *Nat. Genet.* **2021**, *53*, 156–165. [CrossRef] [PubMed]
31. Burgess, S.; Thompson, S.G. Avoiding bias from weak instruments in Mendelian randomization studies. *Int. J. Epidemiol.* **2011**, *40*, 755–764. [CrossRef]

32. Lambert, S.A.; Gil, L.; Jupp, S.; Ritchie, S.C.; Xu, Y.; Buniello, A.; McMahon, A.; Abraham, G.; Chapman, M.; Parkinson, H.; et al. The Polygenic Score Catalog as an open database for reproducibility and systematic evaluation. *Nat. Genet.* **2021**, *53*, 420–425. [CrossRef]
33. Jia, G.; Lu, Y.; Wen, W.; Long, J.; Liu, Y.; Tao, R.; Li, B.; Denny, J.C.; Shu, X.-O.; Zheng, W. Evaluating the Utility of Polygenic Risk Scores in Identifying High-Risk Individuals for Eight Common Cancers. *JNCI Cancer Spectr.* **2020**, *4*, pkaa021. [CrossRef] [PubMed]
34. Hsu, L.; Jeon, J.; Brenner, H.; Gruber, S.B.; Schoen, R.E.; Berndt, S.I.; Chan, A.T.; Chang-Claude, J.; Du, M.; Gong, J.; et al. A Model to Determine Colorectal Cancer Risk Using Common Genetic Susceptibility Loci. *Gastroenterology* **2015**, *148*, 1330–1339.e14. [CrossRef] [PubMed]
35. Fritsche, L.G.; Patil, S.; Beesley, L.J.; VandeHaar, P.; Salvatore, M.; Ma, Y.; Peng, R.B.; Taliun, D.; Zhou, X.; Mukherjee, B. Cancer PRSweb: An Online Repository with Polygenic Risk Scores for Major Cancer Traits and Their Evaluation in Two Independent Biobanks. *Am. J. Hum. Genet.* **2020**, *107*, 815–836. [CrossRef]
36. Huyghe, J.R.; Bien, S.A.; Harrison, T.A.; Kang, H.M.; Chen, S.; Schmit, S.L.; Conti, D.V.; Qu, C.; Jeon, J.; Edlund, C.K.; et al. Discovery of common and rare genetic risk variants for colorectal cancer. *Nat. Genet.* **2019**, *51*, 76–87. [CrossRef]
37. Kachuri, L.; Graff, R.E.; Smith-Byrne, K.; Meyers, T.J.; Rashkin, S.R.; Ziv, E.; Witte, J.S.; Johansson, M. Pan-cancer analysis demonstrates that integrating polygenic risk scores with modifiable risk factors improves risk prediction. *Nat. Commun.* **2020**, *11*, 6084. [CrossRef] [PubMed]
38. Shi, Z.; Yu, H.; Wu, Y.; Lin, X.; Bao, Q.; Jia, H.; Perschon, C.; Duggan, D.; Helfand, B.T.; Zheng, S.L.; et al. Systematic evaluation of cancer-specific genetic risk score for 11 types of cancer in The Cancer Genome Atlas and Electronic Medical Records and Genomics cohorts. *Cancer Med.* **2019**, *8*, 3196–3205. [CrossRef]
39. Graff, R.E.; Cavazos, T.B.; Thai, K.K.; Kachuri, L.; Rashkin, S.R.; Hoffman, J.D.; Alexeeff, S.E.; Blatchins, M.; Meyers, T.J.; Leong, L.; et al. Cross-cancer evaluation of polygenic risk scores for 16 cancer types in two large cohorts. *Nat. Commun.* **2021**, *12*, 970. [CrossRef]
40. Archambault, A.N.; Su, Y.-R.; Jeon, J.; Thomas, M.; Lin, Y.; Conti, D.V.; Win, A.K.; Sakoda, L.C.; Lansdorp-Vogelaar, I.; Peterse, E.F.P.; et al. Cumulative Burden of Colorectal Cancer–Associated Genetic Variants Is More Strongly Associated with Early-Onset vs Late-Onset Cancer. *Gastroenterology* **2020**, *158*, 1274–1286.e12. [CrossRef]
41. Schmit, S.L.; Edlund, C.K.; Schumacher, F.R.; Gong, J.; Harrison, T.A.; Huyghe, J.R.; Qu, C.; Melas, M.; Van Den Berg, D.J.; Wang, H.; et al. Novel Common Genetic Susceptibility Loci for Colorectal Cancer. *JNCI J. Natl. Cancer Inst.* **2019**, *111*, 146–157. [CrossRef]
42. Dvorak, P.; Pesta, M.; Soucek, P. ABC gene expression profiles have clinical importance and possibly form a new hallmark of cancer. *Tumor Biol.* **2017**, *39*, 1010428317699800. [CrossRef]
43. Hlavata, I.; Mohelnikova-Duchonova, B.; Vaclavikova, R.; Liska, V.; Pitule, P.; Novak, P.; Bruha, J.; Vycital, O.; Holubec, L.; Treska, V.; et al. The role of ABC transporters in progression and clinical outcome of colorectal cancer. *Mutagenesis* **2012**, *27*, 187–196. [CrossRef] [PubMed]
44. Wang, B.; Wang, X.; Tseng, Y.; Huang, M.; Luo, F.; Zhang, J.; Liu, J. Distinguishing colorectal adenoma from hyperplastic polyp by WNT2 expression. *J. Clin. Lab. Anal.* **2021**, *35*, 1–10. [CrossRef]
45. Zhang, Z.; Zhu, H.; Li, Q.; Gao, W.; Zang, D.; Su, W.; Yang, R.; Zhong, J. Gene Expression Profiling of Tricarboxylic Acid Cycle and One Carbon Metabolism Related Genes for Prognostic Risk Signature of Colon Carcinoma. *Front. Genet.* **2021**, *12*, 1–14. [CrossRef]
46. Chang, L.C.; Hsu, Y.C.; Chiu, H.M.; Ueda, K.; Wu, M.S.; Kao, C.H.; Shen, T.L. Exploration of the Proteomic Landscape of Small Extracellular Vesicles in Serum as Biomarkers for Early Detection of Colorectal Neoplasia. *Front. Oncol.* **2021**, *11*, 1–12. [CrossRef]
47. Frey, M.R.; Edelblum, K.L.; Mullane, M.T.; Liang, D.; Polk, D.B. The ErbB4 Growth Factor Receptor Is Required for Colon Epithelial Cell Survival in the Presence of TNF. *Gastroenterology* **2009**, *136*, 217–226. [CrossRef] [PubMed]
48. Frey, M.R.; Hilliard, V.C.; Mullane, M.T.; Polk, D.B. ErbB4 promotes cyclooxygenase-2 expression and cell survival in colon epithelial cells. *Lab. Investig.* **2010**, *90*, 1415–1424. [CrossRef]
49. Williams, C.S.; Bernard, J.K.; Beckler, M.D.; Almohazey, D.; Washington, M.K.; Smith, J.J.; Frey, M.R. ERBB4 is over-expressed in human colon cancer and enhances cellular transformation. *Carcinogenesis* **2015**, *36*, 710–718. [CrossRef] [PubMed]
50. Jia, X.; Wang, H.; Li, Z.; Yan, J.; Guo, Y.; Zhao, W.; Gao, L.; Wang, B.; Jia, Y. HER4 promotes the progression of colorectal cancer by promoting epithelial-mesenchymal transition. *Mol. Med. Rep.* **2020**, *21*, 1779–1788. [CrossRef]
51. Baiocchi, G.; Lopes, A.; Coudry, R.A.; Rossi, B.M.; Soares, F.A.; Aguiar, S.; Guimarães, G.C.; Ferreira, F.O.; Nakagawa, W.T. ErbB family immunohistochemical expression in colorectal cancer patients with higher risk of recurrence after radical surgery. *Int. J. Colorectal Dis.* **2009**, *24*, 1059–1068. [CrossRef]
52. Chu, S.; Wang, H.; Yu, M. A putative molecular network associated with colon cancer metastasis constructed from microarray data. *World J. Surg. Oncol.* **2017**, *15*, 115. [CrossRef] [PubMed]
53. Bae, J.A.; Yoon, S.; Park, S.Y.; Lee, J.H.; Hwang, J.E.; Kim, H.; Seo, Y.W.; Cha, Y.J.; Hong, S.P.; Kim, H.; et al. An unconventional KITENIN/ErbB4-mediated downstream signal of EGF upregulates c-Jun and the invasiveness of colorectal cancer cells. *Clin. Cancer Res.* **2014**, *20*, 4115–4128. [CrossRef] [PubMed]
54. Ren, H.; Li, Z.; Tang, Z.; Li, J.; Lang, X. Long noncoding MAGI2-AS3 promotes colorectal cancer progression through regulating miR-3163/TMEM106B axis. *J. Cell. Physiol.* **2020**, *235*, 4824–4833. [CrossRef] [PubMed]
55. Poursheikhani, A.; Abbaszadegan, M.R.; Nokhandani, N.; Kerachian, M.A. Integration analysis of long non-coding RNA (lncRNA) role in tumorigenesis of colon adenocarcinoma. *BMC Med. Genomics* **2020**, *13*, 108. [CrossRef] [PubMed]

56. Yang, X.; Wu, S.; Li, X.; Yin, Y.; Chen, R. MAGI2-AS3 rs7783388 polymorphism contributes to colorectal cancer risk through altering the binding affinity of the transcription factor GR to the MAGI2-AS3 promoter. *J. Clin. Lab. Anal.* **2020**, *34*, 1–11. [CrossRef]
57. Mlecnik, B.; Bindea, G.; Angell, H.K.; Sasso, M.S.; Obenauf, A.C.; Fredriksen, T.; Lafontaine, L.; Bilocq, A.M.; Kirilovsky, A.; Tosolini, M.; et al. Functional network pipeline reveals genetic determinants associated with in situ lymphocyte proliferation and survival of cancer patients. *Sci. Transl. Med.* **2014**, *6*, 228ra37. [CrossRef]
58. Yang, Q.-X.; Liu, T.; Yang, J.-L.; Liu, F.; Chang, L.; Che, G.-L.; Lai, S.-Y.; Jiang, Y.-M. Low expression of NTF3 is associated with unfavorable prognosis in hepatocellular carcinoma. *Int. J. Clin. Exp. Pathol.* **2020**, *13*, 2280–2288.
59. Liu, R.; Li, R.; Yu, H.; Liu, J.; Zheng, S.; Li, Y.; Ye, L. NTF3 Correlates With Prognosis and Immune Infiltration in Hepatocellular Carcinoma. *Front. Med.* **2021**, *8*, 795849. [CrossRef]
60. Zhao, R.; Liu, Y.; Wu, C.; Li, M.; Wei, Y.; Niu, W.; Yang, J.; Fan, S.; Xie, Y.; Li, H.; et al. BRD7 Promotes Cell Proliferation and Tumor Growth Through Stabilization of c-Myc in Colorectal Cancer. *Front. Cell Dev. Biol.* **2021**, *9*, 659392. [CrossRef]
61. Pelullo, M.; Nardozza, F.; Zema, S.; Quaranta, R.; Nicoletti, C.; Besharat, Z.M.; Felli, M.P.; Cerbelli, B.; d'Amati, G.; Palermo, R.; et al. Kras/ADAM17-Dependent Jag1-ICD Reverse Signaling Sustains Colorectal Cancer Progression and Chemoresistance. *Cancer Res.* **2019**, *79*, 5575–5586. [CrossRef]
62. Kiriyama, K.; Hirohashi, Y.; Torigoe, T.; Kubo, T.; Tamura, Y.; Kanaseki, T.; Takahashi, A.; Nakazawa, E.; Saka, E.; Ragnarsson, C.; et al. Expression and function of FERMT genes in colon carcinoma cells. *Anticancer Res.* **2013**, *33*, 167–173.
63. Huyghe, J.R.; Harrison, T.A.; Bien, S.A.; Hampel, H.; Figueiredo, J.C.; Schmit, S.L.; Conti, D.V.; Chen, S.; Qu, C.; Lin, Y.; et al. Genetic architectures of proximal and distal colorectal cancer are partly distinct. *Gut* **2021**, *70*, 1325–1334. [CrossRef] [PubMed]
64. Passarelli, M.N.; Newcomb, P.A.; Makar, K.W.; Burnett-Hartman, A.N.; Potter, J.D.; Upton, M.P.; Zhu, L.C.; Rosenfeld, M.E.; Schwartz, S.M.; Rutter, C.M. Blood lipids and colorectal polyps: Testing an etiologic hypothesis using phenotypic measurements and Mendelian randomization. *Cancer Causes Control* **2015**, *26*, 467–473. [CrossRef] [PubMed]
65. Ma, Y.; Yang, Y.; Wang, F.; Zhang, P.; Shi, C.; Zou, Y.; Qin, H. Obesity and risk of colorectal cancer: A systematic review of prospective studies. *PLoS ONE* **2013**, *8*, e53916. [CrossRef]
66. Saus, E.; Iraola-Guzmán, S.; Willis, J.R.; Brunet-Vega, A.; Gabaldón, T. Microbiome and colorectal cancer: Roles in carcinogenesis and clinical potential. *Mol. Aspects Med.* **2019**, *69*, 93–106. [CrossRef] [PubMed]
67. Flemer, B.; Lynch, D.B.; Brown, J.M.R.; Jeffery, I.B.; Ryan, F.J.; Claesson, M.J.; O'Riordain, M.; Shanahan, F.; O'Toole, P.W. Tumour-associated and non-tumour-associated microbiota in colorectal cancer. *Gut* **2017**, *66*, 633–643. [CrossRef] [PubMed]
68. Miyake, T.; Mori, H.; Yasukawa, D.; Hexun, Z.; Maehira, H.; Ueki, T.; Kojima, M.; Kaida, S.; Iida, H.; Shimizu, T.; et al. The Comparison of Fecal Microbiota in Left-Side and Right-Side Human Colorectal Cancer. *Eur. Surg. Res.* **2021**, *62*, 248–254. [CrossRef]
69. Phipps, O.; Quraishi, M.N.; Dickson, E.A.; Steed, H.; Kumar, A.; Acheson, A.G.; Beggs, A.D.; Brookes, M.J.; Al-Hassi, H.O. Differences in the on-and off-tumor microbiota between right-and left-sided colorectal cancer. *Microorganisms* **2021**, *9*, 1108. [CrossRef]
70. Yang, X.; Pan, Y.; Wu, W.; Qi, Q.; Zhuang, J.; Xu, J.; Han, S. Analysis of prognosis, genome, microbiome, and microbial metabolome in different sites of colorectal cancer. *J. Transl. Med.* **2019**, *17*, 1–22. [CrossRef]
71. Clos-Garcia, M.; Garcia, K.; Alonso, C.; Iruarrizaga-Lejarreta, M.; D'amato, M.; Crespo, A.; Iglesias, A.; Cubiella, J.; Bujanda, L.; Falcón-Pérez, J.M. Integrative analysis of fecal metagenomics and metabolomics in colorectal cancer. *Cancers* **2020**, *12*, 1142. [CrossRef]
72. Lu, Y.; Chen, J.; Zheng, J.; Hu, G.; Wang, J.; Huang, C.; Lou, L.; Wang, X.; Zeng, Y. Mucosal adherent bacterial dysbiosis in patients with colorectal adenomas. *Sci. Rep.* **2016**, *6*, 26337. [CrossRef]
73. Luo, H.; Liu, L.; Zhao, J.-J.; Mi, X.-F.; Wang, Q.-J.; Yu, M. Effects of oxaliplatin on inflammation and intestinal floras in rats with colorectal cancer. *Eur. Rev. Med. Pharmacol. Sci.* **2020**, *24*, 10542–10549. [CrossRef] [PubMed]

MDPI

Article

Clinical Score to Predict Recurrence in Patients with Stage II and Stage III Colon Cancer

David Viñal [1], Sergio Martinez-Recio [1], Daniel Martinez-Perez [1], Iciar Ruiz-Gutierrez [1], Diego Jimenez-Bou [1], Jesús Peña-Lopez [1], Maria Alameda-Guijarro [1], Gema Martin-Montalvo [1], Antonio Rueda-Lara [1], Laura Gutierrez-Sainz [1], Maria Elena Palacios [2], Ana Belén Custodio [1], Ismael Ghanem [1], Jaime Feliu [3,*] and Nuria Rodríguez-Salas [4]

1 Department of Medical Oncology, Hospital Universitario La Paz, 28046 Madrid, Spain
2 Department of Pathology, Hospital Universitario La Paz, 28046 Madrid, Spain
3 Department of Medical Oncology, Hospital Universitario La Paz, IdiPAZ, Catedra UAM-AMGEN, CIBERONC, 28046 Madrid, Spain
4 Department of Medical Oncology, Hospital Universitario La Paz, IdiPAZ, CIBERONC, 28046 Madrid, Spain
* Correspondence: jaime.feliu@salud.madrid.org

Simple Summary: The prognosis of patients with stage II and stage III colon cancer is heterogeneous. Clinical and pathological characteristics may help to further refine the recurrence risk. We built a prognostic score and categorized patients into two risk groups in a training and validation cohort. We assigned two points to T4 and one point to N2 and high tumor budding based on the multivariate cox regression analysis for time to recurrence (TTR) in the training cohort. Forty-five percent of the patients were assigned to the low-risk group and compared to the high-risk group, had a significantly longer TTR. These results were confirmed in the validation cohort.

Abstract: Background: The prognosis of patients with stage II and stage III colon cancer is heterogeneous. Clinical and pathological characteristics, such as tumor budding, may help to further refine the recurrence risk. Methods: We included all the patients with localized colon cancer at Hospital Universitario La Paz from October 2016 to October 2021. We built a prognostic score for recurrence in the training cohort based on multivariate cox regression analysis and categorized the patients into two risk groups. Results: A total of 440 patients were included in the training cohort. After a median follow-up of 45 months, 81 (18%) patients had a first tumor recurrence. T4, N2, and high tumor budding remained with a p value <0.05 at the last step of the multivariate cox regression model for time to recurrence (TTR). We assigned 2 points to T4 and 1 point to N2 and high tumor budding. Forty-five percent of the patients were assigned to the low-risk group (score = 0). Compared to the high-risk group (score 1–4), patients in the low-risk group had a significantly longer TTR (hazard ratio for disease recurrence of 0.14 (95%CI: 0.00 to 0.90; $p < 0.045$)). The results were confirmed in the validation cohort. Conclusions: In our study, we built a simple score to predict tumor recurrence based on T4, N2, and high tumor budding. Patients in the low-risk group, that comprised 44% of the cohort, had an excellent prognosis.

Keywords: colonic neoplasms; chemotherapy; adjuvant; tumor budding

Citation: Viñal, D.; Martinez-Recio, S.; Martinez-Perez, D.; Ruiz-Gutierrez, I.; Jimenez-Bou, D.; Peña-Lopez, J.; Alameda-Guijarro, M.; Martin-Montalvo, G.; Rueda-Lara, A.; Gutierrez-Sainz, L.; et al. Clinical Score to Predict Recurrence in Patients with Stage II and Stage III Colon Cancer. *Cancers* **2022**, *14*, 5891. https://doi.org/10.3390/cancers14235891

Academic Editor: Stephane Dedieu

Received: 10 June 2022
Accepted: 21 November 2022
Published: 29 November 2022

1. Introduction

Colorectal cancer is the third most common tumor and the second cause of cancer-related cause of death globally [1]. In patients with stage II and stage III colon cancer, the prognosis is heterogeneous, and survival varies depending on numerous factors. Classically, for the pathologic stage at diagnosis, according to the American Joint Committee on Cancer (AJCC)/Union for International Cancer Control (UICC), the tumor, node, and metastasis (TNM) staging classification was considered the most important indicator of outcome [2]. However, patients with stage IIIa disease may have a more favorable prognosis than patients with IIb stage, which may indicate that other factors contribute significantly to the prognosis of the patient. Globally, it is estimated that 35% of the patients will eventually recur [3]. To further improve the outcome, chemotherapy has been established as a standard of care for stage III colon cancer with a 10 to 20% of survival benefit (depending on the regimen of chemotherapy) and is an option for patients with intermediate- or high-risk stage II [4,5]. The latest European Society for Medical Oncology (ESMO) guidelines include lymph node sampling <12 and T4 stage including perforation as major prognostic factors and high-grade tumor, vascular invasion, lymphatic invasion, perineural invasion, tumor presentation with obstruction, and high preoperative carcinoembryonic antigen (CEA) levels as minor prognostic factors. The National Comprehensive Cancer Network (NCCN) also includes high tumor budding and close, indeterminate, or positive margins as risk factors for recurrence. To better define the prognosis and recurrence risk of patients with resected colon cancer, several nomograms have been published [6–8]. One of the most widely used is the Memorial Sloan Kettering Cancer Center (MSKCC) colon cancer recurrence nomogram, which predicts freedom from recurrence based on nine clinicopathological features including age, tumor size, preoperative carcinoembryonic antigen (CEA), use of adjuvant chemotherapy, and other indicators of tumor invasiveness [6]. A recently published update simplified the score to five items; however, tumor-infiltrating lymphocytes were included in the nomogram, a feature not available in many centers [8].

The aim of this study is to create a simple clinical score to predict recurrence using clinical and pathological variables available in routine clinical practice and to select a subgroup of patients with excellent prognosis according to this score.

2. Materials and Methods

This is a single-institution retrospective observational study. We included all patients who underwent curative surgery for stage II and stage III colon cancer between October 2016 and October 2021 at Hospital Universitario La Paz (HULP), Madrid (Spain). The study protocol specified the inclusion criteria as follows: age above 18 years and completely resected colon adenocarcinoma located at >15 cm of the anal verge as determined by endoscopy or above the peritoneal reflection in the surgical resection without any evidence of metastatic disease. Main exclusion criteria were as follows: macroscopic evidence of residual tumor in the surgical specimen; no chemotherapy or radiotherapy were allowed before surgery; severe renal or hepatic disorder; bone marrow suppression; or disabling peripheral neuropathy. This study was approved by the Ethics Committee of HULP and was conducted in accordance with ethical standards of the Helsinki Declaration of the World Medical Association. Baseline disease, demographics, clinical data, treatment characteristics, and outcomes were analyzed from the medical record of each patient. Adjuvant chemotherapy was administered according to ESMO guidelines [4,9]. Patients were followed every 3 months with CT scan and CEA for the first 2 years from the surgery and every 6 months with CT scan and CEA from years 3 to 5. Colonoscopy was performed every 3 years starting 1 year after surgery.

The primary objective of the study was the identification of factors associated with time to recurrence (TTR). We chose TTR as the primary endpoint based on previous reports by other groups [8]. The sample was divided into a training cohort (patients diagnosed between October 2016 and September 2020, n = 440) and a validation cohort (patients diagnosed from October 2020 to September 2021, n = 100). TTR was calculated from the date of the surgery until the date of tumor recurrence or last follow-up. OS was defined as the time between the date of diagnosis and the date of death or last follow-up. The analysis was performed with a data cut-off of 15 September 2022. The relation between TTR and OS with each of the variables was analyzed using the log-rank test. Survival analysis was performed using the Kaplan–Meier method. Univariate cox regression analyses and multivariate proportional hazards regression model were carried out in the training cohort to identify independent prognostic factors for disease recurrence. We performed a correlation assessment using the Spearman's rho test. Multicollinearity among variables was defined as a rho test value ≥ 0.50. In fact, we excluded adjuvant chemotherapy treatment as it positively correlates with the presence of high-risk features (Spearman's rho test = 0.533; $p < 0.001$). In the multivariate analysis, we included the variables significantly associated with TTR in the univariate analysis. Multivariate analysis was performed with backward elimination. Prognostic factors that yielded a p value < 0.05 at the last step of multivariate cox regression analysis were included in the score. For the development of the score, each factor was assigned a particular score based on its β coefficient. The β coefficient for each risk factor was divided by the lowest β coefficient and rounded to the nearest whole number. Model calibration and discrimination were assessed in the training cohort by the area under the receiver operating characteristic (ROC) curve [10,11]. The final score of each patient was the sum of the points. The prognostic score was then applied to each patient. Survival by prognostic group was represented by Kaplan–Meier curves, and p values were calculated using the log-rank test. The training sample was divided into two risk strata (low-risk group and high-risk group) based on the approximate median of risk score. Hazard ratios (HRs) were calculated using cox proportional hazard regression, with p values calculated using the Wald statistics. The performance of the two-risk group strategy was tested for TTR in the validation cohort. All statistical analyses were carried out using SPSS v.25.

3. Results

A total of 440 patients with stage II and stage III colon cancer underwent curative surgery between October 2016 and October 2020 and were included in the training cohort. The baseline characteristics are depicted in Table 1. The median age at diagnosis was 74 years (range 35–95), and 44% of the patients were female. The primary tumor was distributed equally in the right and left colon. Stage II and stage III were observed in 50% percent of the patients each. Of note, preoperative CEA was available in 219 patients and was high in 19% of them. Twenty-five percent of the patients had high tumor budding. A total of 225 (51%) patients received adjuvant chemotherapy: 61 patients with stage II (27%) and 164 patients with stage III (75%).

After a median follow-up of 45 months (range, 0,1 to 66 months), 81 (18%) patients had a first tumor recurrence: 27 (12%) patients with stage II and 54 (24%) patients with stage III. Ninety-six (17%) patients died: 39 (17%) patients with stage II and 57 (26%) patients with stage III. The median TTR and OS were not reached for the whole cohort. Univariate cox regression analysis showed that T4 (tumor invades the visceral peritoneum or invades or adheres to the adjacent organ or structure), N2 (four or more regional nodes are positive) [12], R1 (incomplete tumor resection with microscopic surgical resection margin involvement) [13], bowel obstruction and perforation at diagnosis, lymphovascular and perineural invasion, high tumor budding (defined as ≥10 buds) [14], grade 3, and deficient mismatch repair were significantly associated with TTR. Only T4 (hazard ratio (HR), 3.46 [95% confidence interval (CI): 1.68 to 7.13], $p < 0.01$), N2 (HR, 2.29 (95%CI, 1.19 to 4.38), $p = 0.01$), and high tumor budding (HR, 1.91 (95%CI, 1.02 to 3.54), $p = 0.04$) remained with a p value <0.05 at the last step of the multivariate cox regression model, and were selected to create the clinical score (see Table 2).

Based on the β coefficient of each feature (see Table 2), we assigned 2 points to T4, and 1 point to N2 and high tumor budding. Therefore, patients were assigned from 0 to 4 points (score 0 = 138, score 1 = 44, score 2 = 57, score 3 = 52, and score 4 = 13 patients). The area under the ROC curve for tumor recurrence at 36 months was 0.77 (95%CI, 0.70 to 0.84), $p < 0.01$ (Figure 1).

Table 1. Baseline characteristics of the patients.

Characteristic (n = Training Cohort)	Training Cohort (n = 440)	Validation Cohort (n = 100)
Sex (female)	193 (44)	50 (50)
Age	74 (35–95)	75 (45–97)
Age < 50	22 (5)	4 (4)
Location		
Right	212 (48)	56 (56)
Left	228 (52)	44 (44)
Stage at diagnosis		
II	222 (50)	54 (54)
III	218 (50)	46 (46)
T		
1	5 (1)	1 (1)
2	17 (4)	3 (3)
3	252 (58)	50 (50)
4	163 (37)	46 (46)
N		
0	222 (50)	58 (58)
1	150 (34)	33 (33)
2	68 (16)	9 (9)
R0	413 (94)	93 (93)
Preoperative CEA >5 ng/ml	41 (19)	10 (20)
Bowel obstruction at diagnosis	45 (10)	15 (15)
Bowel perforation at diagnosis	37 (8)	5 (5)
Lymphovascular invasion	184 (43)	47 (47)
Perineural invasion	85 (20)	31 (31)
Budding		
Low	142 (47)	49 (49)
Medium	84 (28)	30 (30)
High	78 (25)	21 (21)
Grade		
1	18 (4)	1 (1)
2	367 (88)	91 (91)
3	33 (8)	8 (8)
Mucinous	82 (16)	19 (19)
Rignet cell	13 (3)	5(5)
≥12 resected lymph nodes	382 (90)	96 (96)
dMMR	68 (17)	14 (14)

CEA, carcinoembryonic antigen; dMMR, deficient mismatch repair; R0, complete tumor resection with all margins histologically uninvolved.

Table 2. Multivariate Cox Regression Analysis.

Characteristic	β Coefficient	HR (95%CI)	p Value
T4	1.243	3.46 (1.68–7.13)	0.001
N2	0.829	2.29 (1.19–4.38)	0.012
High tumor budding	0.647	1.91 (1.02–3.54)	0.041

At 36 months, 95%, 83%, 73%, 60%, and 19% of the patients with scores 0, 1, 2, 3, and 4 were recurrence-free, respectively. The median TTR was not reached in patients with scores 0–3. Patients with score 4 had a median TTR of 29 months (95% confidence interval (CI): 0.1 to 60.23). Significant differences were observed between the groups ($p < 0.001$), see Figure 2. Patients were divided into a low-risk group (score = 0; n = 138; 45% of the patients) and a high-risk group (score = 1–4; n = 166; 55% of the patients). At 36 months, 95% and 67% of the patients in the low-risk and high-risk groups were recurrence-free, respectively. Patients assigned to the low-risk group had a significantly longer TTR than patients assigned to the high-risk group. The median TTR was not reached in either group, with a HR for disease recurrence of 0.13 (95%CI: 0.05 to 0.31; p <0.001), see Figure 3.

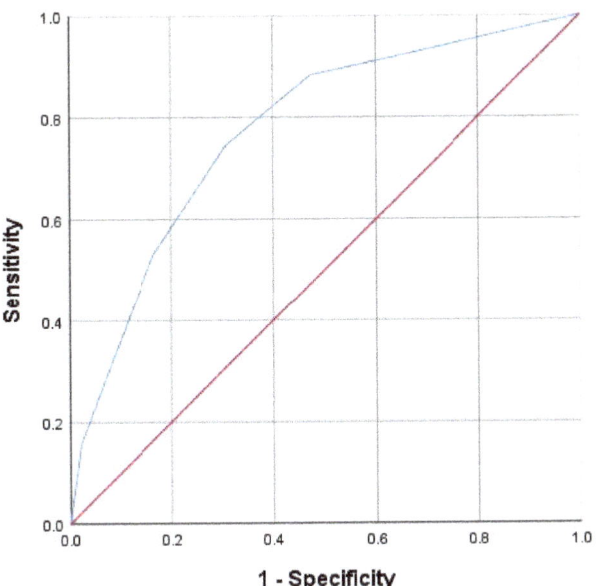

Figure 1. ROC curve of prognostic score (0 to 4 points) for recurrence at 24 months.

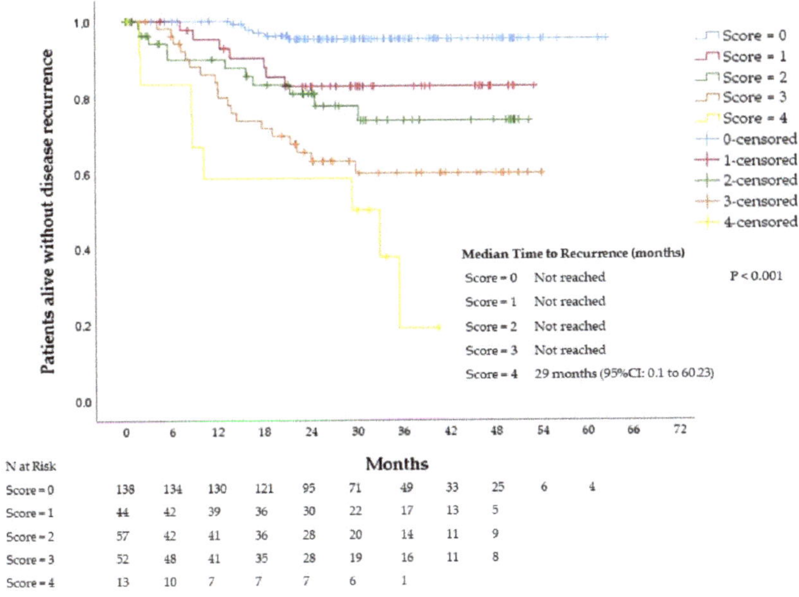

Figure 2. Time to recurrence according to the score (0–4) in the training cohort.

A total of 100 patients were included in the validation cohort. The baseline characteristics are depicted in Table 1. The median age at diagnosis was 75 years (range 45–97), and 50% of the patients were female. The primary tumor was distributed equally in the right and left colon. Stage II was observed in 54% percent of the patients each. Twenty-one percent of the patients had high tumor budding. A total of 43 patients received adjuvant chemotherapy. Patients were assigned to the low-risk (n = 46; 46%) and high-risk (n = 54;

54%) groups. After a median follow-up of 15 months (range, 2 to 25 months), 15 (15%) of the patients had a first tumor recurrence. Recurrences were observed in five (9%) patients with stage II and 10 (22%) patients with stage III. According to our score, all the recurrences were observed in the high-risk group. At 12 months, 100% and 79% of the patients in the low-risk and high-risk groups were recurrence-free, respectively. Patients assigned to the low-risk group had significantly longer TTR than patients assigned to the high-risk group. The median TTR was not reached in either group. HR for disease recurrence of 0.14 (95%CI: 0.00 to 0.90; p <0.045), see Figure 4.

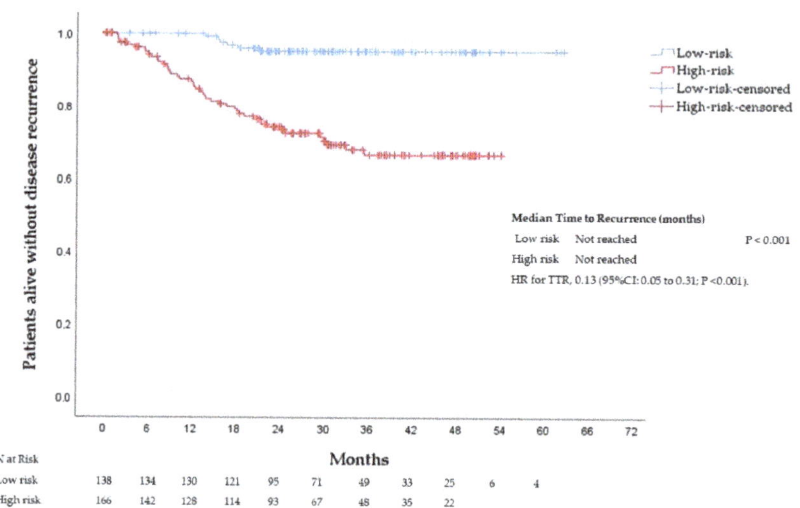

Figure 3. Time to recurrence according to risk groups (low-risk vs. high-risk) in the training cohort. HR, hazard ratio; CI, confidence interval.

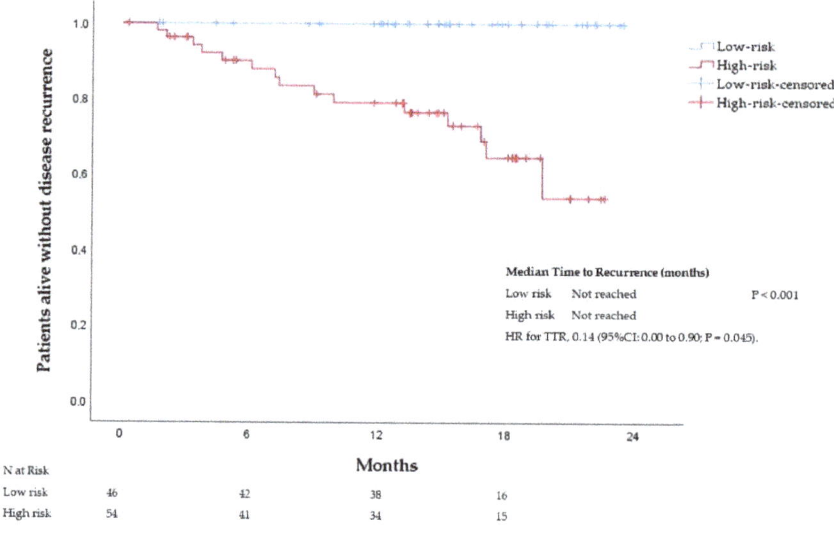

Figure 4. Time to recurrence according to risk groups in the validation cohort.

4. Discussion

In this study, we created a simple score using three clinicopathological parameters available in routine clinical practice to better estimate the recurrence risk in patients with stage II and stage III colon cancer. This score shows that the probability of recurrence ranges from 5% in patients with a score = 0 to 81% in patients with a score = 4, with an AUC of 0.77. More importantly, the score can discriminate a subgroup of patients (low-risk group, score = 0), so that even with locally advanced disease, they will have an excellent prognosis after completing the standard treatment recommendations according to their stage. This low-risk group comprises the 45% of the training cohort included in the multivariate analysis and 46% of the patients in the validation cohorts.

Multiple scores and nomograms have attempted to overcome the aforementioned limitations of the AJCC's TNM staging system for the prediction of outcomes. One of the most relevant is the MSKCC nomogram published in 2008 for the estimation of the recurrence risk of patients with stages I to III colon cancer after a complete resection (R0) of the tumor [6]. The nomogram was based on nine variables including patient age, tumor location, preoperative carcinoembryonic antigen, T stage, number of positive and negative lymph nodes, lymphovascular invasion, perineural invasion, and use of postoperative chemotherapy. The nomogram successfully predicted relapse with a concordance index of 0.77, improving the stratification provided by the AJCC staging scheme and was externally validated in multiple cohorts [15–17]. However, the high number of features and its complexity may prevent it from being used as a practical tool in clinical practice. The MSKCC clinical calculator was updated in 2019 [8]. The nomogram was simplified to six variables and incorporated recently validated molecular and histologic factors, including microsatellite genomic phenotype; AJCC T category; number of tumors involved; lymph nodes; presence of high-risk pathologic features, such as venous, lymphatic, or perineural invasion; presence of tumor-infiltrating lymphocytes; and use of adjuvant chemotherapy. The concordance index was 0.792, and external validation confirmed the utility for the prediction of recurrence. Unfortunately, the generalization of this nomogram was hampered because tumor-infiltrating lymphocytes are not reflexively measured in many centers, including ours.

Our score was built with variables that showed a p value < 0.05 in the multivariate cox regression model and included T4, N2, and budding. Primary colon cancer is classified as T4 per the AJCC TNM staging 8th edition when it invades the visceral peritoneum or invades or adheres to an adjacent organ or structure [12]. T4 has classically been considered a negative prognostic factor. In fact, patients with T4 stage II disease have worse outcomes than patients with stage IIIa disease. A recent subanalysis of patients with stage II colon cancer included in the IDEA collaboration showed that high-risk stage II patients with T4 disease have a worse outcome than those with T3 disease [18]. The IDEA collaboration also highlighted that those patients with stage III with T4 and/or N2 are a different population with a worse prognosis than the other patients with stage III (T1−3 and N1) and suggested the use of these risk groups as stratification categories in randomized trials [19]. Tumor budding refers to isolated or clusters of up to four cancer cells located at the invasive tumor front [20]. A growing amount of evidence has confirmed its prognostic value in localized colon cancer, independent of the tumor grade [21,22]. A recently published subanalysis from the IDEA-France phase III trial [23] showed that tumor budding is an independent prognostic factor in stage III colon cancer patients. The DFS at 3 years was 79% vs 67% ($p = 0.001$) in patients with budding grade 1 vs 2−3 with a HR for recurrence or death of 1.41 (95% CI, 1.12 to 1.77), $p = 0.003$, after adjustment for relevant clinicopathological features. Interestingly, high tumor budding was associated with perineural ($p < 0.01$) and vascular ($p = 0.002$) invasions, which may explain that these well-known adverse prognostic features are not present in the last step of our multivariate analysis. The role of tumor budding in predicting benefit from adjuvant chemotherapy is still controversial. In a subanalysis of the SACURA trial [24], a nonsignificant improvement of 5% in the 5-year recurrence rate was observed in patients with stage II and stage III colon cancer treated with adjuvant chemotherapy vs surgery alone. In patients with pT1, tumor budding currently influences decision making. More recently, the ASCO guidelines were updated and added high tumor budding (≥10 buds, high grade) to the list of adverse prognostic factors to classify patients in the high-risk subpopulation that may derive more benefit from chemotherapy [25]. However, the ESMO guidelines for the management of localized stage II colon cancer still do not consider tumor budding in the decision making. In light of the results of our group and those of previous groups, high tumor budding might be considered as a risk factor.

Prognostic characterization and subgroup categorization in patients with localized colon cancer have more implications than providing the patient a tailored risk of recurrence. Some authors suggest that stratification categories based on T and N should be included in randomized trials for localized colon cancer. Assessing the risk of recurrence may also have implications for the follow-up. The ESMO guidelines recommend a CT scan of the chest and abdomen every 6 to 12 months for the first 3 years in patients who are at higher risk of recurrence according to the TNM classification. Other authors suggest that the preferred approach should be performing two CT scans at 12 and 36 months independent of the stage and risk groups due to the lack of survival benefit of a more intensive approach [26]. We suggest that due to the significantly different risk of recurrence according to subgroups, and the possible benefit of early treatment of oligometastatic disease, the follow-up should be tailored accordingly, or at least taken into account in future follow-up trials.

The limitations to our study are mostly due to its retrospective and unicentric nature. A significant amount of data are missing including tumor budding and preoperative CEA, mainly in the training cohort. Preoperative CEA should be performed before surgery; however, data are missing in half of the patients due to multiple reasons including emergency surgery or even human error. The advantage of our score is that it is based on three features that should be available in every patient with a colorectal cancer diagnosis. Adjuvant treatment was given to the patients following indications by the ESMO guidelines [4,9]. This feature was not considered in the analysis because the benefit of therapy may be masked by the administration in the high-risk subgroup. In fact, we found a correlation between the presence of high-risk features as defined by the ESMO guidelines and the administration of adjuvant chemotherapy (Spearman's rho test = 0.533; $p < 0.001$). Therefore, this score should not be interpreted as a predictive marker of benefit for adjuvant chemotherapy but rather as a predictive marker for recurrence in patients that have followed the standard treatment strategy for localized colon cancer. Nevertheless, we consider that tumor budding is such a strong predictive marker for recurrence that should also be considered as a risk factor and should be included in the guidelines for adjuvant chemotherapy. Only patients with complete (R0) resection were included in the initial MSKCC nomogram [6]; however, approximately 10% of patients have involved resection margins at the pathological report of the surgery. We therefore consider that this feature should be included in a real-world analysis. Finally, although the results of the internal validation cohort seem to confirm the performance of our score, the sample size was small, and the cohort was still immature. We did not perform an external validation and thus an accurate determination of the AUC and calibration of the model was not possible.

5. Conclusions

In conclusion, defining subgroups of patients with localized colon cancer at a high risk of recurrence has implications in the treatment strategy, trial designs, and follow-up. Although the traditional AJCC TNM staging provides adequate prognostic estimation, a more personalized approach using high-risk clinicopathological features may be more precise and practical. In our study, we built a simple score to accurately predict tumor recurrence based on T4, N2, and high tumor budding. Patients with a score = 0, that comprises 44% of the cohort, had an excellent prognosis. A longer follow-up is needed, and an external validation is recommended to confirm our results.

Author Contributions: Conceptualization, D.V., N.R.-S. and J.F.; methodology, D.V., L.G.-S. and J.F.; formal analysis, D.V.; investigation, D.V., S.M.-R., D.M.-P., I.R.-G., D.J.-B., J.P.-L., M.A.-G., G.M.-M. and A.R.-L.; data curation, D.V., S.M.-R. and D.M.-P.; writing—original draft preparation, D.V.; writing—review and editing, L.G.-S., M.E.P., A.B.C., I.G. and J.F.; visualization, D.V. and J.F.; supervision, J.F. and N.R.-S. All authors have read and agreed to the published version of the manuscript.

Funding: This research received no external funding.

Institutional Review Board Statement: The study was conducted in accordance with the Declaration of Helsinki and approved by the Institutional Review Board (or Ethics Committee) of Hospital Universitario La Paz (PI-3607).

Informed Consent Statement: The local ethical committee approved the use of anonymized historic samples and data for the study and waived informed consent from the patients.

Data Availability Statement: The data presented in this study are available on request from the corresponding author.

Conflicts of Interest: The authors declare no conflict of interest.

References

1. Global Cancer Observatory (GLOBOCAN). Available online: https://gco.iarc.fr/ (accessed on 28 March 2022).
2. O'Connell, J.B.; Maggard, M.A.; Ko, C.Y. Colon Cancer Survival Rates with the New American Joint Committee on Cancer Sixth Edition Staging. *J. Natl. Cancer Inst.* **2004**, *96*, 1420–1425. [CrossRef] [PubMed]
3. Sargent, D.; Sobrero, A.; Grothey, A.; O'Connell, M.J.; Buyse, M.; André, T.; Zheng, Y.; Green, E.; Labianca, R.; O'Callaghan, C.; et al. Evidence for Cure by Adjuvant Therapy in Colon Cancer: Observations Based on Individual Patient Data from 20,898 Patients on 18 Randomized Trials. *J. Clin. Oncol.* **2009**, *27*, 872–877. [CrossRef] [PubMed]
4. Argilés, G.; Tabernero, J.; Labianca, R.; Hochhauser, D.; Salazar, R.; Iveson, T.; Laurent-Puig, P.; Quirke, P.; Yoshino, T.; Taieb, J.; et al. Localised colon cancer: ESMO Clinical Practice Guidelines for diagnosis, treatment and follow-up. *Ann. Oncol.* **2020**, *31*, 1291–1305. [CrossRef] [PubMed]
5. Available online: https://www.nccn.org/professionals/physician_gls/pdf/colon.pdf (accessed on 28 March 2022).
6. Weiser, M.R.; Landmann, R.G.; Kattan, M.W.; Gonen, M.; Shia, J.; Chou, J.; Paty, P.B.; Guillem, J.G.; Temple, L.K.; Schrag, D.; et al. Individualized Prediction of Colon Cancer Recurrence Using a Nomogram. *J. Clin. Oncol.* **2008**, *26*, 380–385. [CrossRef] [PubMed]
7. Konishi, T.; Shimada, Y.; Hsu, M.; Wei, I.H.; Pappou, E.; Smith, J.J.; Nash, G.M.; Guillem, J.G.; Paty, P.B.; Garcia-Aguilar, J.; et al. Contemporary Validation of a Nomogram Predicting Colon Cancer Recurrence, Revealing All-Stage Improved Outcomes. [published correction appears in JNCI Cancer Spectr. 2019 Nov 26;3(4):pkz089]. *JNCI Cancer Spectr.* **2019**, *3*, pkz015. [CrossRef]
8. Weiser, M.R.; Hsu, M.; Bauer, P.S.; Chapman, W.C., Jr.; González, I.A.; Chatterjee, D.; Lingam, D.; Mutch, M.G.; Keshinro, A.; Shia, J.; et al. Clinical Calculator Based on Molecular and Clinicopathologic Characteristics Predicts Recurrence Following Resection of Stage I-III Colon Cancer. *J. Clin. Oncol.* **2021**, *39*, 911–919. [CrossRef]
9. Labianca, R.; Nordlinger, B.; Beretta, G.D.; Mosconi, S.; Mandalà, M.; Cervantes, A.; Arnold, D.; ESMO Guidelines Working Group. Early colon cancer: ESMO Clinical Practice Guidelines for diagnosis, treatment and follow-up. *Ann. Oncol.* **2013**, *24* (Suppl. S6), vi64–vi72. [CrossRef]
10. Feliu, J.; Pinto, A.; Basterretxea, L.; Vicente, B.L.-S.; Paredero, I.; Llabrés, E.; Jiménez-Munárriz, B.; Antonio-Rebollo, M.; Losada, B.; Espinosa, E.; et al. Development and Validation of an Early Mortality Risk Score for Older Patients Treated with Chemotherapy for Cancer. *J. Clin. Med.* **2021**, *10*, 1615. [CrossRef]
11. Concato, J.; Feinstein, A.R.; Holford, T.R. The Risk of Determining Risk with Multivariable Models. *Ann. Intern. Med.* **1993**, *118*, 201–210. [CrossRef]
12. Jessup, J.M.; Goldberg, R.M.; Asare, E.A.; Benson, A.; Brierley, J.; Chang, G.; Chen, V.; Compton, C.; De Nardi, P.; Goodman, K. Colon and Rectum. In *AJCC Cancer Staging Manual*, 8th ed.; Amin, M.B., Ed.; AJCC: Chicago, IL, USA, 2017; p. 251.
13. Gress, D.M.; Edge, S.B.; Greene, F.L.; Washington, M.K.; Asare, E.A.; Brierley, J.D.; Byrd, D.R.; Compton, C.C.; Jessup, J.M.; Winchester, D.P.; et al. Principles of Cancer Staging. In *AJCC Cancer Staging Manual*, 8th ed.; Amin, M.B., Ed.; AJCC: Chicago, IL, USA, 2017; p. 29.
14. Lugli, A.; Kirsch, R.; Ajioka, Y.; Bosman, F.; Cathomas, G.; Dawson, H.; El Zimaity, H.; Fléjou, J.-F.; Hansen, T.P.; Hartmann, A.; et al. Recommendations for reporting tumor budding in colorectal cancer based on the International Tumor Budding Consensus Conference (ITBCC) 2016. *Mod. Pathol.* **2017**, *30*, 1299–1311. [CrossRef]
15. Kazem, M.; Khan, A.; Selvasekar, C. Validation of nomogram for disease free survival for colon cancer in UK population: A prospective cohort study. *Int. J. Surg.* **2016**, *27*, 58–65. [CrossRef] [PubMed]
16. Liu, M.; Qu, H.; Bu, Z.; Chen, D.; Jiang, B.; Cui, M.; Maoxing, L.; Yang, H.; Wang, Z.; Jiadi, X.; et al. Validation of the Memorial Sloan-Kettering Cancer Center Nomogram to Predict Overall Survival After Curative Colectomy in a Chinese Colon Cancer Population. *Ann. Surg. Oncol.* **2015**, *22*, 3881–3887. [CrossRef]

17. Collins, I.M.; Kelleher, F.; Stuart, C.; Collins, M.; Kennedy, J. Clinical Decision Aids in Colon Cancer: A Comparison of Two Predictive Nomograms. *Clin. Color. Cancer* **2012**, *11*, 138–142. [CrossRef] [PubMed]
18. Iveson, T.J.; Sobrero, A.F.; Yoshino, T.; Souglakos, I.; Ou, F.-S.; Meyers, J.P.; Shi, Q.; Grothey, A.; Saunders, M.P.; Labianca, R.; et al. Duration of Adjuvant Doublet Chemotherapy (3 or 6 months) in Patients With High-Risk Stage II Colorectal Cancer. *J. Clin. Oncol.* **2021**, *39*, 631–641. [CrossRef] [PubMed]
19. André, T.; Meyerhardt, J.; Iveson, T.; Sobrero, A.; Yoshino, T.; Souglakos, I.; Grothey, A.; Niedzwiecki, D.; Saunders, M.; Labianca, R.; et al. Effect of duration of adjuvant chemotherapy for patients with stage III colon cancer (IDEA collaboration): Final results from a prospective, pooled analysis of six randomised, phase 3 trials. *Lancet Oncol.* **2020**, *21*, 1620–1629. [CrossRef] [PubMed]
20. Lugli, A.; Zlobec, I.; Berger, M.D.; Kirsch, R.; Nagtegaal, I.D. Tumour budding in solid cancers. *Nat. Rev. Clin. Oncol.* **2021**, *18*, 101–115. [CrossRef]
21. Hase, K.; Shatney, C.; Johnson, D.; Trollope, M.; Vierra, M. Prognostic value of tumor "budding" in patients with colorectal cancer. *Dis. Colon Rectum* **1993**, *36*, 627–635. [CrossRef]
22. Goldstein, N.S.; Hart, J. Histologic Features Associated with Lymph Node Metastasis in Stage T1 and Superficial T2 Rectal Adenocarcinomas in Abdominoperineal Resection Specimens: Identifying a Subset of Patients for Whom Treatment with Adjuvant Therapy or Completion Abdominoperineal Resection should be Considered After Local Excision. *Am. J. Clin. Pathol.* **1999**, *111*, 51–58. [CrossRef]
23. Basile, D.; Broudin, C.; Emile, J.; Falcoz, A.; Pagès, F.; Mineur, L.; Bennouna, J.; Louvet, C.; Artru, P.; Fratte, S.; et al. Tumor budding is an independent prognostic factor in stage III colon cancer patients: A post-hoc analysis of the IDEA-France phase III trial (PRODIGE-GERCOR). published online ahead of print, 2022 Mar 16. *Ann. Oncol.* **2022**, *33*, 628–637. [CrossRef]
24. Ueno, H.; Ishiguro, M.; Nakatani, E.; Ishikawa, T.; Uetake, H.; Matsuda, C.; Nakamoto, Y.; Kotake, M.; Kurachi, K.; Egawa, T.; et al. Prospective Multicenter Study on the Prognostic and Predictive Impact of Tumor Budding in Stage II Colon Cancer: Results From the SACURA Trial. *J. Clin. Oncol.* **2019**, *37*, 1886–1894. [CrossRef]
25. Baxter, N.N.; Kennedy, E.B.; Bergsland, E.; Berlin, J.; George, T.J.; Gill, S.; Gold, P.J.; Hantel, A.; Jones, L.; Lieu, C.; et al. Adjuvant Therapy for Stage II Colon Cancer: ASCO Guideline Update. *J. Clin. Oncol.* **2022**, *40*, 892–910. [CrossRef] [PubMed]
26. Jain, A.; Sjoquist, K.; Yip, D. ESMO localised colon cancer guidelines: 'can we improve on our surveillance protocols? *Ann. Oncol.* **2020**, *31*, 1778–1779. [CrossRef] [PubMed]

Article

Routine Immunohistochemical Analysis of Mismatch Repair Proteins in Colorectal Cancer—A Prospective Analysis

Joana Lemos Garcia [1,*], Isadora Rosa [1,2], Sofia Saraiva [1], Inês Marques [1], Ricardo Fonseca [3], Pedro Lage [1,2], Inês Francisco [2,4], Patrícia Silva [2,4], Bruno Filipe [2,4], Cristina Albuquerque [2,4] and Isabel Claro [1,2]

1 Gastroenterology Department, Instituto Português de Oncologia de Lisboa Francisco Gentil, 1099-023 Lisbon, Portugal; isarosa@ipolisboa.min-saude.pt (I.R.); amenezes@ipolisboa.min-saude.pt (S.S.); inesnmarques3@gmail.com (I.M.); plage@ipolisboa.min-saude.pt (P.L.); iclaro@ipolisboa.min-saude.pt (I.C.)
2 Familial Cancer Clinic, Instituto Português de Oncologia de Lisboa Francisco Gentil, 1099-023 Lisbon, Portugal; mfrancisco@ipolisboa.min-saude.pt (I.F.); palsilva@ipolisboa.min-saude.pt (P.S.); bfilipe@ipolisboa.min-saude.pt (B.F.); calbuque@ipolisboa.min-saude.pt (C.A.)
3 Pathology Department, Instituto Português de Oncologia de Lisboa Francisco Gentil, 1099-023 Lisbon, Portugal; rifonseca@ipolisboa.min-saude.pt
4 Molecular Pathobiology Investigation Unit, Instituto Português de Oncologia de Lisboa Francisco Gentil, 1099-023 Lisbon, Portugal
* Correspondence: jgarcia@ipolisboa.min-saude.pt

Simple Summary: Recognition of a hereditary colorectal cancer (CRC) syndrome is crucial. Our aim was to assess the value of routine immunohistochemistry screening for mismatch repair proteins deficiency in CRC patients under 70 years-old. In our cohort, this inclusive strategy allowed the identification of Lynch Syndrome patients that could otherwise be missed using a restrictive approach that relies only on Amsterdam and Bethesda criteria. This study strengthens current recommendations and highlights the role of universal CRC screening for MMR protein status.

Abstract: Recognition of a hereditary colorectal cancer (CRC) syndrome is crucial and Lynch Syndrome (LS) is the most frequent immunohistochemistry (IHC)—screening for mismatch repair proteins (MMR) deficiency in CRC is therefore advocated. An unicentric cohort study was conducted in a central Oncological Hospital to assess its results. All patients under 70 years-old admitted between July 2017–June 2019 and submitted to surgery for CRC were included. Of 275 patients, 56.0% were male, median age 61.0 (IQR:54.5–65.0), with synchronous tumors in six. Histology revealed high grade adenocarcinoma in 8.4%; mucinous and/or signet ring differentiation in 11.3%; and lymphocytic infiltration in 29.8%. Amsterdam (AC) and Bethesda (BC) Criteria were fulfilled in 11 and 74 patients, respectively. IHC revealed loss of expression of MMR proteins in 24 (8.7%), mostly MLH1 and PMS2 (n = 15) and PMS2 (n = 4). Among these, no patients fulfilled AC and 13 fulfilled BC. BRAF mutation or MLH1 promoter hypermethylation was found in four patients with MLH1 loss of expression. Genetic diagnosis was performed in 51 patients, 11 of them with altered IHC. LS was diagnosed in four, and BC was present in three. One patient would not have been diagnosed without routine IHC screening. These results strengthen the important role of IHC screening for MMR proteins loss of expression in CRC.

Keywords: colorectal cancer; Lynch Syndrome; mismatch repair proteins

Citation: Lemos Garcia, J.; Rosa, I.; Saraiva, S.; Marques, I.; Fonseca, R.; Lage, P.; Francisco, I.; Silva, P.; Filipe, B.; Albuquerque, C.; et al. Routine Immunohistochemical Analysis of Mismatch Repair Proteins in Colorectal Cancer—A Prospective Analysis. *Cancers* **2022**, *14*, 3730. https://doi.org/10.3390/cancers14153730

Academic Editor: Stephane Dedieu

Received: 10 June 2022
Accepted: 29 July 2022
Published: 31 July 2022

1. Introduction

Colorectal cancer (CRC) is the third most common cancer type [1–3] and its incidence in some developed countries is increasing among the young (less than 50 years-old) [4–8]. Hereditary syndromes may be responsible for 15–22% of CRC cases [7,9,10].

Recognition of a hereditary CRC syndrome is of paramount importance, since it impacts on patients' surgical management and surveillance as well as on their families screening and surveillance programs [11]. Lynch Syndrome (LS) is the most frequent hereditary CRC syndrome, accounting for 1–3% of all CRC. It occurs due to autosomal dominant mutations in the mismatch repair (MMR) genes *MLH1*, *MSH2*, *MSH6* and *PMS2* or deletions on the cell adhesion molecule (*EPCAM*) gene, which is located upstream of *MSH2*. The MMR defect (which may also be somatic, mostly due to MLH1 promoter hypermethylation) will lead to failure to correct DNA replication errors with accumulation of mutations, resulting in a microsatellite instability (MSI) phenotype. Diagnosis of MSI is via polymerase chain reaction (PCR) amplification of specific microsatellite repeats. Alternatively, immunohistochemistry (IHC) can show absence of expression of MMR proteins in the tumor [12].

Lynch Syndrome can be suspected through family history and clinical data collection, considering the Amsterdam criteria and the revised Bethesda guidelines (Table 1), or using computer-based calculators [12]. However, this strategy lacks sensitivity and specificity. Clinical criteria limitations are overcome by routine IHC staining for MMR proteins in all CRC samples [13–15] in a cost-effective manner [16–18]. International guidelines recommend tumor screening for MMR deficiency for all colorectal cancers regardless of age at diagnosis [19,20] or in patients bellow 70 years-old [21]. In case of MSH2, MSH6 or PMS2 loss of expression, germline testing should ensue. If there is loss of MLH1 or MLH1/PMS2 expression, somatic tumor mutations should be ruled-out first, by searching for *BRAF V600E* mutation and/or *MLH1* promoter hypermethylation [12].

Table 1. Clinical Criteria for Lynch Syndrome Screening (adapted from [22,23]).

Amsterdam II
At least 3 relatives with an HNPCC—associated cancer (CRC, endometrial, stomach, ovary, ureter/renal pelvis, brain, small bowel, hepatobiliary tract and skin (sebaceous) tumors) 1. One is a first degree relative of the other two 2. At least two successive generations affected 3. At least one of the syndrome-associated cancers should be diagnosed at <50 years of age 4. FAP should be excluded in any CRC cases5. Tumors should be verified whenever possible
Revised Bethesda Guidelines
Colorectal tumors from individuals should be tested for MSI in the following situations 1. CRC diagnosed in a patient who is <50 years of age 2. Presence of synchronous or metachronous CRC, or other HNPCC-associated tumors regardless of age. 3. CRC with MSI-H histology diagnosed in a patient who is <60 years of age. 4. CRC diagnosed in one or more first-degree relatives with an HNPCC-related tumor, with one of the cancers being diagnosed under age 50 years. 5. CRC diagnosed in two or more first- or second-degree relatives with HNPCC-related tumors, regardless of age.

HNPCC—Hereditary Non-polyposis Colorectal Cancer, CRC—Colorectal cancer, FAP—Familial Adenomatous Polyposis, MSI-H—Microsatellite Instability-High.

Currently, MMR defects' identification in CRC and has a role beyond LS identification—selection of stage II patients for chemotherapy (CT), choice of the type of adjuvant CT and selection of stage IV patients for immunotherapy all depend on MSI status.

The goal of this study was to assess the importance of routine IHC screening for MMR defects in CRC patients in the identification of Lynch Syndrome patients, in a real-world setting.

2. Materials and Methods

A unicentric cohort study was conducted at the Portuguese Oncological Institute of Lisbon, Portugal, which integrates a Familial Risk Clinic. In this hospital, around 290 new colorectal cancer patients are admitted per year by the Multidisciplinary Colorectal Cancer Group. In their first appointment, relevant personal and clinical data are collected, including family history of neoplasia. All CRC cases are reviewed in a weekly multidisciplinary meeting. All tumors are classified according to the World Health Organization (WHO) Classification of Tumors (2019) [24] and staged using the American Joint Committee on Cancer (AJCC) (8th edition) [25] TNM system.

2.1. Patient Selection

All patients reviewed in the multidisciplinary CRC meeting from 01-07-2016 to 30-06-2019 who were 70 years-old or younger and underwent primary tumor resection surgery were included, in a total of 275 patients.

2.2. Data Collection

Data collected included demographic information, tumor location, radiological and pathological staging, therapeutic modalities performed, family history of CRC and other LS-spectrum cancers, MMR protein status, *BRAF V600E* mutation status, MMR gene promotor methylation and germline mutation analysis. For stage at diagnosis classification, pathological staging was the gold standard, except in patients who underwent neoadjuvant treatment, for whom radiological staging at diagnosis was preferred.

2.3. Hospital Standard Procedures

2.3.1. CRC Sample Processing

In our institution, until 2021, according to the 2009 Jerusalem Workshop recommendations [21], in all patients 70 years old or younger who underwent surgery for CRC, the tumor was screened for loss of expression of MMR proteins by immunohistochemistry. To assess the expression of MLH1, PMS2, MSH2 and MSH6 proteins, IHC analysis is performed using Ventana CC1 equipment (sample in 10% formalin buffer, using thermal recuperation method) and monoclonal antibodies anti-MLH1 (clone ES05), anti-PMS2 (clone EP51), anti-MSH2 (clone G219-1129) and anti-MSH6 (clone EP49) (Figure 1).

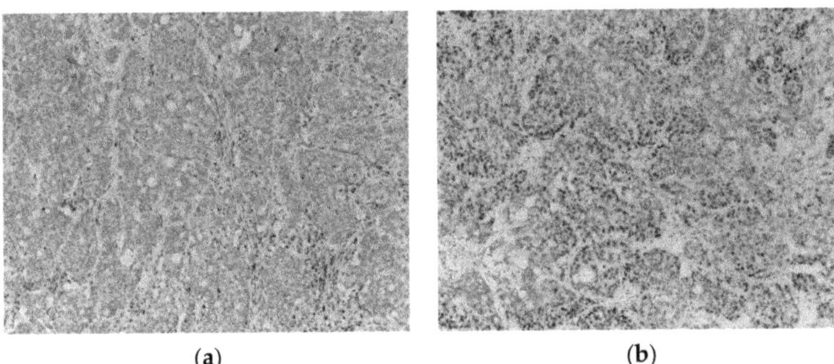

Figure 1. Immunohistochemistry showing loss of MLH1 (**a**) and maintained MSH2 (**b**) staining (10×).

To exclude somatic mutations that lead to MLH1-defective cases, since 2019, tumors with MLH1 loss of expression are further investigated for *BRAF V600E* mutation: DNA from samples of tumor tissue is amplified by PCR using primers for BRAF exon 15 and the product is sequenced using Sanger sequencing on Big Dye terminator v1.1 sequencing kit (Applied Biosystems) on an automatic ABI PrismTM 3130 Genetic Analyzer (Applied Biosystems).

BRAF V600E mutation analysis results were also available in some stage IV (at diagnosis or during follow-up) patients, in whom the test was performed for chemotherapy selection, regardless of IHC results.

2.3.2. Family Risk Clinic Referral

In case Amsterdam or revised Bethesda criteria are fulfilled or when germline MMR genes' mutations are suspected after IHC analysis, the patients are referred to the Familial

Risk Clinic. All patients with 10 or more adenomas or those who fulfil the World Health Organization criteria for Serrated Polyposis Syndrome are also referred.

In cases referred for evaluation in the Familial Risk Clinic, additional tumor testing before genetic diagnosis may be done, at physician's discretion, according to available evidence and international recommendations.

2.3.3. Molecular and Genetic Testing

Microsatellite Instability Analysis

Between 2016 and 2017, this was carried out using the Bethesda microsatellite markers: BAT26, BAT25, D17S250, D2S123 and D5S346 [26–28]. In tumor samples exhibiting microsatellite instability (MSI) in only one marker, or without a conclusive result in at least one marker, two additional markers were analyzed (BAT40 and MYCL1). From 2017 onwards, the MSI analysis was performed with 10 microsatellite markers (the above mentioned and 3 additional mononucleotide repeat marker—NR21, NR24 and NR27).

Between 2016 and 2017, DNA was isolated from CRC-PDEs samples using the KAPA Express Extract Kit (KAPABIOSYSTEMS, Potters Bar, United Kingdom) and from paraffin-embedded tissue (FFPE) colorectal cancer and normal colonic mucosa using proteinase K digestion, which was followed by phenol/chloroform extraction and ethanol precipitation [29]. From 2017, the Maxwell® RSC DNA FFPE Kit (Promega, Madison, WI, USA) was used to isolate DNA from FFPE samples in the Maxwell® RSC Instrument (Promega). Each tumor and paired normal DNA were amplified by PCR for each of the microsatellite markers, using fluorescent labelled primers (Applied Biosystems, Foster City, CA USA), specific for each locus [30,31]. PCR products were analyzed in the ABI PrismTM 3130 Genetic Analyzer using the GeneMapper software (Applied Biosystems). Tumors presenting MSI in >40% of the markers analyzed were classified as MSI-High (MSI-H); otherwise they were classified as MSI-Low (MSI-L) [32]. Tumors without MSI in any of the markers were considered to be microsatellite stable (MSS).

MMR Gene Promoter Methylation Analysis

The analysis of MMR gene promoters methylation was performed by methylation-specific multiplex ligation-dependent probe amplification (MS_MLPA) [33], using the MS-MLPA kits ME011 MMR (MRC-Holland, Amsterdam, the Netherlands). MS-MLPA reactions were performed as described by the manufacturer. MS-MLPA fragments were analyzed on the ABI Prism 3130TM Genetic Analyzer (Applied Biosystems) and normalized using the Coffalyser. NET software (MRC-Holland, Amsterdam, the Netherlands). A baseline for positive methylation was calculated for each gene as described previously [34]. A ratio of 0.15 or higher, corresponding to 15% of methylated DNA, was indicative of *MLH1* promoter methylation.

Germline Mutation Analysis

In case of MMR proteins' deficiency in IHC analysis, mutations in *MLH1*, *MSH2*, *MSH6*, *PMS2* and *EPCAM* were investigated. In other cases, Next Generation Sequencing (NGS) multigene panels were used, according to clinical data and family history.

Germline mutation analysis was performed after signed informed consent, by NGS using multigene panels (TruSight Cancer kit (Illumina, San Diego, CA, USA)) and MLPA (multiplex ligation-dependent probe amplification) analysis (MRC-Holland, Amsterdam, the Netherlands). All pathogenic, probably pathogenic or of uncertain pathogenicity mutations (frequency less than 1% in the population) are confirmed by Sanger sequencing, from an independent DNA sample. The interpretation of the variants is performed according to the rules established by LOVD-InSIGHT (International Society for Gastrointestinal Hereditary Tumors—http://www.insight-group.org/criteria last accessed on the 1 June 2022).

2.4. Statistical Analysis

For statistical analysis, SPSS Statistics 26 (IBM) was used. Demographic and clinical characteristics were presented as frequencies. Continuous variables were expressed as median and standard deviation or as median and interquartile range, according to data distribution, and were compared using t-Student or Wilcoxon tests, respectively. Qualitative variables were compared using chi-square or Fisher Exact tests. Multiple variables were analyzed using logistic regression models. A p value lower than 0.05 was considered statistically significant.

3. Results

3.1. Clinical Characterization

A total of 275 patients were included, 56.0% males, with a median age at diagnosis of 61.0 (IQR 54.5–65.0) years old. Tumors were mostly (53.1%) stage III at diagnosis and histological report revealed high grade (G3) tumors in 8.4%, mucinous and/or signed ring morphology in 11.3% and lymphocytic infiltrate in 29.8%. Population and tumor characteristics are depicted in Tables 2 and 3. Mean follow-up time was 40.6 ± 15.6 months. After personal and family history investigation, 11 (4.0%) patients fulfilled Amsterdam criteria (AC) and 74 (26.9%) revised Bethesda criteria (BC).

Table 2. Clinical characteristics.

Variable	Frequency
Gender	
Female	121 (44.0%)
Male	154 (56.0%)
Age at CRC diagnosis (median, IQR)	61.0 (54.5–65.0)
Tumor location	
Right colon	60 (21.8%)
Left colon	77 (28.0%)
Rectum	138 (50.2%)
Synchronous CRC	6
Stage (AJCC 8th edition)	
I	50 (18.2%)
II	63 (22.9%)
III	146 (53.1%)
IV	16 (5.8%)
Neoadjuvant treatment	
None	162 (58.9%)
Radiotherapy	12 (4.4%)
Chemoradiotherapy	98 (35.6%)
Chemotherapy	3 (1.1%)
Resection technique	
Right hemicolectomy	55 (20.0%)
Left hemicolectomy	13 (4.7%)
Sigmoidectomy	49 (17.8%)
Anterior rectal resection	116 (42.2%)
Abdominoperineal resection	23 (8.4%)
Total colectomy/proctocolectomy	8/3 (2.9/1.1%)
Trans-anal minimally invasive surgery	2 (0.7%)
Endoscopic	7 (2.3%)
Urgent surgery for occlusion	9 (3.3%)
Intraoperatively perforated tumor	2 (0.7%)

CRC—colorectal cancer. In case of synchronous CRC, location and staging of the more advanced neoplasia was selected to present in the table.

Table 3. Tumor characteristics.

Variable	Frequency
Differentiation grade	
Low-grade (G1–G2)	204 (74.2%)
High-grade (G3)	23 (8.4%)
N/A	48
Histological subtype	
Mucinous	26 (9.5%)
Signet ring	2 (0.7%)
Mucinous and signed ring	3 (1.1%)
Tubular and cribiform	2 (0.7%)
Serrated	1 (0.3%)
NOS	241 (87.6%)
Lympho-vascular invasion	69 (25.1%)
Perineural invasion	37 (13.5%)
Lymphocytic infiltrate	82 (29.8%)
Tumor budding	64 (23.3%)

N/A—not available. NOS—no other specification.

3.2. *Immunohistochemical Analysis*

IHC evaluation revealed loss of MMR proteins' expression in 24 cases (8.7%)– MLH1 and PMS2 (n = 15) (Figure 1); PMS2 (n = 4); MSH2 and MSH6 (n = 1); MSH2 (n = 1); MSH6 (n = 2); MLH1, PMS2 and MSH6 (n = 1). AC and BC were fulfilled in 0 and 13 of such cases, respectively (Table 4).

Altered IHC analysis showed a significant association with tumor location in the right colon ($p < 0.001$), poor differentiation ($p = 0.015$) and mucinous histology ($p = 0.016$), but not with gender ($p = 0.157$), age ($p = 0.709$), stage ($p = 0.44$), lympho-vascular ($p = 0.279$) or perineural invasion ($p = 0.567$), lymphocytic infiltrate ($p = 0.052$) or tumor budding ($p = 0.499$).

3.3. *Analysis of MMR Deficient Cases—BRAFV600E Mutation Status, MMR Gene Methylation and Germline Mutation Analysis*

From the 16 patients with MLH1 loss of expression (15 with MLH1/PMS2 loss of expression, one with MLH1/PMS2/MSH6 loss of expression), somatic BRAF V600E mutation testing was carried out in seven, and found in one patient—the IHC alteration was considered somatic and the patient was not referred for genetic testing. From the remaining six patients, three had MLH1 promoter hypermethylation and three did not show either of the somatic alterations. Genetic testing was performed in these last three patients, of whom one had confirmed LS; in the other two, no germline mutation was detected (Table 4).

BRAF V600E mutation testing results were also available in three other patients in whom the analysis was requested by oncologists, for chemotherapy selection (Table 4).

Five patients with altered IHC died before the Family Risk Clinic appointment/germline mutation analysis and one refused genetic testing. Family Risk Clinic appointment is pending or genetic testing is still ongoing in six patients.

Therefore, in total, genetic test results were available in 11 of the 24 patients with altered IHC and in one with artifacts, and Lynch Syndrome was diagnosed in four of them.

Patients with Lynch Syndrome were men in three cases, and aged less than 50 years-old in three (median age 37.0 (IQR 27.5–51.8)). AC were not fulfilled in any of the patients, and three met BC; IHC was altered in three and unavailable in one due to artifacts (Table 4).

Tumor was in the right colon in three and rectum in one, stage I in one and III in three cases. Histology report revealed low-grade (G1/G2) tumors with no other specification, no lymphocytic infiltrate and no unfavorable invasions in all cases.

All patients were alive without evidence of cancer relapse at last follow-up (median follow-up = 33.0 months (IQR: 26.8–54.3)).

The presence of Lynch Syndrome had a significant association with younger age at diagnosis ($p < 0.001$) and right-sided tumors ($p = 0.037$), but not with gender ($p = 0.634$), stage ($p = 0.718$), differentiation ($p = 1.000$), histological subtype ($p = 1.000$), lympho-vascular invasion ($p = 0.575$), perineural invasion ($p = 1.000$), lymphocytic infiltrate ($p = 0.323$) or tumor budding ($p = 1.000$).

Table 4. Clinical and molecular characterization of cases with altered MMR status by immunohistochemical analysis.

ID	Age (years)	Gender	CRC Location at Diagnosis	CRC Stage	CRC Histopathology					AC	BC	Immunohistochemistry–Unexpressed Proteins	*BRAF V600E*	*MLH1* Promoter Hyper-Methylation	Genetic Diagnosis' Results
					G3	LV/P	Mucinous	LI	Bd						
1	64	Male	Right colon	III	+	+	–	+	+	No	No	MLH1 and PMS2	N/A	N/A	N/A
2	61	Female	Left colon	II	–	–	–	–	–	No	Yes	MLH1 and PMS2	N/A	N/A	No mutation detected
3	29	Female	Left colon	III	N/A	+	+	+	–	No	Yes	MLH1 and PMS2	No	Yes	No mutation detected
4	39	Male	Right colon	III	–	+	–	+	–	No	Yes	MLH1 and PMS2	N/A	N/A	No mutation detected
5	54	Female	Right colon	II	+	+	–	–	–	No	No	MLH1 and PMS2	No	Yes	No mutation detected
6	32	Male	Right colon	III	–	–	–	–	–	No	Yes	MSH2 and MSH6	N/A	N/A	LS-*MSH2* Frameshift mutation c.388_389delp.Gln130ValfsTer2
7	66	Male	Rectum	III	–	–	–	+	–	No	No	PMS2	N/A	N/A	N/A
8	60	Female	Rectum	III	–	–	–	–	–	No	No	MLH1 and PMS2	N/A	N/A	N/A
9	67	Female	Right colon	III	+	–	–	+	–	No	No	MLH1 and PMS2	No	Yes	N/A
10	63	Male	Rectum	III	–	–	–	+	–	No	No	MSH6	N/A	N/A	N/A
11	64	Male	Rectum	III	–	–	–	–	–	No	Yes	MLH1 and PMS2	N/A	N/A	N/A
12	62	Male	Left colon	I	–	–	–	–	–	No	No	MSH6	N/A	N/A	N/A
13	65	Male	Left colon	I	–	–	–	–	–	No	Yes	MSH2	N/A	N/A	N/A
14	51	Male	Right colon	II	–	+	–	+	–	No	Yes	MLH1 and PMS2	N/A	N/A	N/A
15	54	Male	Right colon	II	–	–	–	+	–	No	Yes	MLH1 and PMS2	No	No	No mutation detected
16	67	Male	Right colon	II	–	–	–	–	–	No	No	MLH1 and PMS2	N/A	N/A	N/A
17	67	Male	Right colon	II	–	–	–	+	–	No	No	MLH1 and PMS2	N/A	N/A	No mutation detected
18	67	Male	Right colon	III	+	+	–	+	+	No	Yes	PMS2	Yes	N/A	N/A
19	67	Male	Right colon	IV	+	+	–	+	+	No	Yes	PMS2	Yes	N/A	N/A
20	63	Male	Right colon	II	–	–	–	+	–	No	No	MLH1 and PMS2	Yes	N/A	N/A
21	56	Male	Rectum	IV	–	+	–	–	+	No	Yes	MLH1 and PMS2	N/A	N/A	No mutation detected
22	42	Male	Right colon	I	–	–	–	–	–	No	Yes	MLH1 and PMS2	No	No	LS-*MLH1* Missense mutation c.2041G > Ap.(Ala681Thr)
23	55	Male	Right colon	III	–	–	–	–	+	No	No	PMS2	No	N/A	LS-*PMS2* Deletion exons 1 to 14 (c.(?-87)_(2445+1_2446-1)del)
24	64	Male	Right colon	IV	+	+	–	–	+	No	Yes	MLH1, PMS2 and MSH6	No	No	No mutation detected
25	26	Male	Rectum	III	–	–	–	–	–	No	Yes	PMS2 N/A (artifacts), MLH1, MSH2 and MSH6 expressed	No	N/A	LS-*PMS2* Deletion exons 12 to 14 (c.(2006+1_2007-1)_(2445+1_2446-1)del)

ID–identification; CRC–colorectal cancer; G3–poorly differentiated; LV/P–lympho-vascular and/or perineural invasion; Bd–Budding; LI–lymphocytic infiltrate; AC–Amsterdam criteria; BC–Revised Bethesda criteria; N/A–non applicable/ non available; LS–Lynch Syndrome.

3.4. Germline Mutation Analysis in MMR Proficient Cases

In 10 patients with altered IHC and in one with artifacts, germline MMR mutation analysis was performed and in 40 patients a multigene panel was used. From these, one Familial Adenomatous Polyposis and one *MUTYH*-associated Polyposis were diagnosed (both in patients with multiple adenomas). A *MUTYH* heterozygote mutation was found in a patient with CRC at the age of 47 with family history of colonic adenomas. Familial Colorectal Cancer Type X) was diagnosed in a patient in whom no mutation was found after multigene panel testing.

4. Discussion

This study presents the clinical picture of CRC in an adult population under 70 years old. As expected, most cases were sporadic cancers. Nevertheless, the use of IHC, combined with personal and familial data, allowed the attending physicians to diagnose Lynch Syndrome in four (1.5%) cases. It is important to notice that one of these patients did not fulfill Amsterdam II or Bethesda criteria and genetic diagnosis would have been missed if IHC analysis had not been performed.

Accurate and timely identification of Lynch Syndrome patients is extremely important, since surveillance for colonic and extra-colonic malignancies can increase survival and improve quality of live. This is relevant both for the patients and for at-risk relatives that may benefit from genetic study [13,35,36]. Even if LS is a rare entity, the cost of missing this diagnosis is significant.

Altered IHC was detected in 9.6% of the cases, a rate that is lower than expected, given that deficient-MMR protein status can be found in 15–30% of sporadic CRC [37]. The rates found may be due to the young population studied, where all CRC in patients aged more than 70 years old were excluded. Indeed, microsatellite instability in sporadic cases is frequently associated to MLH1 promoter methylation and these features are more frequently detected in older female patients, some of them often older than 70 years old [38].

In 16 patients, there was MLH1 $\pm$ PMS2 loss of expression in the tumor. A major limitation of our study was the fact that somatic *BRAF V600E* mutation/MLH1 promoter hypermethylation analysis' results were available in only a minority of these cases. Routine *BRAF* testing after a MLH1 loss of expression result has only been implemented in our hospital in the last year of the study. Nevertheless, from seven patients with available results, one had *BRAF V600E* mutation and three others had *MLH1* promoter hypermethylation. These findings highlight the benefits of a step-up approach [20,39], that prevents a significant proportion of patients from undergoing most likely inconclusive genetic testing. This strategy makes sense not only in an economic standpoint, but also considering the psychological burden associated with genetic testing [39].

Further advantages of MMR status investigation are the possibility of personalized therapies. MSI tumors may have a reduced response to 5-FU chemotherapy and a better overall prognosis in early stages. Therefore, most stage II MMR deficient CRC patients do not seem to benefit from adjuvant chemotherapy, namely, with 5-FU [40]. Another scenario is metastatic MSI CRC, where therapy with immune checkpoint inhibitors may be proposed, since these patients often show sustained responses to this class of drugs. This is explained by the increased expression of several immune checkpoints in MMR deficient tumors, resulting from the production of abnormal proteins which elicit antigen-driven immune responses [40,41].

Although IHC analysis, molecular and genetic studies' results were prospectively recorded, clinical data collection was retrospective, which is a limitation of the study, which may be relevant in details such as family history that may not have been carefully reported in all cases. However, the study was conducted in an oncological center which integrates a Family RiskClinic, in strict interaction with a Molecular Biology Laboratory and therefore has the means and expertise to pursue genetic investigation when indicated, limiting bias due to unrecognized hereditary cancer patients. Furthermore, this is a sequential series of patients with a relevant number of cases included, reflecting real-life practice.

5. Conclusions

This study strengthens current recommendations and highlights the role of universal CRC screening for MMR protein status. This inclusive strategy allows the identification of Lynch Syndrome patients that could otherwise be missed using a restrictive approach that relies only on Amsterdam and Bethesda criteria.

Author Contributions: Conceptualization, J.L.G. and I.R.; methodology, J.L.G. and I.R.; validation, J.L.G. and I.R.; formal analysis, J.L.G.; investigation, J.L.G. and S.S.; resources, C.A., I.F., P.S., B.F. and R.F.; writing—original draft preparation, J.L.G., I.R, C.A., I.F., P.S., B.F.; writing—review and editing, S.S., C.A., R.F., P.L., I.M. and I.C.; supervision, I.C.; project administration, I.R. All authors have read and agreed to the published version of the manuscript.

Funding: This research received no external funding.

Institutional Review Board Statement: The study was conducted in accordance with the Declaration of Helsinki. Approval was obtained from the Ethical Committee of Instituto Português de Oncologia de Lisboa Francisco Gentil (Document Number: UIC/1538; date of approval 26 July 2022).

Informed Consent Statement: Patients gave informed consent for all genetic studies.

Data Availability Statement: Data not shared due to confidentiality.

Acknowledgments: The authors would like to thank David Carvalho Fiel who was responsible for proofreading the article.

Conflicts of Interest: The authors have no conflicts of interest to declare.

References

1. Siegel, R.L.; Miller, K.D.; Fuchs, H.E.; Jemal, A. Cancer Statistics, 2021. *CA Cancer J. Clin.* **2021**, *71*, 7–33. [CrossRef] [PubMed]
2. ECIS-European Cancer Information System. Available online: https://ecis.jrc.ec.europa.eu (accessed on 2 January 2022).
3. Bray, F.; Ferlay, J.; Soerjomataram, I.; Siegel, R.L.; Torre, L.A.; Jemal, A. Global cancer statistics 2018: GLOBOCAN estimates of incidence and mortality worldwide for 36 cancers in 185 countries. *CA Cancer J. Clin.* **2018**, *68*, 394–424. [CrossRef] [PubMed]
4. Meyer, J.E.; Narang, T.; Schnoll-Sussman, F.H.; Pochapin, M.B.; Christos, P.J.; Sherr, D.L. Increasing incidence of rectal cancer in patients aged younger than 40 years: An analysis of the surveillance, epidemiology, and end results database. *Cancer* **2010**, *116*, 4354–4359. [CrossRef] [PubMed]
5. Meester, R.G.S.; Doubeni, C.A.; Lansdorp-Vogelaar, I.; Goede, S.L.; Levin, T.R.; Quinn, V.P.; van Ballegooijen, M.; Corley, D.A.; Zauber, A.G. Colorectal cancer deaths attributable to nonuse of screening in the United States. *Ann. Epidemiol.* **2015**, *25*, 208–213.e1. [CrossRef] [PubMed]
6. Siegel, R.L.; Jemal, A.; Ward, E.M. Increase in incidence of colorectal cancer among young men and women in the United States. *Cancer Epidemiol. Biomark. Prev.* **2009**, *18*, 1695–1698. [CrossRef] [PubMed]
7. Silla, I.O.; Rueda, D.; Rodriguez, Y.; Garcia, J.L.; de La Cruz Vigo, F.; Perea, J. Early-onset colorectal cancer: A separate subset of colorectal cancer. *World J. Gastroenterol.* **2014**, *20*, 17288–17296. [CrossRef]
8. Siegel, R.L.; Torre, L.A.; Soerjomataram, I.; Hayes, R.B.; Bray, F.; Weber, T.K.; Jemal, A. Global patterns and trends in colorectal cancer incidence in young adults. *Gut* **2019**, *68*, 2179–2185. [CrossRef] [PubMed]
9. Chang, D.T.; Pai, R.K.; Rybicki, L.A.; Dimaio, M.A.; Limaye, M.; Jayachandran, P.; Koong, A.C.; Kunz, P.A.; Fisher, G.A.; Ford, J.A.; et al. Clinicopathologic and molecular features of sporadic early-onset colorectal adenocarcinoma: An adenocarcinoma with frequent signet ring cell differentiation, rectal and sigmoid involvement, and adverse morphologic features. *Mod. Pathol.* **2012**, *25*, 1128–1139. [CrossRef] [PubMed]
10. Levine, O.; Zbuk, K. Colorectal cancer in adolescents and young adults: Defining a growing threat. *Pediatr. Blood Cancer* **2019**, *66*, e27941. [CrossRef] [PubMed]
11. Boardman, L.A.; Vilar, E.; You, Y.N.; Samadder, J. AGA Clinical Practice Update on Young Adult–Onset Colorectal Cancer Diagnosis and Management: Expert Review. *Clin. Gastroenterol. Hepatol.* **2020**, *18*, 2415–2424. [CrossRef]
12. Ryan, E.; Sheahan, K.; Creavin, B.; Mohan, H.M.; Winter, D.C. The current value of determining the mismatch repair status of colorectal cancer: A rationale for routine testing. *Crit. Rev. Oncol. Hematol.* **2017**, *116*, 38–57. [CrossRef] [PubMed]
13. Chiaravalli, A.M.; Carnevali, I.; Sahnane, N.; Leoni, E.; Furlan, D.; Berselli, M.; Sessa, F.; Tibiletti, M.G. Universal screening to identify Lynch syndrome: Two years of experience in a Northern Italian Center. *Eur. J. Cancer Prev.* **2020**, *29*, 281–288. [CrossRef] [PubMed]
14. Moreira, L.; Balaguer, F.; Lindor, N.; de la Chapelle, A.; Hampel, H.; Aaltonen, L.A.; Hopper, J.L.; Le Marchand, L.; Gallinger, S.; Newcomb, P.A.; et al. Identification of Lynch Syndrome Among Patients with Colorectal Cancer. *JAMA* **2012**, *308*, 1555–1565. [CrossRef] [PubMed]

15. Pérez-Carbonell, L.; Ruiz-Ponte, C.; Guarinos, C.; Alenda, C.; Payá, A.; Brea, A.; Egoavil, C.M.; Castillejo, A.; Barberá, V.M.; Bessa, X.; et al. Comparison between universal molecular screening for Lynch syndrome and revised Bethesda guidelines in a large population-based cohort of patients with colorectal cancer. *Gut* **2012**, *61*, 865–872. [CrossRef]
16. Mvundura, M.; Grosse, S.D.; Hampel, H.; Palomaki, G.E. The cost-effectiveness of genetic testing strategies for Lynch syndrome among newly diagnosed patients with colorectal cancer. *Genet. Med.* **2010**, *12*, 93–104. [CrossRef] [PubMed]
17. Ladabaum, U.; Wang, G.; Terdiman, J.; Blanco, A.; Kuppermann, M.; Boland, C.R.; Ford, J.; Elkin, E.; Phillips, K.A. Strategies to Identify the Lynch Syndrome Among Patients with Colorectal Cancer. *Ann. Intern. Med.* **2011**, *155*, 69–79. [CrossRef] [PubMed]
18. Kang, Y.J.; Killen, J.; Caruana, M.; Simms, K.; Taylor, N.; Frayling, I.M.; Snowsill, T.; Huxley, N.; Coupe, V.M.; Hughes, S.; et al. The predicted impact and cost-effectiveness of systematic testing of people with incident colorectal cancer for Lynch syndrome. *Med. J. Aust.* **2020**, *212*, 72–81. [CrossRef] [PubMed]
19. Weiss, J.M.; Gupta, S.; Burke, C.A.; Axell, L.; Chen, L.M.; Chung, D.C.; Clayback, K.M.; Dallas, S.; Felder, S.; Gbolahan, O.; et al. Genetic/familial high-risk assessment: Colorectal, version 1.2021 featured updates to the NCCN Guidelines. *J. Natl. Compr. Cancer Netw.* **2021**, *19*, 1122–1132.
20. Molecular Testing Strategies for Lynch Syndrome in People with Colorectal Cancer Diagnostics Guidance. 2017. Available online: www.nice.org.uk/guidance/dg27 (accessed on 2 January 2022).
21. Boland, C.R.; Shike, M. Report from the Jerusalem Workshop on Lynch Syndrome-Hereditary Nonpolyposis Colorectal Cancer. *Gastroenterology* **2010**, *138*, 2197.e1–2197.e7. [CrossRef] [PubMed]
22. Umar, A.; Boland, C.R.; Terdiman, J.P.; Syngal, S.; de la Chapelle, A.; Rüschoff, J.; Fishel, R.; Lindor, N.M.; Burgart, L.J.; Hamelin, R.; et al. Revised Bethesda Guidelines for Hereditary Nonpolyposis Colorectal Cancer (Lynch Syndrome) and Microsatellite Instability. *J. Natl. Cancer Inst.* **2004**, *96*, 261–268. [CrossRef]
23. Vasen, H.F.; Watson, P.; Mecklin, J.P.; Lynch, H. New clinical criteria for hereditary nonpolyposis colorectal cancer (HNPCC, Lynch syndrome) proposed by the International Collaborative Group on HNPCC. *Gastroenterology* **1999**, *116*, 1453–1456. [CrossRef]
24. Nagtegaal, I.D.; Odze, R.D.; Klimstra, D.; Paradis, V.; Rugge, M.; Schirmacher, P.; Washington, K.M.; Carneiro, F.; Cree, I.A.; WHO Classification of Tumours Editorial Board. The 2019 WHO classification of tumours of the digestive system. *Histopathology* **2020**, *76*, 182–188. [CrossRef] [PubMed]
25. Weiser, M.R. AJCC 8th Edition: Colorectal Cancer. *Ann. Surg. Oncol.* **2018**, *25*, 1454–1455. [CrossRef] [PubMed]
26. Francisco, I.; Albuquerque, C.; Lage, P.; Belo, H.; Vitoriano, I.; Filipe, B.; Claro, I.; Ferreira, S.; Rodrigues, P.; Chaves, P.; et al. Familial colorectal cancer type X syndrome: Two distinct molecular entities? *Fam. Cancer* **2011**, *10*, 623–631. [CrossRef] [PubMed]
27. Silva, P.; Albuquerque, C.; Lage, P.; Fontes, V.; Fonseca, R.; Vitoriano, I.; Filipe, B.; Rodrigues, P.; Moita, S.; Ferreira, S.; et al. Serrated polyposis associated with a family history of colorectal cancer and/or polyps: The preferential location of polyps in the colon and rectum defines two molecular entities. *Int. J. Mol. Med.* **2016**, *38*, 687–702. [CrossRef] [PubMed]
28. Cortez-Pinto, J.; Claro, I.; Francisco, I.; Lage, P.; Filipe, B.; Rodrigues, P.; Chaves, P.; Albuquerque, C.; Dias Pereira, A. Pediatric Colorectal Cancer: A Heterogenous Entity. *J. Pediatr. Hematol. Oncol.* **2020**, *42*, 131–135. [CrossRef] [PubMed]
29. Albuquerque, C.; Breukel, C.; van der Luijt, R.; Fidalgo, P.; Lage, P.; Slors, F.J.; Leitão, C.N.; Fodde, R.; Smits, R. The 'just-right' signaling model: APC somatic mutations are selected based on a specific level of activation of the beta-catenin signaling cascade. *Hum. Mol. Genet.* **2002**, *11*, 1549–1560. [CrossRef]
30. Boland, C.R.; Thibodeau, S.N.; Hamilton, S.R.; Sidransky, D.; Eshleman, J.R.; Burt, R.W.; Meltzer, S.J.; Rodriguez-Bigas, M.A.; Fodde, R.; Ranzani, G.N.; et al. A National Cancer Institute Workshop on Microsatellite Instability for cancer detection and familial predisposition: Development of international criteria for the determination of microsatellite instability in colorectal cancer. *Cancer Res.* **1998**, *58*, 5248–5257. [PubMed]
31. Buhard, O.; Cattaneo, F.; Wong, Y.F.; Yim, S.F.; Friedman, E.; Flejou, J.F.; Duval, A.; Hamelin, R. Multipopulation analysis of polymorphisms in five mononucleotide repeats used to determine the microsatellite instability status of human tumors. *J. Clin. Oncol.* **2006**, *24*, 241–251. [CrossRef] [PubMed]
32. Umar, A. Lynch syndrome (HNPCC) and microsatellite instability. *Dis. Markers* **2004**, *20*, 179–180. [CrossRef] [PubMed]
33. Nygren, A.O.; Ameziane, N.; Duarte, H.M.; Vijzelaar, R.N.; Waisfisz, Q.; Hess, C.J.; Schouten, J.P.; Errami, A. Methylation-specific MLPA (MS-MLPA): Simultaneous detection of CpG methylation and copy number changes of up to 40 sequences. *Nucleic Acids Res.* **2005**, *33*, e128. [CrossRef] [PubMed]
34. Rosa, I.; Silva, P.; da Mata, S.; Magro, F.; Carneiro, F.; Peixoto, A.; Silva, M.; Sousa, H.T.; Roseira, J.; Parra, J.; et al. Methylation patterns in dysplasia in inflammatory bowel disease patients. *Scand. J. Gastroenterol.* **2020**, *55*, 646–655. [CrossRef] [PubMed]
35. Gupta, S.; Provenzale, D.; Llor, X.; Halverson, A.L.; Grady, W.; Chung, D.C.; Haraldsdottir, S.; Markowitz, A.J.; Slavin, T.P.; Hampel, H.; et al. NCCN Guidelines Insights: Genetic/Familial High-Risk Assessment: Colorectal, Version 2.2019. *J. Natl. Compr. Cancer Netw.* **2019**, *17*, 1032–1041. [CrossRef] [PubMed]
36. Stjepanovic, N.; Moreira, L.; Carneiro, F.; Balaguer, F.; Cervantes, A.; Balmaña, J.; Martinelli, E.; ESMO Guidelines Committee. Hereditary gastrointestinal cancers: ESMO Clinical Practice Guidelines for diagnosis, treatment and follow-up. *Ann. Oncol.* **2019**, *30*, 1558–1571. [CrossRef]
37. Poulogiannis, G.; Frayling, I.M.; Arends, M.J. DNA mismatch repair deficiency in sporadic colorectal cancer and Lynch syndrome. *Histopathology* **2010**, *56*, 167–179. [CrossRef] [PubMed]

38. Goshayeshi, L.; Ghaffarzadegan, K.; Khooei, A.; Esmaeilzadeh, A.; Khorram, M.R.; Mozaffari, H.M.; Kiani, B.; Hoseini, B. Prevalence and clinicopathological characteristics of mismatch repair-deficient colorectal carcinoma in early onset cases as compared with late-onset cases: A retrospective cross-sectional study in Northeastern Iran. *BMJ Open* **2018**, *30*, e023102. [CrossRef]
39. Raymond, V.; Everett, J.N. Genetic counselling and genetic testing in hereditary gastrointestinal cancer syndromes. *Best Pract. Res. Clin. Gastroenterol.* **2009**, *23*, 275–283. [CrossRef] [PubMed]
40. Currais, P.; Rosa, I.; Claro, I. Colorectal cancer carcinogenesis: From bench to bedside. *World J. Gastrointest. Oncol.* **2022**, *14*, 654–663. [CrossRef] [PubMed]
41. Lizardo, D.Y.; Kuang, C.; Hao, S.; Yu, J.; Huang, Y.; Zhang, L. Immunotherapy efficacy on mismatch repair-deficient colorectal cancer: From bench to bedside. *Biochim. Biophys. Acta (BBA)-Rev. Cancer* **2020**, *1874*, 188447. [CrossRef] [PubMed]

 cancers

MDPI

Review

Precision Medicine in Metastatic Colorectal Cancer: Targeting ERBB2 (HER-2) Oncogene

Javier Torres-Jiménez [1,2,*,†], Jorge Esteban-Villarrubia [2,3,†] and Reyes Ferreiro-Monteagudo [2]

1 Medical Oncology Department, MD Anderson Cancer Center Madrid, 28033 Madrid, Spain
2 Medical Oncology Department, University Hospital Ramon y Cajal, 28034 Madrid, Spain; jorge.esteban@salud.madrid.org (J.E.-V.); mariareyes.ferreiro@salud.madrid.org (R.F.-M.)
3 Medical Oncology Department, University Hospital 12 de Octubre, 28041 Madrid, Spain
* Correspondence: jtorresj@mdanderson.es or javier.torres.jim@gmail.com
† These authors contributed equally to this work.

Simple Summary: Colorectal cancer (CRC) is the third most common cancer in terms of incidence rate in adults and the second most common cause of cancer-related death in Europe. The treatment of metastatic CRC (mCRC) is based on the use of chemotherapy, anti-vascular endothelial growth factor (VEGF), and anti-epidermal growth factor receptor (EGFR) for RAS wild-type tumors. Precision medicine tries to identify molecular alterations that could be treated with targeted therapies. Although ERBB2 (also known as HER-2) has an important therapeutic role in breast and esophagogastric cancer, there are no approved ERBB2-targeted therapies for mCRC. The purpose of this review is to describe the landscape of ERBB2-positive mCRC.

Abstract: Colorectal cancer (CRC) is the third most common cancer in terms of incidence rate in adults and the second most common cause of cancer-related death in Europe. The treatment of metastatic CRC (mCRC) is based on the use of chemotherapy, anti-vascular endothelial growth factor (VEGF), and anti-epidermal growth factor receptor (EGFR) for RAS wild-type tumors. Precision medicine tries to identify molecular alterations that could be treated with targeted therapies. ERBB2 amplification (also known as HER-2) has been identified in 2–3% of patients with mCRC, but there are currently no approved ERBB2-targeted therapies for mCRC. The purpose of this review is to describe the molecular structure of ERBB2, clinical features of these patients, diagnosis of ERBB2 alterations, and the most relevant clinical trials with ERBB2-targeted therapies in mCRC.

Keywords: colorectal cancer; precision medicine; ERBB2; HER-2

Citation: Torres-Jiménez, J.; Esteban-Villarrubia, J.; Ferreiro-Monteagudo, R. Precision Medicine in Metastatic Colorectal Cancer: Targeting ERBB2 (HER-2) Oncogene. *Cancers* **2022**, *14*, 3718. https://doi.org/10.3390/cancers14153718

Academic Editor: Stephane Dedieu

Received: 29 June 2022
Accepted: 28 July 2022
Published: 30 July 2022

1. Introduction

Colorectal cancer (CRC) is the third cancer in terms of incidence rate in adults and the second most common cause of cancer-related deaths in Europe [1–3]. A total of 25% of CRC patients have metastatic lesions at diagnosis, and almost 50% of patients with early-stage CRC will develop disseminated or metastatic disease. The median overall survival (mOS) for patients with metastatic colorectal cancer (mCRC) is approximately 30 months (m) with current standard-of-care-therapies, according to phase III clinical trials and real-world data [4].

Most patients with mCRC have incurable disease, and treatment is based on systemic therapy with palliative intent. Different combinations of chemotherapeutic agents ((5-fluorouracil 40 (5-FU)/leucovorin (LV)/oxaliplatin (FOLFOX), 5-FU/LV/irinotecan (FOLFIRI), 5- 41 FU/LV/oxaliplatin/irinotecan (FOLFOXIRI)) and anti-vascular endothelial growth factor (VEGF), such as bevacizumab and aflibercept, have been developed recently, and they are used in the first- and second-line of treatment of mCRC [5,6]. Other therapies (trifluridine/tipiracil, regorafenib, raltitrexed) are used in third-line and successive lines of treatment [7,8].

Precision medicine includes the integration of molecular tumor profiles into clinical decision-making in cancer treatment. In other words, it consists in the identification of molecular targets, which would allow starting treatment with targeted therapies. Precision medicine is a challenge in oncology and it is changing the routine clinical practice [9,10].

EGFR (epidermal growth factor receptor, also known as ERBB1) is one of the first oncogenic targets in mCRC. KRAS and NRAS (RAS, rat sarcoma virus) mutations are associated with primary resistance to anti-EGFR therapies, so cetuximab and panitumumab are indicated only for RAS wild-type tumors [5,6,11].

Several target molecular biomarkers have changed the landscape of treatment of mCRC. These targeted therapies have demonstrated their effectiveness in clinical trials, obtaining the approval of the regulatory agencies: encorafenib and cetuximab in v-raf murine sarcoma viral oncogene homolog B1 (BRAF) V600E mutations, larotrectinib or entrectinib in neurotrophic tyrosine receptor kinase (NTRK) fusions, nivolumab/ipilimumab or pembrolizumab in deficient mismatch repair/microsatellite instability-high and high tumor mutation burden [12–17].

HER-2 (human epidermal growth factor receptor 2, also known as ERBB2) is a predictive biomarker that allows for the use of targeted therapies in breast and esophagogastric cancer in routine clinical practice [18–21]. ERBB2 activation by ERBB2 gene amplification or mutations is associated with anti-EGFR resistance in patients with mCRC [22,23]. ERBB2 is now under investigation for precision medicine in patients with mCRC [24–27]. Several clinical trials have evaluated the function of ERBB2-targeted therapies in patients with ERBB2-positive mCRC [28–30]. Although these clinical trials have promising results, ERBB2-targeted therapies have not been approved for patients with ERBB2-positive mCRC.

This review focuses on the knowledge of targeting ERBB2 oncogene in mCRC in the era of precision medicine: ERBB2 receptor biology, clinical features, and diagnosis of patients with ERBB2-positive mCRC and clinical trials that evaluated targeting therapies in patients with ERBB2-positive mCRC.

2. Molecular Biology of HER2 Receptor

ERBB2 is part of the family of epidermal growth factor receptors (ERBB). This family represents a group of receptor tyrosine kinases (RTK). The other members of this family are EGFR (ERBB1), HER3 (ERBB3), and HER4 (ERBB4). After binding with diverse ligands such as epidermal growth factor (EGF) or epiregulin (EREG), these receptors are able to heterodimerize, which leads to autophosphorylation. This allows the binding of diverse downstream signaling molecules, resulting in the activation of multiple pathways such as mitogen-activated protein kinase (MAPK), phosphoinositide 3-kinase (PI3K), Src pathways, and signal transducer and activator of transcription (STAT) transcription factors. ERBB2 is the only receptor capable of heterodimerizing with other ERBB receptors without binding any ligand and has an important role in the transphosphorylation of their dimerization partner [31].

The best-known pathogenic mechanisms involved in ERBB2 aberrant activation are overexpression of ERBB2 and activating mutations, both leading to constitutive activation of the receptor. Overexpression of ERBB2 in the membrane can lead to ERBB2 homodimerization and ligand-independent activation. In CRC, both mechanisms have been described. Traditionally, ERBB2 alterations have been considered to be mutually exclusive with KRAS/NRAS/BRAF alterations, although rare exceptions have been reported [32]. Modern series report that ERBB2 alterations are present in approximately 5% of CRC patients [32]. ERBB2 amplification would be present in approximately 3% of patients [33–36], while ERBB2-activating mutations in less than 2%. Co-existing amplifications and mutations would represent less than 1% of patients [32]. ERBB2-activating mutations are located in diverse regions of the receptor, such as extracellular domain II (S310F), juxtamembrane region (R678Q), and kinase domain (L775S, L866M, V777L, V842I, R868W). ERBB2 inhibition with Neratinib and Afatinib (two EGFR tyrosine kinase inhibitors) resulted in diminished cell growth in transfected cell lines [37]. Consistent with these findings, expression of

ERBB2 by IHC (immunochemistry, membrane, and cytoplasmic staining) was significantly higher in adenomas compared to normal colorectal mucosa, and was significantly higher in adenocarcinomas compared to adenomas, suggesting a role in tumorigenesis [38].

Figure 1 represents an overview of HER2 signaling and different mechanisms of action of targeted therapies that will be further discussed in the article.

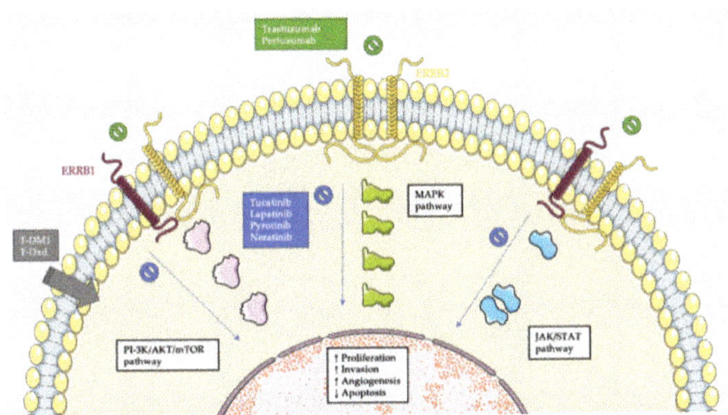

Figure 1. Molecular biology of HER2 receptor and mechanisms of action of main available drugs. Activation of HER2 by overexpression (enabling uncontrolled homo- or heterodimerization) or by activating mutations leads to constitutive activation of MAPK, PI-3K/AKT/mTOR, Src, and JAK/STAT pathways. Available drugs block this activation by inhibition of the dimerization or by inhibition of the tyrosine kinase domain of the receptor. T-DM1 and T-Dxd exert their cytopathic effects by liberation of chemotherapy in high concentrations in tumors expressing HER2.

3. Diagnosis of HER2-Positive in mCRC

Reported rates of ERBB2 positivity have varied widely in earlier studies, due to differences in antibody clone selection, scoring criteria, staining platform, and cohort composition. Scoring criteria used in other carcinomas in which ERBB2 has a pathogenic role (breast and gastroesophageal) can produce false results, as some differences in ERBB2 expression have been noted. For example, ERBB2 expression in CRC cells is often restricted to the basolateral membranes of tumor cells and stains uniformly across the tumor. These patterns are different from breast (uniform staining across the membrane) and gastric patterns (basolateral staining but patchy pattern) [39]. These observations led to the development of a validated scoring system, the HERACLES, used in the HERACLES-A trial that is discussed later.

In the HERACLES diagnostic criteria, the pattern of expression, intensity of staining, and percentage of positive cells are used to define positivity. This is defined by intense (3+) expression in ≥50% of cells. Equivocal cases are defined by moderate (2+) expression in ≥50% or 3+ ERBB2 in more than 10% but less than 50% of tumor cells. These equivocal cases require in situ hybridization (FISH) to elucidate ERBB2 overexpression. If FISH testing confirmed an ERBB2/CEP17 (centromere enumeration probe for chromosome 17) ratio of 2 or higher in 50% or more cells, this is considered a positive result. 0+ and 1+ staining intensity are considered negative [36]. Authors from Japan, Korea, and the USA recently published a harmonization broadening provisional diagnostic criteria for ERBB2-positive mCRC. These criteria differ from those previously described in that the membrane staining positivity of a lower percentage of cells (10%) is taken into account. For example, a complete, lateral, or circumferential membrane staining with strong intensity and within >10% of tumor cells would be considered as IHC 3+, while an incomplete, lateral, or circumferential membrane staining with weak/moderate intensity and within >10% of tumor cells; or complete, lateral, or circumferential membrane staining with strong intensity and within ≤ 10% of tumor cells would be considered as IHC 2+. ERBB2 positivity was defined as IHC 3+

or IHC 2+/FISH positive [40]. ERBB2-positive, low expression has recently gained interest due to recent encouraging published results in breast cancer. Some authors suggest that IHC 2+/FISH-negative cases should be considered as ERBB2-low in CRC [41]. Implications for practice will be discussed afterward.

More recently, next-generation sequencing (NGS) has gained interest as an alternative technique to assess ERBB2 positivity as it can also provide information on ERBB2 and other oncogenic drivers' mutational status. In a recent study, applying HERACLES criteria, IHC and NGS showed 92% concordance at the positive ERBB2 cutpoint and 99% concordance if equivocal cases were also considered positive. On the other hand, if ERBB2 IHC is treated as a screening tool, HERACLES-defined positive HER2 staining is 47% sensitive and 100% specific, whereas HERACLES-defined equivocal staining or greater is 93% sensitive and 100% specific for amplification by NGS [42]. Trying to harmonize IHC/FISH criteria, some authors suggest that CRC can be diagnosed as ERBB2+ with NGS if a copy number variant (CNV) of $\geq$ 5.0 is found in NGS, while CNV of 4.0 and 4.9 should be confirmed by IHC/FISH. This suggestion was validated in a retrospective cohort [40]. However, in a translational exploratory analysis in the HERACLES trial, the authors found that an ERBB2 copy number superior to 9.45 was predictive of response and progression-free survival [43]. Thus, more research is needed to find an optimal cut-off value for both diagnosis and prediction of benefit.

ctDNA (circulating tumor DNA) is becoming an attractive detection technique as it is less invasive than a conventional biopsy. This would allow repeating determinations during disease to track response and progression and to early detect the emergence of cellular clones resistant to therapy. This idea is also supported by evidence from a longitudinal tracking of ctDNA in the blood of patients included in the HERACLES trial. In this study, the dynamics of the presence of ERBB2 alleles increased in patients that were not responding to treatment and decreased in patients who had tumors that were responsive to treatment. Moreover, emerging KRAS mutant clones, BRAF amplification, mutations in ERBB2, and alterations in PI3KCA and PTEN appeared after progression to treatment with anti-ERBB2 agents. Some of these mutations had been previously linked to anti-ERBB2 resistance [44]. However, an adequate concordance between techniques is of capital importance before implementing the use of ctDNA to detect ERBB2 alterations. In an analysis of the HERACLES trial, ctDNA sequencing by Guardant360 assay correctly identified 96.6% of samples as ERBB2 amplified. Moreover, this study suggests a plasma copy number (pCN) $\geq$2.4 copies as a possible threshold representative of those patients whose HER2 amplification is the primary driver of malignancy. However, to improve the diagnostic performance of ctDNA, the authors developed an adjusted plasma copy number (apCN) in order to correct for variation in plasma tumor fraction between samples that can affect the tumor contribution to the circulating DNA pool. This apCN showed a stronger correlation than pCN ($r = 0.86$ vs. $r = 0.52$) between tissue HER2 copy number [45]. This approach was used in a later substudy of the TRIUMPH trial (discussed later) where apCN's association with clinical benefit was similar between tissue and ctDNA NGS [46]. These observations suggest that serial determinations of ctDNA of patients treated with anti-ERBB2 therapies could be useful to monitor response to treatment and to elucidate resistance mechanisms and alternative therapeutic approaches.

However, there are points of debate regarding the diagnosis of ERBB2 positivity in CRC. The first one is the concordance between ERBB2 positivity between primary and metastatic lesions. Discordance rates seem to be relatively high, as a recent study using the HERACLES system suggests. In this study, the primary positivity rate was 11.2%, while in corresponding lymph nodes was 10.1% and 31.8% in liver metastases, showing a low concordance. However, this study has its limitations as no FISH was performed to confirm equivocal samples, and no information was given about the treatments the patient had received and their temporal relationship with sample collection. This could be relevant as ERBB2 could represent an acquired resistance mechanism of anti-EGFR treatments, as will be discussed afterward. There is also evidence that ERBB2 expression could be dynamic, with changes not only after treatment with anti-EGFR drugs but also with changes after anti-ERBB2 exposure as a resistance mechanism. In a patient included in the HERACLES trial, a warm autopsy protocol was applied, which allowed for the analysis of progressing hepatic lesions after treatment with trastuzumab and lapatinib. Two of the three progressive lesions were ERBB2 negative after the treatment, providing a biological rationale for the progression [44]. It is also not clear as to how chemotherapy (QT) treatment alone could influence ERBB2 expression. There is little retrospective evidence in this regard, with a study showing only 2.2% of 139 patients after chemoradiotherapy having ERBB2 overexpression in surgical specimens, lower than usually reported in the literature [34]. However, no information about ERBB2 status in previous biopsy specimens has been reported in this study, so we cannot conclude this low prevalence was only due to treatment.

4. Clinical Features of Patients with HER2-Positive mCRC

Evidence suggests that ERBB2 tumors are more common in the left side of the colon (including the rectum), although they may not be confined to the left side. This may be related to differences in organogenesis during embryonic development [47]. There is also evidence that canonical molecular subtype (CMS2) is enriched in ERBB2-positive tumors. CMS2 represents 37% of cases, with a greater prevalence of left-sided tumors with epithelial differentiation, alterations in WNT and MYC signaling, and more frequent copy number gains in oncogenes (including ERBB2) [48,49]. Preclinical data may suggest that CMS2 tumors are more responsive to EGFR and ERBB2 blockade by tyrosine kinase inhibitors than the other subtypes [50].

There is also evidence of a different pattern of dissemination in patients with ERBB2-positive disease. In a retrospective cohort of CRC patients with resected brain metastases, up to 12% had IHC 3+ for ERBB2, which is higher than expected according to the reported prevalence of ERBB2-positive primaries [51]. The development of central nervous system (CNS) metastases has also been linked with treatment with trastuzumab and lapatinib in the HERACLES trial. CNS progression appeared in up to 19% of patients treated in this trial, a high prevalence compared to historical series [52]. There is also evidence linking ERBB2 positivity with a higher probability of developing lung metastases [22] and ovarian metastases [53]. ERBB2 positivity in ovarian metastases was also correlated with the presence of peritoneal metastases [53].

Regarding ERBB2 as a prognostic factor, evidence is conflicting. Older studies found associations of ERBB2 positivity with worse overall survival (OS) and worse stage at diagnosis; however, these studies considered cytoplasmic staining as well as membranous staining. These methods contrast with modern diagnostic criteria, so these results are difficult to interpret [38,54]. In this regard, in a modern and large (1654 patients) primary colorectal cancer study, ERBB2 positivity (1.6%; 26 patients) was associated with advanced stages and a non-significant tendency towards worse OS [55]. Furthermore, a post-hoc analysis of the PETACC-8 trial (1795 patients) found that stage-III ERBB2 positive and ERBB2 exon 19–21 mutated patients (2.9%; 49 patients, and 1%; 17 patients, respectively) had a shorter time to recurrence and worse OS, and that observation was maintained after adjusting for other adverse prognostic factors as KRAS mutation [56]. In another study, ERBB2-low patients were found to be more frequent than ERBB2 positive, with a significantly better prognosis in terms of OS than ERBB2-positive patients (33.3 months vs. 18.2 months), and in terms of PFS (2.2 months vs. 7.8 months) [41]. Another study found that nuclear staining of cyclooxygenase-2 (COX-2) correlated with high ERBB2 staining in colorectal patients. Negative nuclear staining of COX-2 and low ERBB2 staining correlated with a better prognosis than high nuclear staining of COX-2 and high ERBB2 membrane staining. However, this study did not use the above-mentioned diagnostic criteria for ERBB2 positivity, and hence its results are difficult to extrapolate [57].

On the other hand, a large cohort of 3256 patients were included in the QUASAR (adjuvant trial, stage I, II, and III patients) PICCOLO and FOCUS (metastatic patients) trial. In this cohort, 2.2% ($n = 29$) of stage IV patients and 1.3% ($n = 25$) of stage II and III patients were found to have ERBB2 positivity by IHC. There was no significant correlation between ERBB2 and recurrence or overall survival [35]. Furthermore, a German study that included 264 patients found that ERBB2 positivity (26–7%; 60 patients) was associated with better disease-free survival (DFS). This study used diagnostic criteria with a low cut-off value, which explains the high proportion of ERBB2-positive patients [58]. As different methodologies were used in the aforementioned studies, as well as in the inconsistent results found, there is no current consensus on the role of ERBB2 as a prognostic factor in CRC. Table 1 summarizes the main studies about the prognostic significance of ERBB2 in CRC.

Table 1. Main studies on the prognostic significance of ERBB2 positivity.

Study (Year)	Patients	Stage	ERBB2 Positivity Criteria	Prognostic Significance
Yagisawa et al. (2021) [41]	370	IV	International harmonization	Better prognosis of ERBB-low patients
Sawada et al. (2018) [33]	359	I–IV	HERACLES	No differences in OS
Park et al. (2018) [34]	145	I–III	Modified HERACLES	No differences in survival
Richman et al. (2016) [35]	3256	I–IV	Gastric cancer scoring	No differences in OS or PFS
Laurent-Puig et al. (2016) [56]	1804	III	HERACLES + NGS	Lower DFS and OS
Heppner et al. (2014) [55]	1645	I–IV	Gastric cancer scoring	No significant trend to poorer OS
Conradi et al. (2013) [58]	264	II–IV	Gastric cancer scoring	Better DFS
Kruszewsky et al. (2010) [59]	202	I–IV	Membranous + cytoplasmic staining	No association with OS
Osako et al. (1998) [38]	146	Dukes A-D	Membranous + cytoplasmic staining	Poorer survival in cytoplasmic staining
Kapitanovic et al. (1997) [54]	221	Bening, premalignant and malignant lesions	Membranous staining	Strong staining correlates with poorer survival

Abbreviations: DFS: disease-free survival; NGS: next-generation sequencing; OS: overall survival.

ERBB2 has also been proposed as a marker of resistance to anti-EGFR therapies. A preclinical study has suggested that ERBB2 amplification could mediate anti-EGFR primary resistance in xenograft models, particularly in KRAS/NRAS/BRAF/PI3KCA wild-type patients. Importantly, this resistance to cetuximab could be reversed with a combined inhibition of ERBB2 and EGFR [60]. Another study found evidence of ERBB2 amplification in ctDNA in patients primarily resistant to anti-EGFR therapy [61]. A retrospective study suggested that ERBB2 patients were less likely to respond to anti-EGFR therapies. However, this reduction in response rates was not directly correlated with survival. This study found a non-significant trend to worse progression-free survival and no significant differences in OS [22]. A study focused on ERBB2-low patients found a significant difference in progression-free survival (PFS) between ERBB2-low and ERBB2-positive patients treated with anti-EGFR agents (7.8 m vs 2.2 m) [41]. Other experiments and studies have also suggested that HER2 could represent a mechanism of acquired resistance to antiEGFR therapies. The introduction of ERBB2 in cells that were previously sensitive to cetuximab conferred resistance to this drug by causing abnormal activation of ERK1/2 [62]. Another study of ctDNA in patients previously treated with anti-EGFR therapies showed amplification of the ERBB2 gene upon progression in 22% of patients. In this study, a patient who progressed on cetuximab had a progressive lesion rebiopsied, showing evidence of HER2 overexpression. Notably, ERBB2 amplification was absent in primary tumor [63]. However, to date, there are no specific recommendations regarding the role of ERBB2 to guide therapeutic decisions on anti-EGFR therapies or the role of rebiopsy in progression.

5. Clinical Trials for Patients with ERBB2-Positive mCRC

There are several types of therapies that target ERBB2: monoclonal antibodies, tyrosine kinase inhibitors (TKIs), and antibody–drug conjugates (ADCs). Table 2 summarizes clinical trials for patients with ERBB2-positive mCRC and the presented results.

Table 2. Clinical trials targeting ERBB2-positive mCRC.

Trial	Reference	Treatment	n	Prior Lines of Treatment	Mutational Status	mPFS (m)	ORR (%)
Trastuzumab + QT							
Clark et al.	[64]	Trastuzumab + FOLFOX		<2	NS	NR	24
Ramanathan et al.	[65]	Trastuzumab + irinotecan	9	≤1	NS	NR	71
Monoclonal antibodies							
MyPathway	[66]	Trastuzumab + pertuzumab	57	≥1	RAS WT	2.9	32
TAPUR	[67]	Trastuzumab + pertuzumab	28	≥0	NS	NR	14
TRIUMPH	[68]	Trastuzumab + pertuzumab	27 (Tissue)	≥1	RAS WT	4.0	30
			25 (ctDNA)			3.1	25
Monoclonal antibody + TKI							
HERACLES-A	[43,69]	Trastuzumab + lapatinib	35	≥2	KRAS WT	4.7	28
Yuan et al.	[70]	Trastuzumab + pyrotinib	11	≥2	RAS WT and mutated	NR	27
MOUNTAINEER	[71]	Trastuzumab + tucatinib	23	≥2	RAS WT	8.1	52
ADCs							
HERACLES-B	[72]	Pertuzumab + T-DM1	31	≥2	RAS/BRAF WT	4.1	10
DESTINY-CRC01	[73]	TD	53 (Cohort A)	≥2	RAS/BRAF WT	6.9	45

Abbreviations: ADCs: antibody–drug conjugates, ctDNA: circulating tumor DNA, m: month, mPFS: median progression-free survival, NR: not reported, NS: not specified, ORR: overall response rates, QT: chemotherapy, T-DM1: trastuzumab emtansine, TD: trastuzumab deruxtecan, TKI: tyrosine kinase inhibitor, WT: wild type.

5.1. Monoclonal Antibodies

Trastuzumab and pertuzumab are monoclonal antibodies that target ERBB2. They bind to the extracellular domains of the receptor, inhibiting dimerization and promoting antibody-dependent cellular cytotoxic effects [74,75].

Initial trials investigated the combination of trastuzumab with QT. A phase II clinical trial evaluated trastuzumab and FOLFOX as the second or third line of treatment in ERBB2-positive patients. The ORR (overall response rate) was 24%, and the median duration of response was 4.5 m (2.7–11) [64]. Ramanathan et al. led a phase II clinical trial that evaluated trastuzumab plus irinotecan in ERBB2-positive mCRC patients. ERBB2 overexpression was detected in 8% of screened patients by IHC. Nine patients were included, and partial responses were seen in five of seven evaluable patients. However, they concluded that the low rate of ERBB2 overexpression limited more investigations in mCRC [65].

Several clinical trials have evaluated the effectiveness of the combination of two ERBB2-directed monoclonal antibodies. MyPathway was a phase IIa, multiple-basket, clinical trial that evaluated the combination of trastuzumab and pertuzumab in 57 patients with pretreated ERBB2-positive mCRC. mPFS was 2.9 m, mOS was 11.5 m, and ORR was 32%. The most common treatment-emergent adverse events were diarrhea (33%), fatigue (32%), and nausea (30%). ORR was 40% in patients with KRAS WT (wild-type) and 8% in patients with KRAS mutated tumors, so KRAS status was associated with anti-ERBB2 therapeutic efficacy [66].

TAPUR was a phase II basket clinical trial that investigated the addition of trastuzumab to pertuzumab in 28 pretreated patients with mCRC and ERBB2 overexpression/amplification. ORR was 14%. Differences in ORR compared to other studies might be explained by the inclusion of patients with concomitant RAS variations (additional analyses by RAS mutation status are pending). Two patients had at least one grade III adverse event related to trastuzumab and pertuzumab, including anemia, infusional reactions, and left ventricular dysfunction [67].

TRIUMPH was a phase II clinical trial that enrolled patients with RAS WT mCRC and ERBB2 amplification detected in tissue or ctDNA. mPFSs were 4.0 m and 3.1 m in patients with ERBB2-positive tissue and ctDNA (Guardant360), respectively. ORRs were 30% and 28% in patients with ERBB2-positive tissue and ctDNA, respectively. Patients without ctDNA variations in RAS/BRAF/PIK3CA/ERBB2 had a better response than those with a ctDNA variation in one of these genes: ORR was 44% vs. 0% in ERBB2-positive tissue and 37% vs. 0% in ERBB2-positive tissue ctDNA, respectively. mPFSs were 4.0 m (1.4–5.6) and 3.1 m (1.4.5.6) in patients with ERBB2-positive tissue and ctDNA, respectively, whereas mOSs were 10.1 (4.5–16.5) and 8.8 m (4.3–12.9) in patients with ERBB2-positive tissue and ctDNA, respectively. TRIUMPH demonstrated that decreased ctDNA fraction and ERBB2 plasma copy number three weeks after treatment initiation correlated with treatment response [46,68].

5.2. *Tyrosine Kinase Inhibitors (TKIs)*

Lapatinib, pyrotinib, tucatinib, and neratinib are oral TKIs. They inhibit the intracellular tyrosine kinase domain and phosphorylation of the ERBB2 receptor, inhibiting cell growth [76]. Several clinical trials have evaluated the efficacy of dual ERBB2 inhibition through the combination of trastuzumab and TKI.

HERACLES-A was a non-randomized, open-label, phase II clinical trial where treatment-refractory KRAS WT ERBB2-positive mCRC patients were treated with trastuzumab and lapatinib. A total of 914 patients were screened, and 48 patients (5%) were identified as ERBB2-positive (IHC 3+ in $\geq$ 50% of cells or IHC 2+ and an ERBB2:CEP17 ratio > 2 in more than 50% of cells by FISH). However, only 27 patients were eligible for the trial. In total, 74% of patients previously received at least four lines of treatment. None of the 15 patients who were evaluable for prior response to anti-EGFR therapy had obtained an objective response with cetuximab or panitumumab. Six patients (22%) had grade III adverse events: fatigue in four patients, skin rash in one patient, and increased bilirubin concentration in one patient [43].

Long-term follow-up analysis of the HERACLES-A study shows that 35 patients received trastuzumab and lapatinib, ORR was 28%, mPFS was 4.7 m (95% CI: 3.7–1), and mOS was 10.0 m (95% CI: 7.9–15.8). Progression in CNS occurred in 19% of patients [69].

Yuan et al. led a phase II clinical trial of 11 patients with ERBB2-positive mCRC treated with trastuzumab and pyrotinib. The ORR was 27%, 50% in patients with KRAS wild-type mCRC, and 60% in patients with RAS wild-type disease. Diarrhea was the most common grade III adverse event (73%), causing dose interruption and reduction in 64% and 45% of patients, respectively [70].

The MOUNTAINEER study was an open-label, single-arm phase II clinical trial when 23 pre-treated RAS WT ERBB2-positive mCRC received trastuzumab and tucatinib. The ORR was 55%, mPFS was 6.2 m (95% CI: 3.5–NE), and mOS 17.3 m (95% CI: 12.3–NE). The grade III adverse events were low (8%) [71].

Jacobs et al. reported the results of a phase Ib clinical trial involving 11 patients with RAS/BRAF/PIK3CA WT; ERRB2-positive tumors were treated with neratinib and cetuximab. However, it did not show responses: seven received stable disease, four of whom had ERBB2 amplification either in the primary tumor or the enrolment biopsy [77].

5.3. Antibody–Drug Conjugates (ADCs)

Trastuzumab emtansine (T-DM1) and trastuzumab deruxtecan (TD) are ADCs. Whereas trastuzumab is linked to a microtubule inhibitor in T-DM1, trastuzumab is joined to topoisomerase inhibitor. If the trastuzumab binds ERBB2, the ADC is internalized, the linker is cleaved, and a cytotoxic effect is made [78].

The HERACLES-B trial was a single-arm, phase II clinical trial that investigated the combination of pertuzumab and T-DM1 in RAS/RAF WT ERBB2-positive mCRC patients refractory to standard treatments. A total of 31 patients were enrolled. The primary endpoint of the study was ORR, being negative for this endpoint (9.7%, 95% CI: 0–28). However, 21 patients (67.7%) had stable disease resulting in a disease control rate of 77.4%. mPFS was 4.1 m (95% CI: 3.6–5.9); this result was similar to the HERACLES-A study. Grade III adverse events were observed in two patients (thrombocytopenia), and the most frequent grade II adverse events were nausea and fatigue [72].

The DESTINY-CRC01 trial was a phase II clinical trial that evaluated TD in treatment-refractory patients with RAS/BRAF V600E WT ERBB2-positive mCRC. Patients were enrolled in one of three cohorts on the basis of the level of ERBB2 amplification to explore the association of ERBB2 expression with the activity of TD in mCRC: cohort A (ICH3-positive or ICH2-positive and FISH-positive), cohort B (IHC2-positive and ISH-negative), and cohort C (IHC1-positive). All patients received TD 6.3 mg/kg every three weeks [73]. This dose was the same as recommended for gastric cancer and higher than breast cancer, and it was chosen because of previous studies of pharmacokinetics and antitumor activity [79–82].

A total of 78 patients were enrolled in the DESTINY-CRC01: 53 in cohort A, 7 in cohort B, and 18 in cohort C. The ORR was 45.3% in cohort A: 57.5% in patients that were ERBB2 ICH3 positive and 7.7 in patients that were ERBB2 ICH2 positive/ISH-positive. ORR was 0% in cohorts B and C. mPFS was 6.9 m, 2.1 m, and 1.4 in cohort A, cohort B, and cohort C, respectively. mOS was 15.5 m, 7.3 m, and 7.7 m in cohort A, cohort B, and cohort C, respectively. However, grade III or worse adverse events that occurred in at least 10% of all patients were decreased neutrophil count and anemia. Five patients had interstitial lung disease or pneumonitis (two grade 2; one grade 3; two grade 5, the only treatment-related deaths). A higher clinical response was detected with higher plasma ctDNA ERBB2 copy number. Antitumor activity was observed in patients regardless of ctDNA-detected activating RAS or PIK3CA mutations [83].

5.4. Ongoing Clinical Trials and Novel Anti-ERBB2 Therapies

Several ongoing clinical trials are exploring anti-ERBB2 therapies that evaluate the efficacy of small molecule inhibitors, ADCs, and their combination with established therapies [84].

The MATCH trial is a clinical trial of targeted therapy diagnosed by genetic testing in solid tumors or lymphomas after progression of at least one line of treatment. Two cohorts of patients with ERBB2-amplified tumors are treated with trastuzumab plus pertuzumab (cohort J) or T-DM1 (cohort Q) [85].

The MOUNTAINEER trial has been expanded to include a cohort of tucatinib monotherapy (NCT03943313) [86]. NSABP FC-11 is a three-cohort, phase II clinical trial in patients with RAS/BRAF/PIK3CA WT ERBB2-positive mCRC. This study compares the efficacy of neratinib and trastuzumab (Arm-1: patients who have ERBB2 amplification and prior anti-EGFR treatment or ERBB2 mutation with or without prior anti-EGFR treatment) vs. neratinib plus cetuximab (Arm-2: patients who are ERBB2 non-amplificated or ERBB2 amplification without prior anti-EGFR treatment) (NCT03457896) [87]. The first results of NSABP FC-11 were presented at the ASCO 2022 meeting [88]. Arm-1 closed due to poor accrual, and those patients have been excluded from further analysis. Arm 2 enrolled 21 patients with 15 evaluable for response by imaging. Of the 15 evaluable patients, there were 6 PR, 5 of 13 ERBB2 non-amplification, 1 of 2 ERBB2 amplification, and 5 SD. The ORR in all patients who were treated with at least one dose was 33%. Common grade 3–4 were diarrhea (24%), rash (8%), and abdominal pain/distension (8%).

Following the results of DESTINY-CRC01, DESTINY-CRC02 is a phase II clinical trial that is going to determine the efficacy and safety of TD in patients ERBB2-positive at 5.4 mg/kg and 6.4 mg/kg doses [89]. The dose of 5.4 mg/kg has not been tested in ERBB2-positive mCRC patients, but this dose has shown efficacy in other tumors [79–81].

Several trials explore the role of anti-ERBB2 therapies in earlier lines of treatment compared to QT. The MODUL trial is a randomized, open-label, parallel-group study that evaluates the efficacy and safety of biomarker-driven maintenance treatment in the first line of treatment in mCRC, including an ERBB2-positive cohort (capecitabine, trastuzumab, and pertuzumab) (NCT02291289). SWOG study (S1613) is a multicenter, randomized, phase II clinical trial that tries to compare the efficacy of trastuzumab plus pertuzumab vs. cetuximab plus irinotecan in patients with RAS/RAF WT ERBB2-positive mCRC (NCT03365882). Patients have to have been treated with at least one prior line of therapy for mCRC that did not include anti-EGFR or anti-ERBB2 agents.Zanidatamab (ZW25) is a bispecific antibody that binds to two different regions on the ERBB2 receptor, increasing antibody binding density and improving receptor internalization and downregulation. It is used in phase I and II clinical trials in patients with ERBB2-positive gastrointestinal cancers, including mCRC (NCT02892123, NCT03929666) [90]. A166 uses an antibody with the same amino acid sequence as trastuzumab and it is linked to duostatin-5. Safety profile of A166 has been observed in a phase I clinical trial. ZW49 has an auristatin payload conjugated to the antibody ZW25, which binds to the same ERBB2 domains as trastuzumab and pertuzumab. ZW49 is being evaluated in a phase I clinical trial (NCT03602079).

A phase I clinical trial will investigate the efficacy and safety of two chimeric (trastuzumab-like and pertuzumab-like) ERBB2 vaccines in patients with various metastatic solid tumors, including mCRC (NCT01376505). Another phase I trial uses an allogenic-donor-derived natural killer (NK) cell cancer immunotherapy (FATE-NK100) as monotherapy or in combination with trastuzumab or cetuximab in multiple ERBB2-positive tumors (NCT03319459). An anti-ERBB2 chimeric antigen receptor (CAR)-modified T cell therapy is evaluated in several ERBB2-positive solid tumors, including mCRC (NCT02713984) [91]. Moreover, HER2-AdVST (ERBB2 chimeric antigen receptor-modified adenovirus-specific cytotoxic T lymphocytes) joined to an intra-tumor injection of CAdVEC (an oncolytic adenovirus that helps the immune system) is being evaluated in an ongoing clinical trial (NCT03740256). Other clinical trials are trying to show the effectiveness of peptide vaccines (NCT01376505). Patients receive an ERBB2/neu peptide vaccine comprising measles virus epitope MVF-ERBB2-2 (266–296) and MVF-ERBB2 (597–626) emulsified with nor-MDP in ISA 720 intramuscularly.

6. Discussion

The results of clinical trials targeting ERBB2 positivity in mCRC have shown promising results in ORR and PFS, especially when standard treatments have been administrated. This demonstrates the importance of the diagnosis of target molecular biomarkers in the era of precision medicine. Whereas the ORR of these clinical trials are 10–40%, the trifluridine/tipiracil (TAS-102) and regorafenib have ORR of 2% and 1%, respectively [7,8]. Table 3 shows the salient points of this review.

Table 3. Salient points of the review.

Molecular biology
• The best-known pathogenic mechanisms involved in ERBB2 aberrant activation are overexpression of ERBB2 and activating mutations.
Diagnosis of HER2-positivity in mCRC
• The HERACLES diagnostic criteria are nowadays the diagnostic criteria most commonly used, although not the only ones described in the literature: ○ Positive: intense (3+) expression in ≥50% of cells. ○ Equivocal: moderate (2+) expression in ≥50% or 3+ ERBB2 in more than 10% but less than 50% of tumor cells. FISH must be performed, with an ERBB2/CEP17 ratio of 2 or higher in 50% or more cells, considered a positive result. ○ Negative: 0+ and 1+ staining. • NGS could represent an alternative diagnostic technique, but adequate threshold positivity must be defined. • ctDNA is a promising less-invasive diagnostic technique but needs to be validated.
Clinical features of patients with HER2-positive mCRC
• ERBB2-positive tumors are more common in the left side of the colon. CMS2 is enriched in ERBB2-positive tumors. • Regarding ERBB2 as a prognostic factor, evidence is conflicting. • ERBB2 has also been proposed as a marker of resistance to anti-EGFR therapies, innate or acquired.
Clinical trials for patients with ERBB2-positive mCRC
• MyPathway, TAPUR, and TRIUMP were phase II clinical trials that have evaluated the effectiveness of the combination of two ERBB2-directed monoclonal antibodies (trastuzumab and pertuzumab). • Several clinical trials have evaluated the paper of dual ERBB2 inhibition by the combination of trastuzumab and TKI: HERACLES-A (lapatinib) and MOUNTAINEER (tucatinib), showing promising ORR. • The HERACLES-B clinical trial used the combination of pertuzumab and T-DM1, and the DESTINY-CRC01 clinical trial used trastuzumab-deruxtecan. They showed an important ORR. • Several ongoing clinical trials are exploring the efficacy of small molecule inhibitors, ADCs, and their combination with established therapies or the role of anti-ERBB2 therapies in earlier lines of treatment compared to QT. • However, therapies are currently not approved for these patients, so the enrollment of patients in a clinical trial is recommended.

Although some results of the clinical trials shown above are not definitive, they illustrate the necessity of the development of phase III clinical trials in ERBB2-positive mCRC patients. These phase III clinical trials have to try to answer some important clinical questions: Which is the better sequence of treatment, starting with targeted therapy or standard treatment? Which targeted therapy is better? Is sequential targeted therapy relevant in ERBB2-positive mCRC?

We do not have results that suggest which is the better sequence of treatment in ERBB2-positive mCRC because the majority of the described clinical trials are realized in patients in whom standard treatment fails [92,93]. We do not know which targeted therapy shows better ORR or mPFS. The management of adverse events of targeted therapy is well known—for example, the appearance of left ventricular dysfunction with trastuzumab or interstitial lung disease when TD is used. However, some results suggest that sequential therapy may be relevant in ERRB2-positive mCRC because 30% of patients in DESTINY-CRC01 were previously treated with other anti-ERRB2 therapies.

In an ideal scenario, when a patient is diagnosed with mCRC, a biopsy has been analyzed by a pathologist and IHC and NGS should be realized. Moreover, a liquid biopsy of ctDNA should be realized. These procedures should be repeated when a line of treatment fails. This strategy would give a lot of molecular information, such as the development of mechanisms of resistance. In this way, a medical oncologist would be able to select the best treatment, including the inclusion in a clinical trial. However, this approach is not useful. Firstly, performing multiple biopsies carries certain risks. Secondly, NGS and ctDNA in mCRC are under research at this moment. The information obtained from NGS and ctDNA is sometimes difficult to integrate and understand. Third, making these procedures is expensive, and in a public health system, the government could deny the payment because this strategy may not be efficient enough (we are looking for target rare molecular biomarkers to start expensive targeted therapies). This could also apply for private insurances.

There is controversy about when the medical oncologist must look for rare target molecular biomarkers. Some oncologists think that rare molecular biomarkers should be determined before starting the first line of treatment. Patients have a good clinical status at this moment, and targeted therapies have better ORR and PFS than QT, as in other pathologies (EGFR mutations in non-small cell lung cancer). On the other hand, other oncologists affirm that the determination of rare molecular biomarkers must be performed when patients maintain good clinical status and standard treatments have failed.

The guidelines of treatment in mCRC show that the determination of RAS (KRAS/NRAS) mutations, BRAF mutations, and deficient mismatch repair should be realized before starting the first line of treatment (6,32). The ESMO (European Society of. Medical Oncology) guidelines do not mention ERBB2 amplification/overexpression [6]. The NCCN (National Comprehensive Cancer Network) guidelines state that testing ERBB2 amplification/overexpression should be made in patients with mCRC and absence of RAS or BRAF mutation. ERBB2-targeted therapies are recommended as subsequent therapy options, encouraging enrollment in a clinical trial (32). The actualization of these guidelines should define the optimal timing and technique for testing, the most adequate panel, and whether all RAS WT mCRC should be tested for ERBB2 [94].

The search for target rare molecular biomarkers illustrates the complexity of precision medicine, so it is required that a medical oncologist has to study molecular biology and clinical treatments. Moreover, it shows the necessity of multidisciplinary work. In our center, we have to work with other specialists to obtain tissue for the molecular diagnosis, and we have a fluid relationship with pathologists. If we obtain a target rare molecular biomarker, we must discuss it with other medical oncologists when we want to include a patient in a clinical trial.

7. Conclusions

The management of mCRC in the era of precision medicine is becoming more complex. Amplification of ERBB2 is present in 3% of patients with mCRC and 5% of patients with RAS and BRAF wild type. Several clinical trials have demonstrated that the ERBB2 receptor represents a good option for targeted therapy in mCRC and may represent an option when standard treatments fail to control mCRC. However, therapies are currently not approved for these patients, and the recommendation is the enrollment of patients in a clinical trial.

Author Contributions: J.T.-J., J.E.-V. and R.F.-M. contributed substantially to conceptualization, methodology, validation, investigation, and writing. Supervision: R.F.-M. All authors have read and agreed to the published version of the manuscript.

Funding: This research received no external funding.

Acknowledgments: The authors wish to thank the patients who participated and are currently participating in the studies mentioned and their families.

Conflicts of Interest: The authors declare no conflict of interest.

Abbreviations

5-FU	5-fluorouracil
ADCs	antibody-drug conjugates
BRAF	v-raf murine sarcoma viral oncogene homolog B1
CEP17	centromere enumeration probe for chromosome 17
CNS	central nervous system
CRC	colorectal cancer
ctDNA	circulating tumor DNA
EGF	epidermal growth factor
EGFR	epidermal growth factor receptor
EREG	epiregulin
ESMO	European Society of Medical Oncology
FISH	fluorescence in situ hybridization
FOLFIRI	5-fluouracil/leucovorin/irinotecan
FOLFOX	5-fluouracil/leucovorin/oxaliplatin
FOLFOXIRI	5-fluouracil/leucovorin/oxaliplatin/irinotecan
HER2	human epidermal growth factor receptor 2
IHC	immunochemistry
LV	leucovorin
m	month
MAPK	mitogen-activated protein kinase
mCRC	metastatic colorectal cancer
mPFS	median progression-free survival
NCCN	National Comprehensive Cancer Network
NGS	next-generation sequencing
NK	natural killer
NTRK	neurotrophic tyrosine receptor kinase
ORR	overall response rates
OS	overall survival
PI3K	phosphoinositide 3-kinase
PFS	progression-free survival
PTEN	phosphatase and tensin homolog
QT	chemotherapy
RAS	rat sarcoma virus
RTK	receptor tyrosine kinase
STAT	signal transducer and activator of transcription
T-DM1	trastuzumab emastine
TD	trastuzumab deruxtecan
TKIs	tyrosine kinase inhibitors
VEGF	anti-vascular endothelial growth factor
WT	wild-type

References

1. Keum, N.; Giovannucci, E. Global Burden of Colorectal Cancer: Emerging Trends, Risk Factors and Prevention Strategies. *Nat. Rev. Gastroenterol. Hepatol.* **2019**, *16*, 713–732. [CrossRef] [PubMed]
2. Rawla, P.; Sunkara, T.; Barsouk, A. Epidemiology of Colorectal Cancer: Incidence, Mortality, Survival, and Risk Factors. *Gastroenterol. Rev./Przegląd Gastroenterol.* **2019**, *14*, 89–103. [CrossRef] [PubMed]
3. Sung, H.; Ferlay, J.; Siegel, R.L.; Laversanne, M.; Soerjomataram, I.; Jemal, A.; Bray, F. Global Cancer Statistics 2020: GLOBOCAN Estimates of Incidence and Mortality Worldwide for 36 Cancers in 185 Countries. *CA A Cancer J. Clin.* **2021**, *71*, 209–249. [CrossRef] [PubMed]
4. Biller, L.H.; Schrag, D. Diagnosis and Treatment of Metastatic Colorectal Cancer: A Review. *JAMA* **2021**, *325*, 669. [CrossRef]
5. Benson, A.B.; Venook, A.P.; Al-Hawary, M.M.; Arain, M.A.; Chen, Y.-J.; Ciombor, K.K.; Cohen, S.; Cooper, H.S.; Deming, D.; Farkas, L.; et al. Colon Cancer, Version 2.2021, NCCN Clinical Practice Guidelines in Oncology. *J. Natl. Compr. Cancer Netw.* **2021**, *19*, 329–359. [CrossRef]
6. Van Cutsem, E.; Cervantes, A.; Adam, R.; Sobrero, A.; Van Krieken, J.H.; Aderka, D.; Aranda Aguilar, E.; Bardelli, A.; Benson, A.; Bodoky, G.; et al. ESMO Consensus Guidelines for the Management of Patients with Metastatic Colorectal Cancer. *Ann. Oncol.* **2016**, *27*, 1386–1422. [CrossRef]

7. Mayer, R.J.; Van Cutsem, E.; Falcone, A.; Yoshino, T.; Garcia-Carbonero, R.; Mizunuma, N.; Yamazaki, K.; Shimada, Y.; Tabernero, J.; Komatsu, Y.; et al. Randomized Trial of TAS-102 for Refractory Metastatic Colorectal Cancer. *N. Engl. J. Med.* **2015**, *372*, 1909–1919. [CrossRef]
8. Grothey, A.; Cutsem, E.V.; Sobrero, A.; Siena, S.; Falcone, A.; Ychou, M.; Humblet, Y.; Bouché, O.; Mineur, L.; Barone, C.; et al. Regorafenib Monotherapy for Previously Treated Metastatic Colorectal Cancer (CORRECT): An International, Multicentre, Randomised, Placebo-Controlled, Phase 3 Trial. *Lancet* **2013**, *381*, 303–312. [CrossRef]
9. Di Nicolantonio, F.; Vitiello, P.P.; Marsoni, S.; Siena, S.; Tabernero, J.; Trusolino, L.; Bernards, R.; Bardelli, A. Precision Oncology in Metastatic Colorectal Cancer—From Biology to Medicine. *Nat. Rev. Clin. Oncol.* **2021**, *18*, 506–525. [CrossRef]
10. Mateo, J.; Steuten, L.; Aftimos, P.; André, F.; Davies, M.; Garralda, E.; Geissler, J.; Husereau, D.; Martinez-Lopez, I.; Normanno, N.; et al. Delivering Precision Oncology to Patients with Cancer. *Nat. Med.* **2022**, *28*, 658–665. [CrossRef]
11. Douillard, J.-Y.; Oliner, K.S.; Siena, S.; Tabernero, J.; Burkes, R.; Barugel, M.; Humblet, Y.; Bodoky, G.; Cunningham, D.; Jassem, J.; et al. Panitumumab–FOLFOX4 Treatment and *RAS* Mutations in Colorectal Cancer. *N. Engl. J. Med.* **2013**, *369*, 1023–1034. [CrossRef]
12. Kopetz, S.; Grothey, A.; Yaeger, R.; Van Cutsem, E.; Desai, J.; Yoshino, T.; Wasan, H.; Ciardiello, F.; Loupakis, F.; Hong, Y.S.; et al. Encorafenib, Binimetinib, and Cetuximab in *BRAF* V600E–Mutated Colorectal Cancer. *N. Engl. J. Med.* **2019**, *381*, 1632–1643. [CrossRef]
13. André, T.; Shiu, K.-K.; Kim, T.W.; Jensen, B.V.; Jensen, L.H.; Punt, C.; Smith, D.; Garcia-Carbonero, R.; Benavides, M.; Gibbs, P.; et al. Pembrolizumab in Microsatellite-Instability–High Advanced Colorectal Cancer. *N. Engl. J. Med.* **2020**, *383*, 2207–2218. [CrossRef]
14. Overman, M.J.; McDermott, R.; Leach, J.L.; Lonardi, S.; Lenz, H.-J.; Morse, M.A.; Desai, J.; Hill, A.; Axelson, M.; Moss, R.A.; et al. Nivolumab in Patients with Metastatic DNA Mismatch Repair-Deficient or Microsatellite Instability-High Colorectal Cancer (CheckMate 142): An Open-Label, Multicentre, Phase 2 Study. *Lancet Oncol.* **2017**, *18*, 1182–1191. [CrossRef]
15. Boni, V.; Drilon, A.; Deeken, J.; Garralda, E.; Chung, H.; Kinoshita, I.; Oh, D.; Patel, J.; Xu, R.; Norenberg, R.; et al. SO-29 Efficacy and Safety of Larotrectinib in Patients with Tropomyosin Receptor Kinase Fusion-Positive Gastrointestinal Cancer: An Expanded Dataset. *Ann. Oncol.* **2021**, *32*, S214–S215. [CrossRef]
16. Patel, M.; Siena, S.; Demetri, G.; Doebele, R.; Chae, Y.; Conkling, P.; Garrido-Laguna, I.; Longo, F.; Rolfo, C.; Sigal, D.; et al. O-3 Efficacy and Safety of Entrectinib in NTRK Fusion-Positive Gastrointestinal Cancers: Updated Integrated Analysis of Three Clinical Trials (STARTRK-2, STARTRK-1 and ALKA-372-001). *Ann. Oncol.* **2020**, *31*, 232–233. [CrossRef]
17. Cohen, R.; Pudlarz, T.; Delattre, J.-F.; Colle, R.; André, T. Molecular Targets for the Treatment of Metastatic Colorectal Cancer. *Cancers* **2020**, *12*, 2350. [CrossRef]
18. Nader-Marta, G.; Martins-Branco, D.; de Azambuja, E. How We Treat Patients with Metastatic HER2-Positive Breast Cancer. *ESMO Open* **2022**, *7*, 100343. [CrossRef]
19. Schettini, F.; Prat, A. Dissecting the Biological Heterogeneity of HER2-Positive Breast Cancer. *Breast* **2021**, *59*, 339–350. [CrossRef]
20. Kahraman, S.; Yalcin, S. Recent Advances in Systemic Treatments for HER-2 Positive Advanced Gastric Cancer. *OTT* **2021**, *14*, 4149–4162. [CrossRef]
21. Zhu, Y.; Zhu, X.; Wei, X.; Tang, C.; Zhang, W. HER2-Targeted Therapies in Gastric Cancer. *Biochim. Biophys. Acta (BBA)-Rev. Cancer* **2021**, *1876*, 188549. [CrossRef]
22. Sartore-Bianchi, A.; Amatu, A.; Porcu, L.; Ghezzi, S.; Lonardi, S.; Leone, F.; Bergamo, F.; Fenocchio, E.; Martinelli, E.; Borelli, B.; et al. HER2 Positivity Predicts Unresponsiveness to EGFR-Targeted Treatment in Metastatic Colorectal Cancer. *Oncologist* **2019**, *24*, 1395–1402. [CrossRef]
23. Huang, W.; Chen, Y.; Chang, W.; Ren, L.; Tang, W.; Zheng, P.; Wu, Q.; Liu, T.; Liu, Y.; Wei, Y.; et al. HER2 Positivity as a Biomarker for Poor Prognosis and Unresponsiveness to Anti-EGFR Therapy in Colorectal Cancer. *J. Cancer Res. Clin. Oncol.* **2022**, *148*, 993–1002. [CrossRef]
24. Greally, M.; Kelly, C.M.; Cercek, A. HER2: An Emerging Target in Colorectal Cancer. *Curr. Probl. Cancer* **2018**, *42*, 560–571. [CrossRef] [PubMed]
25. Lam, M.; Lum, C.; Latham, S.; Tipping Smith, S.; Prenen, H.; Segelov, E. Refractory Metastatic Colorectal Cancer: Current Challenges and Future Prospects. *CMAR* **2020**, *12*, 5819–5830. [CrossRef] [PubMed]
26. Siena, S.; Sartore-Bianchi, A.; Marsoni, S.; Hurwitz, H.I.; McCall, S.J.; Penault-Llorca, F.; Srock, S.; Bardelli, A.; Trusolino, L. Targeting the Human Epidermal Growth Factor Receptor 2 (HER2) Oncogene in Colorectal Cancer. *Ann. Oncol.* **2018**, *29*, 1108–1119. [CrossRef] [PubMed]
27. La Salvia, A.; Lopez-Gomez, V.; Garcia-Carbonero, R. HER2-Targeted Therapy: An Emerging Strategy in Advanced Colorectal Cancer. *Expert Opin. Investig. Drugs* **2019**, *28*, 29–38. [CrossRef]
28. Bitar, L.; Zouein, J.; Haddad, F.G.; Eid, R.; Kourie, H.R. HER2 in Metastatic Colorectal Cancer: A New to Target to Remember. *Biomark. Med.* **2021**, *15*, 135–138. [CrossRef]
29. Strickler, J.H.; Yoshino, T.; Graham, R.P.; Siena, S.; Bekaii-Saab, T. Diagnosis and Treatment of ERBB2-Positive Metastatic Colorectal Cancer: A Review. *JAMA Oncol.* **2022**, *8*, 760–769. [CrossRef]
30. Mohamed, A.A.; Lau, D.K.; Chau, I. HER2 Targeted Therapy in Colorectal Cancer: New Horizons. *Cancer Treat. Rev.* **2022**, *105*, 102363. [CrossRef]

31. Yarden, Y.; Pines, G. The ERBB Network: At Last, Cancer Therapy Meets Systems Biology. *Nat. Rev. Cancer* **2012**, *12*, 553–563. [CrossRef]
32. Yaeger, R.; Chatila, W.K.; Lipsyc, M.D.; Hechtman, J.F.; Cercek, A.; Sanchez-Vega, F.; Jayakumaran, G.; Middha, S.; Zehir, A.; Donoghue, M.T.A.; et al. Clinical Sequencing Defines the Genomic Landscape of Metastatic Colorectal Cancer. *Cancer Cell* **2018**, *33*, 125–136.e3. [CrossRef]
33. Sawada, K.; Nakamura, Y.; Yamanaka, T.; Kuboki, Y.; Yamaguchi, D.; Yuki, S.; Yoshino, T.; Komatsu, Y.; Sakamoto, N.; Okamoto, W.; et al. Prognostic and Predictive Value of HER2 Amplification in Patients With Metastatic Colorectal Cancer. *Clin. Colorectal Cancer* **2018**, *17*, 198–205. [CrossRef]
34. Park, J.S.; Yoon, G.; Kim, H.J.; Park, S.Y.; Choi, G.S.; Kang, M.K.; Kim, J.G.; Jang, J.-S.; Seo, A.N. HER2 Status in Patients with Residual Rectal Cancer after Preoperative Chemoradiotherapy: The Relationship with Molecular Results and Clinicopathologic Features. *Virchows Arch.* **2018**, *473*, 413–423. [CrossRef]
35. Richman, S.D.; Southward, K.; Chambers, P.; Cross, D.; Barrett, J.; Hemmings, G.; Taylor, M.; Wood, H.; Hutchins, G.; Foster, J.M.; et al. HER2 Overexpression and Amplification as a Potential Therapeutic Target in Colorectal Cancer: Analysis of 3256 Patients Enrolled in the QUASAR, FOCUS and PICCOLO Colorectal Cancer Trials. *J. Pathol.* **2016**, *238*, 562–570. [CrossRef]
36. Valtorta, E.; Martino, C.; Sartore-Bianchi, A.; Penaullt-Llorca, F.; Viale, G.; Risio, M.; Rugge, M.; Grigioni, W.; Bencardino, K.; Lonardi, S.; et al. Assessment of a HER2 Scoring System for Colorectal Cancer: Results from a Validation Study. *Mod. Pathol.* **2015**, *28*, 1481–1491. [CrossRef]
37. Kavuri, S.M.; Jain, N.; Galimi, F.; Cottino, F.; Leto, S.M.; Migliardi, G.; Searleman, A.C.; Shen, W.; Monsey, J.; Trusolino, L.; et al. HER2 Activating Mutations Are Targets for Colorectal Cancer Treatment. *Cancer Discov.* **2015**, *5*, 832–841. [CrossRef]
38. Osako, T.; Miyahara, M.; Uchino, S.; Inomata, M.; Kitano, S.; Kobayashi, M. Immunohistochemical Study of C-ErbB-2 Protein in Colorectal Cancer and the Correlation with Patient Survival. *Oncology* **1998**, *55*, 548–555. [CrossRef]
39. Nowak, J.A. HER2 in Colorectal Carcinoma. *Surg. Pathol. Clin.* **2020**, *13*, 485–502. [CrossRef]
40. Fujii, S.; Magliocco, A.M.; Kim, J.; Okamoto, W.; Kim, J.E.; Sawada, K.; Nakamura, Y.; Kopetz, S.; Park, W.-Y.; Tsuchihara, K.; et al. International Harmonization of Provisional Diagnostic Criteria for *ERBB2*-Amplified Metastatic Colorectal Cancer Allowing for Screening by Next-Generation Sequencing Panel. *JCO Precis. Oncol.* **2020**, *4*, 6–19. [CrossRef]
41. Yagisawa, M.; Sawada, K.; Nakamura, Y.; Fujii, S.; Yuki, S.; Komatsu, Y.; Yoshino, T.; Sakamoto, N.; Taniguchi, H. Prognostic Value and Molecular Landscape of HER2 Low-Expressing Metastatic Colorectal Cancer. *Clin. Colorectal Cancer* **2021**, *20*, 113–120.e1. [CrossRef]
42. Cenaj, O.; Ligon, A.H.; Hornick, J.L.; Sholl, L.M. Detection of ERBB2 Amplification by Next-Generation Sequencing Predicts HER2 Expression in Colorectal Carcinoma. *Am. J. Clin. Pathol.* **2019**, *152*, 97–108. [CrossRef]
43. Sartore-Bianchi, A.; Trusolino, L.; Martino, C.; Bencardino, K.; Lonardi, S.; Bergamo, F.; Zagonel, V.; Leone, F.; Depetris, I.; Martinelli, E.; et al. Dual-Targeted Therapy with Trastuzumab and Lapatinib in Treatment-Refractory, KRAS Codon 12/13 Wild-Type, HER2-Positive Metastatic Colorectal Cancer (HERACLES): A Proof-of-Concept, Multicentre, Open-Label, Phase 2 Trial. *Lancet Oncol.* **2016**, *17*, 738–746. [CrossRef]
44. Siravegna, G.; Lazzari, L.; Crisafulli, G.; Sartore-Bianchi, A.; Mussolin, B.; Cassingena, A.; Martino, C.; Lanman, R.B.; Nagy, R.J.; Fairclough, S.; et al. Radiologic and Genomic Evolution of Individual Metastases during HER2 Blockade in Colorectal Cancer. *Cancer Cell* **2018**, *34*, 148–162.e7. [CrossRef]
45. Siravegna, G.; Sartore-Bianchi, A.; Nagy, R.J.; Raghav, K.; Odegaard, J.I.; Lanman, R.B.; Trusolino, L.; Marsoni, S.; Siena, S.; Bardelli, A. Plasma HER2 (*ERBB2*) Copy Number Predicts Response to HER2-Targeted Therapy in Metastatic Colorectal Cancer. *Clin. Cancer Res.* **2019**, *25*, 3046–3053. [CrossRef]
46. Nakamura, Y.; Okamoto, W.; Kato, T.; Esaki, T.; Kato, K.; Komatsu, Y.; Yuki, S.; Masuishi, T.; Nishina, T.; Ebi, H.; et al. Circulating Tumor DNA-Guided Treatment with Pertuzumab plus Trastuzumab for HER2-Amplified Metastatic Colorectal Cancer: A Phase 2 Trial. *Nat. Med.* **2021**, *27*, 1899–1903. [CrossRef]
47. Salem, M.E.; Weinberg, B.A.; Xiu, J.; El-Deiry, W.S.; Hwang, J.J.; Gatalica, Z.; Philip, P.A.; Shields, A.F.; Lenz, H.-J.; Marshall, J.L. Comparative Molecular Analyses of Left-Sided Colon, Right-Sided Colon, and Rectal Cancers. *Oncotarget* **2017**, *8*, 86356–86368. [CrossRef]
48. Guinney, J.; Dienstmann, R.; Wang, X.; de Reyniès, A.; Schlicker, A.; Soneson, C.; Marisa, L.; Roepman, P.; Nyamundanda, G.; Angelino, P.; et al. The Consensus Molecular Subtypes of Colorectal Cancer. *Nat. Med.* **2015**, *21*, 1350–1356. [CrossRef] [PubMed]
49. Berg, K.C.G.; Sveen, A.; Høland, M.; Alagaratnam, S.; Berg, M.; Danielsen, S.A.; Nesbakken, A.; Søreide, K.; Lothe, R.A. Gene Expression Profiles of CMS2-Epithelial/Canonical Colorectal Cancers Are Largely Driven by DNA Copy Number Gains. *Oncogene* **2019**, *38*, 6109–6122. [CrossRef] [PubMed]
50. Sveen, A.; Bruun, J.; Eide, P.W.; Eilertsen, I.A.; Ramirez, L.; Murumägi, A.; Arjama, M.; Danielsen, S.A.; Kryeziu, K.; Elez, E.; et al. Colorectal Cancer Consensus Molecular Subtypes Translated to Preclinical Models Uncover Potentially Targetable Cancer Cell Dependencies. *Clin. Cancer Res.* **2018**, *24*, 794–806. [CrossRef] [PubMed]
51. Aprile, G.; De Maglio, G.; Menis, J.; Casagrande, M.; Tuniz, F.; Pisa, E.; Fontanella, C.; Skrap, M.; Beltrami, A.; Fasola, G.; et al. HER-2 Expression in Brain Metastases from Colorectal Cancer and Corresponding Primary Tumors: A Case Cohort Series. *IJMS* **2013**, *14*, 2370–2387. [CrossRef]

52. Sartore-Bianchi, A.; Lonardi, S.; Aglietta, M.; Martino, C.; Ciardiello, F.; Marsoni, S.; Siena, S. Central Nervous System as Possible Site of Relapse in *ERBB2* -Positive Metastatic Colorectal Cancer: Long-Term Results of Treatment With Trastuzumab and Lapatinib. *JAMA Oncol.* **2020**, *6*, 927. [CrossRef]
53. Li, J.-L.; Lin, S.-H.; Chen, H.-Q.; Liang, L.-S.; Mo, X.-W.; Lai, H.; Zhang, J.; Xu, J.; Gao, B.-Q.; Feng, Y.; et al. Clinical Significance of HER2 and EGFR Expression in Colorectal Cancer Patients with Ovarian Metastasis. *BMC Clin. Pathol.* **2019**, *19*, 3. [CrossRef]
54. Kapitanović, S.; Radosević, S.; Kapitanović, M.; Andelinović, S.; Ferencić, Z.; Tavassoli, M.; Primorać, D.; Sonicki, Z.; Spaventi, S.; Pavelic, K.; et al. The Expression of P185(HER-2/Neu) Correlates with the Stage of Disease and Survival in Colorectal Cancer. *Gastroenterology* **1997**, *112*, 1103–1113. [CrossRef]
55. Ingold Heppner, B.; Behrens, H.-M.; Balschun, K.; Haag, J.; Krüger, S.; Becker, T.; Röcken, C. HER2/Neu Testing in Primary Colorectal Carcinoma. *Br. J. Cancer* **2014**, *111*, 1977–1984. [CrossRef]
56. Laurent-Puig, P.; Balogoun, R.; Cayre, A.; Le Malicot, K.; Tabernero, J.; Mini, E.; Folprecht, G.; van Laethem, J.-L.; Thaler, J.; Petersen, L.N.; et al. ERBB2 Alterations a New Prognostic Biomarker in Stage III Colon Cancer from a FOLFOX Based Adjuvant Trial (PETACC8). *Ann. Oncol.* **2016**, *27*, vi151. [CrossRef]
57. Zhou, F.; Huang, R.; Jiang, J.; Zeng, X.; Zou, S. Correlated Non-Nuclear COX2 and Low HER2 Expression Confers a Good Prognosis in Colorectal Cancer. *Saudi J. Gastroenterol.* **2018**, *24*, 301. [CrossRef]
58. Conradi, L.-C.; Styczen, H.; Sprenger, T.; Wolff, H.A.; Rödel, C.; Nietert, M.; Homayounfar, K.; Gaedcke, J.; Kitz, J.; Talaulicar, R.; et al. Frequency of HER-2 Positivity in Rectal Cancer and Prognosis. *Am. J. Surg. Pathol.* **2013**, *37*, 522–531. [CrossRef]
59. Kruszewski, W.J.; Rzepko, R.; Ciesielski, M.; Szefel, J.; Zieliński, J.; Szajewski, M.; Jasiński, W.; Kawecki, K.; Wojtacki, J. Expression of HER2 in Colorectal Cancer Does Not Correlate with Prognosis. *Dis. Markers* **2010**, *29*, 207–212. [CrossRef]
60. Bertotti, A.; Migliardi, G.; Galimi, F.; Sassi, F.; Torti, D.; Isella, C.; Corà, D.; Di Nicolantonio, F.; Buscarino, M.; Petti, C.; et al. A Molecularly Annotated Platform of Patient-Derived Xenografts ("Xenopatients") Identifies HER2 as an Effective Therapeutic Target in Cetuximab-Resistant Colorectal Cancer. *Cancer Discov.* **2011**, *1*, 508–523. [CrossRef]
61. Siravegna, G.; Mussolin, B.; Buscarino, M.; Corti, G.; Cassingena, A.; Crisafulli, G.; Ponzetti, A.; Cremolini, C.; Amatu, A.; Lauricella, C.; et al. Clonal Evolution and Resistance to EGFR Blockade in the Blood of Colorectal Cancer Patients. *Nat. Med.* **2015**, *21*, 795–801. [CrossRef]
62. Yonesaka, K.; Zejnullahu, K.; Okamoto, I.; Satoh, T.; Cappuzzo, F.; Souglakos, J.; Ercan, D.; Rogers, A.; Roncalli, M.; Takeda, M.; et al. Activation of ERBB2 Signaling Causes Resistance to the EGFR-Directed Therapeutic Antibody Cetuximab. *Sci. Transl. Med.* **2011**, *3*, 99ra86. [CrossRef]
63. Takegawa, N.; Yonesaka, K.; Sakai, K.; Ueda, H.; Watanabe, S.; Nonagase, Y.; Okuno, T.; Takeda, M.; Maenishi, O.; Tsurutani, J.; et al. HER2 Genomic Amplification in Circulating Tumor DNA from Patients with Cetuximab-Resistant Colorectal Cancer. *Oncotarget* **2016**, *7*, 3453–3460. [CrossRef] [PubMed]
64. Clark, J.W.; Niedzwiecki, D.; Hollis, D.; Mayer, R. Phase II Trial of 5-Fluorouracil (5-FU), Leucovorin (LV), Oxaliplatin (Ox), and Trastuzumab (T) for Patients with Metastatic Colorectal Cancer (CRC) Refractory to Initial Therapy. *Proc. Am. Soc. Clin. Oncol.* **2003**, *22*, abstr-3584.
65. Ramanathan, R.K.; Hwang, J.J.; Zamboni, W.C.; Sinicrope, F.A.; Safran, H.; Wong, M.K.; Earle, M.; Brufsky, A.; Evans, T.; Troetschel, M.; et al. Low Overexpression of HER-2/Neu in Advanced Colorectal Cancer Limits the Usefulness of Trastuzumab (Herceptin®) and Irinotecan as Therapy. A Phase II Trial. *Cancer Investig.* **2004**, *22*, 858–865. [CrossRef] [PubMed]
66. Meric-Bernstam, F.; Hurwitz, H.; Raghav, K.P.S.; McWilliams, R.R.; Fakih, M.; VanderWalde, A.; Swanton, C.; Kurzrock, R.; Burris, H.; Sweeney, C.; et al. Pertuzumab plus Trastuzumab for HER2-Amplified Metastatic Colorectal Cancer (MyPathway): An Updated Report from a Multicentre, Open-Label, Phase 2a, Multiple Basket Study. *Lancet Oncol.* **2019**, *20*, 518–530. [CrossRef]
67. Gupta, R.; Garrett-Mayer, E.; Halabi, S.; Mangat, P.K.; D'Andre, S.D.; Meiri, E.; Shrestha, S.; Warren, S.L.; Ranasinghe, S.; Schilsky, R.L. Pertuzumab plus Trastuzumab (P+T) in Patients (Pts) with Colorectal Cancer (CRC) with *ERBB2* Amplification or Overexpression: Results from the TAPUR Study. *JCO* **2020**, *38*, 132. [CrossRef]
68. Okamoto, W.; Nakamura, Y.; Kato, T.; Esaki, T.; Komoda, M.; Kato, K.; Komatsu, Y.; Masuishi, T.; Nishina, T.; Sawada, K.; et al. Pertuzumab plus Trastuzumab and Real-World Standard of Care (SOC) for Patients (Pts) with Treatment Refractory Metastatic Colorectal Cancer (MCRC) with *HER2* (*ERBB2*) Amplification (Amp) Confirmed by Tumor Tissue or CtDNA Analysis (TRIUMPH, EPOC1602). *JCO* **2021**, *39*, 3555. [CrossRef]
69. Tosi, F.; Sartore-Bianchi, A.; Lonardi, S.; Amatu, A.; Leone, F.; Ghezzi, S.; Martino, C.; Bencardino, K.; Bonazzina, E.; Bergamo, F.; et al. Long-Term Clinical Outcome of Trastuzumab and Lapatinib for HER2-Positive Metastatic Colorectal Cancer. *Clin. Colorectal Cancer* **2020**, *19*, 256–262.e2. [CrossRef]
70. Yuan, Y.; Fu, X.; Ying, J.; Yang, L.; Fang, W.; Han, W.; Zhang, S. Dual-Targeted Therapy with Pyrotinib and Trastuzumab for HER2-Positive Advanced Colorectal Cancer: Preliminary Results from a Multicenter Phase 2 Trial. *JCO* **2021**, *39*, e15554. [CrossRef]
71. Strickler, J.H.; Zemla, T.; Ou, F.-S.; Cercek, A.; Wu, C.; Sanchez, F.A.; Hubbard, J.; Jaszewski, B.; Bandel, L.; Schweitzer, B.; et al. Trastuzumab and Tucatinib for the Treatment of HER2 Amplified Metastatic Colorectal Cancer (MCRC): Initial Results from the MOUNTAINEER Trial. *Ann. Oncol.* **2019**, *30*, v200. [CrossRef]
72. Sartore-Bianchi, A.; Lonardi, S.; Martino, C.; Fenocchio, E.; Tosi, F.; Ghezzi, S.; Leone, F.; Bergamo, F.; Zagonel, V.; Ciardiello, F.; et al. Pertuzumab and Trastuzumab Emtansine in Patients with HER2-Amplified Metastatic Colorectal Cancer: The Phase II HERACLES-B Trial. *ESMO Open* **2020**, *5*, e000911. [CrossRef]

73. Siena, S.; Di Bartolomeo, M.; Raghav, K.; Masuishi, T.; Loupakis, F.; Kawakami, H.; Yamaguchi, K.; Nishina, T.; Fakih, M.; Elez, E.; et al. Trastuzumab Deruxtecan (DS-8201) in Patients with HER2-Expressing Metastatic Colorectal Cancer (DESTINY-CRC01): A Multicentre, Open-Label, Phase 2 Trial. *Lancet Oncol.* **2021**, *22*, 779–789. [CrossRef]
74. Coussens, L.; Yang-Feng, T.L.; Liao, Y.-C.; Chen, E.; Gray, A.; McGrath, J.; Seeburg, P.H.; Libermann, T.A.; Schlessinger, J.; Francke, U.; et al. Tyrosine Kinase Receptor with Extensive Homology to EGF Receptor Shares Chromosomal Location with *Neu* Oncogene. *Science* **1985**, *230*, 1132–1139. [CrossRef]
75. Richard, S.; Selle, F.; Lotz, J.-P.; Khalil, A.; Gligorov, J.; Soares, D.G. Pertuzumab and Trastuzumab: The Rationale Way to Synergy. *An. Acad. Bras. Ciênc.* **2016**, *88*, 565–577. [CrossRef]
76. Esteban-Villarrubia, J.; Soto-Castillo, J.J.; Pozas, J.; San Román-Gil, M.; Orejana-Martín, I.; Torres-Jiménez, J.; Carrato, A.; Alonso-Gordoa, T.; Molina-Cerrillo, J. Tyrosine Kinase Receptors in Oncology. *IJMS* **2020**, *21*, 8529. [CrossRef]
77. Jacobs, S.A.; Lee, J.J.; George, T.J.; Wade, J.L.; Stella, P.J.; Wang, D.; Sama, A.R.; Piette, F.; Pogue-Geile, K.L.; Kim, R.S.; et al. Neratinib-Plus-Cetuximab in Quadruple-WT (*KRAS, NRAS, BRAF, PIK3CA*) Metastatic Colorectal Cancer Resistant to Cetuximab or Panitumumab: NSABP FC-7, A Phase Ib Study. *Clin. Cancer Res.* **2021**, *27*, 1612–1622. [CrossRef]
78. Ferraro, E.; Drago, J.Z.; Modi, S. Implementing Antibody-Drug Conjugates (ADCs) in HER2-Positive Breast Cancer: State of the Art and Future Directions. *Breast Cancer Res.* **2021**, *23*, 84. [CrossRef]
79. Shitara, K.; Bang, Y.-J.; Iwasa, S.; Sugimoto, N.; Ryu, M.-H.; Sakai, D.; Chung, H.-C.; Kawakami, H.; Yabusaki, H.; Lee, J.; et al. Trastuzumab Deruxtecan in Previously Treated HER2-Positive Gastric Cancer. *N. Engl. J. Med.* **2020**, *382*, 2419–2430. [CrossRef]
80. Modi, S.; Saura, C.; Yamashita, T.; Park, Y.H.; Kim, S.-B.; Tamura, K.; Andre, F.; Iwata, H.; Ito, Y.; Tsurutani, J.; et al. Trastuzumab Deruxtecan in Previously Treated HER2-Positive Breast Cancer. *N. Engl. J. Med.* **2020**, *382*, 610–621. [CrossRef]
81. Cortés, J.; Kim, S.-B.; Chung, W.-P.; Im, S.-A.; Park, Y.H.; Hegg, R.; Kim, M.H.; Tseng, L.-M.; Petry, V.; Chung, C.-F.; et al. Trastuzumab Deruxtecan versus Trastuzumab Emtansine for Breast Cancer. *N. Engl. J. Med.* **2022**, *386*, 1143–1154. [CrossRef] [PubMed]
82. Doi, T.; Shitara, K.; Naito, Y.; Shimomura, A.; Fujiwara, Y.; Yonemori, K.; Shimizu, C.; Shimoi, T.; Kuboki, Y.; Matsubara, N.; et al. Safety, Pharmacokinetics, and Antitumour Activity of Trastuzumab Deruxtecan (DS-8201), a HER2-Targeting Antibody–Drug Conjugate, in Patients with Advanced Breast and Gastric or Gastro-Oesophageal Tumours: A Phase 1 Dose-Escalation Study. *Lancet Oncol.* **2017**, *18*, 1512–1522. [CrossRef]
83. Siena, S.; Raghav, K.; Masuishi, T.; Yamaguchi, K.; Nishina, T.; Elez, E.; Rodriguez, J.; Chau, I.; Di Bartolomeo, M.; Kawakami, H.; et al. 386O Exploratory Biomarker Analysis of DESTINY-CRC01, a Phase II, Multicenter, Open-Label Study of Trastuzumab Deruxtecan (T-DXd, DS-8201) in Patients (Pts) with HER2-Expressing Metastatic Colorectal Cancer (MCRC). *Ann. Oncol.* **2021**, *32*, S532. [CrossRef]
84. Karan, C.; Tan, E.; Sarfraz, H.; Knepper, T.C.; Walko, C.M.; Felder, S.; Kim, R.; Sahin, I.H. Human Epidermal Growth Factor Receptor 2–Targeting Approaches for Colorectal Cancer: Clinical Implications of Novel Treatments and Future Therapeutic Avenues. *JCO Oncol. Pract.* **2022**, OP-21. [CrossRef]
85. Flaherty, K.T.; Gray, R.; Chen, A.; Li, S.; Patton, D.; Hamilton, S.R.; Williams, P.M.; Mitchell, E.P.; Iafrate, A.J.; Sklar, J.; et al. The Molecular Analysis for Therapy Choice (NCI-MATCH) Trial: Lessons for Genomic Trial Design. *JNCI J. Natl. Cancer Inst.* **2020**, *112*, 1021–1029. [CrossRef]
86. Strickler, J.H.; Ng, K.; Cercek, A.; Fountzilas, C.; Sanchez, F.A.; Hubbard, J.M.; Wu, C.; Siena, S.; Tabernero, J.; Van Cutsem, E.; et al. MOUNTAINEER:Open-Label, Phase II Study of Tucatinib Combined with Trastuzumab for HER2-Positive Metastatic Colorectal Cancer (SGNTUC-017, Trial in Progress). *JCO* **2021**, *39*, TPS153. [CrossRef]
87. Jacobs, S.A.; Lee, J.J.; George, T.J.; Yothers, G.; Kolevska, T.; Yost, K.J.; Wade, J.L.; Buchschacher, G.L.; Stella, P.J.; Shipstone, A.; et al. NSABP FC-11: A Phase II Study of Neratinib (N) plus Trastuzumab (T) or n plus Cetuximab (C) in Patients (Pts) with "Quadruple Wild-Type (WT)" (KRAS/NRAS/BRAF/PIK3CA WT) Metastatic Colorectal Cancer (MCRC) Based on HER2 Status—Amplified (Amp), Non-Amplified (Non-Amp), WT, or Mutated (Mt). *JCO* **2019**, *37*, TPS716. [CrossRef]
88. Jacobs, S.A.; George, T.J.; Kolevska, T.; Wade, J.L.; Zera, R.; Buchschacher, G.L.; Al Baghdadi, T.; Shipstone, A.; Lin, D.; Yothers, G.; et al. NSABP FC-11: A Phase II Study of Neratinib (N) plus Trastuzumab (T) or N plus Cetuximab (C) in Patients (Pts) with "Quadruple Wild-Type" Metastatic Colorectal Cancer (MCRC) Based on HER2 Status. *JCO* **2022**, *40*, 3564. [CrossRef]
89. Raghav, K.P.S.; Yoshino, T.; Guimbaud, R.; Chau, I.; Van Den Eynde, M.; Maurel, J.; Tie, J.; Kim, T.W.; Yeh, K.-H.; Barrios, D.; et al. Trastuzumab Deruxtecan in Patients with HER2-Overexpressing Locally Advanced, Unresectable, or Metastatic Colorectal Cancer (MCRC): A Randomized, Multicenter, Phase 2 Study (DESTINY-CRC02). *JCO* **2021**, *39*, TPS3620. [CrossRef]
90. Hausman, D.F.; Hamilton, E.P.; Beeram, M.; Thimmarayappa, J.; Ng, G.; Meric-Bernstam, F. Phase 1 Study of ZW25, a Bispecific Anti-HER2 Antibody, in Patients with Advanced HER2-Expressing Cancers. *JCO* **2017**, *35*, TPS215. [CrossRef]
91. Xu, J.; Meng, Q.; Sun, H.; Zhang, X.; Yun, J.; Li, B.; Wu, S.; Li, X.; Yang, H.; Zhu, H.; et al. HER2-Specific Chimeric Antigen Receptor-T Cells for Targeted Therapy of Metastatic Colorectal Cancer. *Cell Death Dis.* **2021**, *12*, 1109. [CrossRef]
92. Nakamura, Y.; Sawada, K.; Fujii, S.; Yoshino, T. HER2-Targeted Therapy Should Be Shifted towards an Earlier Line for Patients with Anti-EGFR-Therapy Naïve, HER2-Amplified Metastatic Colorectal Cancer. *ESMO Open* **2019**, *4*, e000530. [CrossRef]

93. Raghav, K.P. Sequencing Strategies in the Management of Metastatic Colorectal Cancer with HER2 Amplification. *Clin. Adv. Hematol. Oncol.* **2022**, *20*, 86–88.
94. Baraibar, I.; Ros, J.; Mulet, N.; Salvà, F.; Argilés, G.; Martini, G.; Cuadra, J.L.; Sardo, E.; Ciardiello, D.; Tabernero, J.; et al. Incorporating Traditional and Emerging Biomarkers in the Clinical Management of Metastatic Colorectal Cancer: An Update. *Expert Rev. Mol. Diagn.* **2020**, *20*, 653–664. [CrossRef]

Review

Colorectal Cancer: The Contribution of CXCL12 and Its Receptors CXCR4 and CXCR7

Aïssata Aimée Goïta and Dominique Guenot *

INSERM U1113/Unistra, IRFAC—Interface de Recherche Fondamentale et Appliquée en Cancérologie, 67200 Strasbourg, France; goitaaissata@yahoo.fr
* Correspondence: guenot@unistra.fr; Tel.: +33-388275362

Simple Summary: Many signaling pathways are involved in cancer progression, and among these pathways, the CXCL12 axis and its two receptors CXCR4 and CXCR7 are well described for many cancers. This review presents the current knowledge on the role played by each of the actors of this axis in colorectal cancer and on its consideration in the development of new therapeutic strategies.

Abstract: Colorectal cancer is one of the most common cancers, and diagnosis at late metastatic stages is the main cause of death related to this cancer. This progression to metastasis is complex and involves different molecules such as the chemokine CXCL12 and its two receptors CXCR4 and CXCR7. The high expression of receptors in CRC is often associated with a poor prognosis and aggressiveness of the tumor. The interaction of CXCL12 and its receptors activates signaling pathways that induce chemotaxis, proliferation, migration, and cell invasion. To this end, receptor inhibitors were developed, and their use in preclinical and clinical studies is ongoing. This review provides an overview of studies involving CXCR4 and CXCR7 in CRC with an update on their targeting in anti-cancer therapies.

Keywords: colorectal cancer; chemokine; ACKR3; metastasis; microenvironment; signaling pathways; epigenetics; prognosis; therapy; resistance

Citation: Goïta, A.A.; Guenot, D. Colorectal Cancer: The Contribution of CXCL12 and Its Receptors CXCR4 and CXCR7. *Cancers* **2022**, *14*, 1810. https://doi.org/10.3390/cancers14071810

Academic Editors: David Wong and Stephane Dedieu

Received: 11 February 2022
Accepted: 29 March 2022
Published: 2 April 2022

1. Cancer Colorectal

1.1. Epidemiology

Colorectal cancer (CRC) is the third most common cancer in the world, with an annual estimate in 2020 of 1,148,515 new cases affecting both men and women. Because most patients are diagnosed at metastatic stages of the disease [1], it is the cause of 576,858 deaths per year, making it the second most deadly cancer. Similar to many cancers, the etiology of CRC involves a variety of environmental and individual risk factors, including genetic causes, chronic disease, lifestyle, and age [2].

An average risk is attributed to men and women over 50 years of age with no known predisposing factors. In absence of genetic factors or family history, environmental factors such as diet, a sedentary lifestyle, alcohol, and tobacco abuse influence the development of CRC [3,4]. The high risk, about 20% of the general population, considers family (familial adenomatous polyposis or FAP or Lynch syndrome) and personal history. Thus, this risk is two to five times higher than the average risk of developing CRC for people who have had an adenoma >1 cm, or with at least one first-degree relative who has developed colorectal adenomas or CRC [5]. The risk is also elevated for people affected by a chronic inflammatory bowel disease (IBD) such as ulcerative colitis or Crohn's disease [5].

1.2. Molecular Definition of Colorectal Cancer

CRC occurs and progresses because of an accumulation of sequential mutations and/or genomic abnormalities. Molecular biology techniques have classified CRCs into three major phenotypes according to the abnormalities identified [6]. Tumors with a Chromosome INstability phenotype (CIN) or a MicroSatellite Stability phenotype (MSS) are the most frequently observed (80–85%) [7]. This instability is a result of loss or gain of chromosomes or chromosome fragments leading to loss of tumor suppressor genes or gain of oncogenes [8]. Examples include the loss of chromosomes 5, 17 and 18 on which the APC (5q21), TP53 (17p13) and SMAD2-3 (18q21) genes are located, respectively, and the gains on chromosome 8 for the C-MYC gene on 8q24 [8–12].

The second phenotype represents tumors characterized by MicroSatellite Instability (MSI) and is present in 15–20% of CRCs [13]. This phenotype is characterized by a deficiency in the base MisMatch Repair (MMR) system during replication [14]. This defect results in an accumulation of mutations in microsatellites and repeated sequences of one to twenty nucleotides in the coding region of certain genes involved in colorectal carcinogenesis [15]. Among the genes affected are mainly MLH1 and MSH2, which are also associated with Lynch syndrome [16,17]. The MSI phenotype is also classified into MSI-High and MSI-Low.

A third phenotype has been established by observing methylation/hypermethylation of CG repeat sequences or CpG (cytosine–phosphate–guanine) islands in the promoter regions of some genes, thus repressing their transcriptional expression [18]. These repressions typically affect many tumor suppressor genes such as MLH1, CDKN2A [19]. The latter phenotype can be found associated with either of the previous two phenotypes, as 12% of CIMP cases are found associated with the MSI phenotype and 8% in MSS phenotypes [20].

In general, CRC survival depends directly on the stages. Thus, the overall survival at 5 years for all stages combined is 63%, and the chance of cure is almost total for stage 0 to II cancers (>90%) and 72% for stage III, but it drops to 14% for stage IV, the stage of dissemination to distant organs [21]. Survival also depends on CRC phenotype since patients with MSI tumors have a better prognosis than patients with MSS tumors [22,23]. Other studies have been performed using meta-analyses on transcriptomic data to propose a consensus molecular classification (CMS) of CRCs by defining four subtypes that have been associated with a prognostic value for patient survival [24]. The CMS classification has an important prognostic value and indicates that in non-metastatic CRC (stages 0 to III), the prognosis is favorable for tumors in the CMS-1 subgroup and to a lesser extent for the CMS-2 subgroup. Conversely, in a metastatic situation (stage IV), it is the CMS-1 subgroup that is linked to the worst prognosis since the overall survival of patients with a CMS-1 tumor was 14.8 months against 31.9 months for CMS-2 tumors [25].

Therapeutically, tumor resection remains the primary treatment for all stages of the disease. However, for stages with lymph node involvement or distant metastases (in the liver and lungs), chemotherapy combined or not, with targeted therapy is proposed. Note that therapy using cetuximab or panitumumab, two anti-EGFR (Epidermal Growth Factor Receptor) antibodies, is proposed only to treat CRCs with wild-type KRAS [26]. In recent years, an increasing number of studies have focused on immunotherapy. The basis of immunotherapy is to overcome the mechanisms involved in immune tolerance to tumor self-antigens and to block the immunosuppressive response that occurs in the tumor microenvironment. This process is primarily driven by the inactivation and depletion of T cells via the activation of immune checkpoint inhibitors (ICIs) on the surface of T cells, which prevent them from recognizing tumor neoantigens. Current therapies target the PD-1 receptor (Programmed cell Death protein 1) and its ligand PD-L1 (Programmed cell Death protein Ligand 1), and CTLA-4 (Cytotoxic T lymphocyte Antigen 4).

In the exploratory NICHE study (ClinicalTrials.gov: NCT03026140), patients with early-stage MSS or MSI CRC and neoadjuvant treatment with a single dose of anti-CTLA-4 (ipilimumab) and two doses of anti-PD1 (nivolumab) led to 100% and 27% response in MSI and MMS tumors, respectively [27]. Several phase II and III randomized controlled trials are underway to evaluate the efficacy of immunotherapy in metastatic CRC of both phenotypes (first-line or refractory), with/without chemotherapy [28]. Once it will be validated in larger cohorts and with at least 3 years of recurrence-free survival data, neoadjuvant immunotherapy could potentially become the standard of care for a defined group of patients.

2. CXCL12 and Its Two Receptors CXCR4 and CXCR7

Chemokines are a group of small proteins of 8 to 12 kDa from the family of chemoattractant cytokines [29,30]. To date, about fifty chemokines have been identified, and they are structurally classified into subfamilies of chemokines C, CC, CXC and CX3C according to the presence of the "chemokine domain", represented by the location of four cysteine residues conserved in the N-terminal domain necessary for the formation of disulfide bridges [29,30]. These proteins exert their function by binding to receptors with seven transmembrane domains, which are related to rhodopsin receptors [31]. Thus, there is the CR, CCR, CXCR and CX3CR receptor subfamily. Within each group, several chemokines can bind to several receptors, and inversely, one receptor can bind several chemokines. Because of this redundancy, the absence of chemokines or their receptors by gene invalidation of chemokines or their receptors does not necessarily lead to major effects, except for CXCL12 and its two receptors CXCR4 and CXCR7. Mice invalidated for each of these three proteins die during the embryonic or postnatal period, demonstrating the essential role of these proteins during embryogenesis [32–34].

The chemokine–receptor interaction was initially described to induce lymphocyte migration and recruitment [35,36]. However, it is now clear that their activity extends beyond immune cell migration. Numerous studies have documented that chemokine signaling also guides the migration of neurons, neural crest cells and germ cells during embryonic development and regulates the patterning and remodeling of the vascular system [37–39]. The chemokine–receptor is also a factor in inflammatory diseases [36,40,41], infections [30,40,41] and cancers [42–44]. One of the most studied chemokines is CXCL12, which exerts its biological functions by activating the two receptors CXCR4 and CXCR7.

3. Physiological Roles of CXCL12 and Its Two Receptors CXCR4 and CXCR7

3.1. Chemokine CXCL12

The chemokine CXCL12, also known as stromal-cell-derived factor 1 of the bone marrow (SDF-1), was originally discovered as a factor stimulating the growth of pre-B lymphocyte progenitors CD34+ (pre-B CD34+) [33,35,45,46] and is mainly responsible for the homing and maintenance of hematopoietic stem cells in the bone marrow.

CXCL12 is a homeostatic chemokine whose expression is constitutive in a wide range of tissues and organs such as bone marrow, liver, lung, heart, brain, spleen, and intestine [35,47]; however, its expression can be induced during inflammatory conditions [48,49]. It is expressed in human and mouse with a highly conserved structure, and the gene undergoes splicing that generates six isoforms (CXCL12α to ϕ), the alpha and beta forms being the predominant and ubiquitously expressed forms [50,51].

In the intestinal epithelium, CXCL12 is expressed in an increasing gradient of concentration from the base to the crypt surface [52]. This high expression at the crypt surface contributes to the constant turnover of epithelial tissue as the CXCL12-CXCR4 signaling axis stimulates intestinal epithelial cell migration and enhances the integrity of the innate barrier of the intestinal mucosal epithelium [53].

3.2. CXCR4 Receptor

The CXCR4 receptor (C-X-C motif receptor 4) was originally discovered as a co-receptor for HIV entry into lymphocytes [54]. Human (352 amino acids) and murine (359 amino acids) CXCR4 receptors share 89% homology and are ubiquitously expressed in both embryonic and adult tissue [55]. As the first receptor that can bind CXCL12, it was considered for a long time as its only receptor, since mice deficient in CXCL12 or CXCR4 have similar phenotypes with abnormalities in hematopoiesis, blood vessel formation in the gastrointestinal tract, cerebellar development, cardiac ventricular septum formation and significant embryonic lethality [32,33,56]. The monogamy relationship between CXCL12 and CXCR4 was disproved by the discovery of the orphan receptor RDC1, identified by cDNA cloning in the dog thyroid [57,58].

The interaction of CXCR4 with its ligand CXCL12 activates downstream signaling pathways, including Ras-MAPK, PI3K-AKT-mTOR, Jak2/3-STAT2/4, PLC β and γ2, NF-κB, and JNK/p38 MAPK via interaction with Gβγ subunits, while inhibiting adenylate cyclase and cAMP formation via interaction with Gαi [59]. This signaling pathways activation leads to an alteration in the expression of genes that will modulate different cellular functions such as actin polymerization, cell skeleton rearrangement or cell migration [60,61]. The physiological functions of CXCR4 are not only critical for development and homeostasis but also for the survival of cancer cells.

3.3. *CXCR7 Receptor*

More recently, another receptor CXCR7 (C-X-C motif receptor 7), renamed ACKR3 (Atypical Chemokine Receptor 3) in 2014, has been described to bind CXCL12 with 10-fold higher affinity than CXCR4 [62] and can also bind with CXCL11. A particularity is that the CXCL12-ACKR3 complex does not couple to a G protein but through activation of the β-arrestin pathway [63]. However, a study by Nguyen et al. in HEK293 cells shows that binding of CXCL12 to CXCR7 does not result in activation of signaling pathways via Gαi subunits but activates G-protein-coupled receptor kinase 2 (GRK2) via βγ subunits and phosphorylation of the receptor by recruitment of β-arrestin 2 [64]. In contrast to CXCR4, CXCR7 internalization occurs even in the absence of ligand binding and does not lead to receptor degradation [65].

Similar to CXCR4, CXCR7 can activate many intracellular signaling pathways, including AKT and MAPK pathways, via β-arrestins [63]. CXCR7, which does not activate calcium responses in the presence of CXCL12, is able to modulate CXCR4-activated calcium signaling through the formation of CXCR4/CXCR7 heterodimers [65–67], which can form in the absence of CXCL12 ligand [65]. However, contradictory data indicate that the activation of such heterodimers by CXCL12 leads either to a potentiation of the calcium response with a loss of early activation of ERK kinase [65], or conversely, to a decrease in this calcium response [67].

The physiological implications of the CXCR7 receptor have been demonstrated in CXCR7 knockout mice (CXCR7-/-), which die at birth due to abnormal heart valve development, highlighting the critical role of CXCR7 in cardiogenesis [68]. Other studies have shown that CXCR7 allows for the migration of central nervous system neurons during development by indirectly controlling their migration, through the regulation of the expression level of CXCR4, and the loss of CXCR7 function results in the production of neurons functionally deficient for both receptors [69].

The phenotypic differences described for CXCR4-/- and CXCR7-/- mice [32,34], and recent work examining the role of these receptors in zebrafish development [70,71], support the hypothesis that CXCR7 and CXCR4 have specific and distinct biological roles. In addition, several groups have established that CXCR7 acts as a "scavenger" or "decoy receptor" for extracellular CXCL12 but also for CXCL11, promoting constant cycling between the plasma membrane and the cytoplasm, and thus establishing a CXCL12 gradient. Thus, CXCR7 controls chemokine concentrations in the extracellular space, limiting signaling via other receptors [72,73]. According to a recent study, the balance between intracellular and membrane expression of CXCR7, and thus its scavenger function, is tightly regulated by CXCL12-induced phosphorylation of CXCR7 that ensures its subsequent protection against degradation [74]. This atypical function of CXCR7 is essential for the development of many organs, for the control and coordination of cell migration and positioning [75], and is not only dependent on CXCR7 but requires an intimate interaction between CXCL12, CXCR4 and CXCR7.

4. CXCL12/CXCR4/CXCR7: Pathological Role in CRC

Pathologically, chemokines and their receptors are involved in the development of infectious diseases, in particular the role of CXCR4 as a gateway for the HIV virus in CD4+ T cells [54]. However, recently, the involvement of chemokines has aroused a lot of interest in oncology [43,76,77]. The first evidence emerges from studies in breast cancer, with the involvement of CXCR4 in the control of metastatic dissemination [78].

4.1. Receptor Expression

Numerous studies have investigated the expression level of CXCR4 and CXCR7 receptors in solid cancers and in hematological cancers, given their involvement in the development of the hematopoietic system. These studies show elevated expression of one or both receptors in tumors compared to adjacent healthy tissues [79–81]. Furthermore, in CRC, Romain et al. showed that CXCR4 and CXCR7 expression increases with clinical stages [82]. Several authors have reported that receptor overexpression reflects disease progression and is therefore associated with tumor aggressiveness, decreased survival and poor prognosis [80,81,83–87]. Receptor expression is not only associated with tumor cells but also with endothelial cells of tumor microvessels [88,89] whether in colon, liver, pancreas, prostate, or lung cancers [90,91]. In contrast, Guillemot et al. described CXCR7 expression only in vessels of primary colorectal tumors and in liver and lung metastases [92].

4.2. CXCL12 Expression

In CRC, different expression patterns have been reported. The expression of CXCL12 can be increasing from healthy mucosa to adenomas and adenocarcinomas [93] or, on the contrary, decreasing [94]. Other studies show that CXCL12 expression is higher in tumors compared to healthy tissues [95], and still, others describe heterogeneous tumors since within the same cohort, some tumors express the chemokine strongly while others express it weakly or not at all [96,97]. In tumors, CXCL12 is expressed by epithelial cells but also by vascular endothelial cells [96] and stromal fibroblasts [97]. Finally, some studies observe no difference in expression between healthy mucosa and tumor [98]. In contrast, we showed that CXCL12 expression is strongly decreased in 94% of adenomas and 85% and 75% of MSI and MSS carcinomas, respectively [52]. Similarly, Wendt et al. describe an absence of CXCL12 expression in the CRC epithelium [99]. It is always difficult to understand the reasons underlying different levels of expression of a factor in the same cancer in different studies.

One of the reasons for these discrepancies could be the mixture of colon and rectal tumors in the cohorts and the fact that part of rectal tumors are either irradiated and/or chemically treated before resection, leading to changes in CXCL12 expression level [100,101]. Another reason might be the technique used. For instance, in immunohistochemistry, there may be differences in the reference of the antibody, its dilution, and in the unmasking and revelation technique (enzymatic, immunofluorescence). The heterogeneity of the tumor must also be taken into account, as analyses are usually performed on only a fragment of the tumor. Depending on how the samples are collected, it is possible to be in areas with high, low or no expression of the protein. The number of tissue sections must also be considered; with a limited number of sections, it is possible to be in a tumor area expressing or not the protein. For these reasons, it could be recommended to separately study rectal and colon tumors, as well as to combine the expression of the transcript with that of the protein since these two techniques request separate tumor samples [52].

4.3. CXCL12/CXCR4/CXCR7 Axis in Cellular Interactions

The interaction between tumor cells and the tumor microenvironment, which includes fibroblasts, immune cells and endothelial cells, participates to the development of tumor malignancy through the modulation of a wide variety of proteins in both cancer and stromal cells [102]. For example, there is bidirectional crosstalk between tumor cells and cancer-associated fibroblasts (CAFs). This crosstalk is mediated by cancer cells releasing factors that enhance the ability of fibroblasts to release various tumor-promoting chemokines, which in turn act on malignant cells to promote their proliferative, migratory and invasive properties. In this aspect, the CXCL12-CXCR4 pair plays a fundamental role in a large number of malignancies [103].

More specifically, mesenchymal stromal cells (MSCs) can be recruited to the stroma of developing tumors to enhance metastasis through their ability to secrete growth factors such as CXCL12 to promote tumor cell proliferation and tumor angiogenesis [104,105]. However, MSCs are also able to differentiate into CAFs by enhancing CXCR4 expression and activating the TGF (Tumor Growth Factor) pathway, therefore promoting growth and metastasis by secreting protumor factors [106]. Similarly, Todaro et al. showed that medium conditioned with fibroblasts isolated from primary colon tumors increases the clonogenicity of sphere-cultured colon cells and enhances the migration of CD44 stem cells isolated from CXCR4-expressing human tumors; this medium also converts non-migrating CD44v6-negative cells into migrating CD44v6-positive cells [107]. This phenotype can be mimicked by CXCL12, which also confers metastatic potential and a more aggressive phenotype to these progenitors in vivo.

MSCs present in the tumor stroma may also exert indirect pro-malignant actions by promoting tumor angiogenesis through the recruitment of endothelial progenitor cells and by facilitating the formation and maturation of the tumor vasculature [108]. These patterns are relevant in situations where the primary tumor expresses CXCL12.

In tumors not expressing CXCL12, other chemokines or growth factors (CCL4 or CCL5/CCL1 or CXCL8) released by CRCs have the ability to attract cells of the immune repertoire, angiogenic progenitors, and mesenchymal stem cells, resulting in a metastatic phenotype [109,110]; these molecules can also be produced by stromal cells [111]. In addition, MIF (Macrophage migration Inhibitory Factor) was shown to recruit MSCs to tumors by a physical interaction between MIF and CXCR4 expressing cells observed in vitro and in vivo [112]. Other factors such as fibroblast growth factor (bFGF), VEGF, platelet-derived growth factor (PDGF), insulin-like growth factor (IGF), and TGF-β have been further described for their contribution to tumor growth to MSCs [108].

Conversely, in the liver, hepatic stellate cells (HSCs) constitute the predominant population of CAFs, which are the main components of the tumor microenvironment [113]. Tumor/fibroblast interaction has been involved the progression of cancer, the CXCR4/CXCL12 chemokine axis being a main leader of malignancy [114]. In addition, HSCs, together with liver sinusoidal endothelial cells, are one of the principal sources of CXCL12 secretion in the liver, where they mediate not only the recruitment of CXCR4-expressing tumor cells, but also of CXCR4-expressing immune cells [114]. Immunohistochemical analysis of human liver show that the sinusoidal endothelial cells lining the hepatic vessel wall abundantly express the CXCL12 protein, which is therefore perfectly positioned to interact with circulating tumor cells for the formation of metastases [115].

Therefore, CXCL12 promotes communication between cancer cells and the surrounding non-neoplastic cells in the tumor microenvironment, including endothelial cells and fibroblasts, through activation of CXCR4 and CXCR7. The hypoxic tumor microenvironment can favor the upregulation of CXCR4 and CXCL12 in several cell types such as endothelial cells and cancer cells through mobilization of the hypoxia induced factor 1 (HIF-1α).

Concerning CXCR7-expressing cells, Guillemot et al. found that, in the primary CRC, the presence of the CXCR7 protein was restricted to tumor-associated endothelial cells, whereas it was absent in tumor cells [92]. However, others described CXCR7 expression in tumor-associated blood vessels but also by the malignant cells in CRC [82,116] and other cancer types [117,118].

We could speculate that CXCR7 expression in tumor vessels is a common feature of all cancers, whereas the presence of this receptor in malignant cells would be restricted to a particular type of cancer.

5. Prognostic Value of CXCL121/CXCR4/CXCR7 Axis

5.1. CXCL12 as a Prognostic Factor

Clinically, there are divergent viewpoints on the prognostic value of CXCL12 expression level. High expression is significantly associated with high tumor stage, lymphatic invasion, venous invasion, lymph node and distant metastases, and decreased survival [93,96,97]. Likewise, other studies suggest an association between CXCL12/CXCR4 expression and the induction of adenomas, carcinomas, and the development of metastases [94]. Transcriptomic analysis of a cohort of 49 CRCs and RNA-Seq data from TCGA for 375 CRCs indicate that increased CXCR4/CXCR7+CXCL12 signature expression is the only independent prognostic marker for the presence/occurrence of metastasis and decreased overall survival in both datasets [119].

In contrast, in two cohorts of 290 and 306 patients with stage III CRC, high cytoplasmic expression of CXCL12, assessed by in situ hybridization and immunohistochemistry, is associated with a better 5-year event-free survival [120]. Several studies, conversely, did not find a correlation between high CXCL12 expression levels and clinico-pathological parameters [88,121]. For example, in a meta-analysis of 25 articles published through 2017, increased transcript or protein expression of CXCL12 was not associated with TNM stage, age, gender, or diagnosis, but only with degree of tumor differentiation [121]. In another cohort of 444 CRCs with MSS phenotype [104], the two molecular subgroups C4 and C6 have higher levels of CXCL12 expression than the other four subgroups and are associated with a worse prognosis for patients [122].

Fushimi et al. showed that overexpression of CXCL12 in the CT26 syngeneic colorectal cell line in Balb/C mice resulted in an accumulation of dendritic cells and CD8+ T cells, which significantly slowed tumor growth after subcutaneous implantation [123]. A significant number of studies have shown that CD8+ T-cell infiltration of a tumor is associated with a better prognosis in CRC [124–127]. In a study of 613 stage III CRC specimens, high CD8+ T cell infiltration combined with high CXCL12 expression is associated with superior 5-year overall survival compared to patients with tumors with high CD8+ T cell expression alone [128].

In order to address these conflicting results, it can be hypothesized that during the early stages of carcinogenesis, CXCL12 production might participate in the transformation of the colonic mucosa at the beginning of the carcinogenesis process, whereas at later stages, a lower expression would avoid the recruitment of cytotoxic lymphocytes and facilitate the development of metastasis. Wendt et al. reported that tumor cells that do not express endogenous CXCL12 respond better to exogenous CXCL12 produced by distant organs, leading to metastasis in mice [99].

5.2. CXCR4 as a Prognostic Factor

Regarding the prognostic implication of CXCR4, the literature agrees that high CXCR4 expression in CRC patients is unfavorable, as it correlates with advanced tumor stage and increased risk of recurrence and distant metastasis [96,121,122]. Several meta-analyses conclude that there is a significant association between high CXCR4 expression and poor overall survival [80,121,129–131]. In a similar way, a recent study indicated a particularly poor prognosis for patients having CRCs jointly and strongly express CXCR4 and VEGF (Vascular Endothelial Growth Factor) in more than 50% of cells, and this combination of high expression is a strong and independent predictor of early distant relapse [132]. In another cohort, the CXCR4+CXCR7+CXCL12-β+ signature stratifies patients with risk of metastasis and in a TCGA dataset (n = 375), this signature predicts the presence of metastasis and overall survival [119]. Consistent with these observations, low CXCR4 expression in resections of CRC liver metastases is independently associated with a lower overall recurrence rate and thus improved disease-related survival [133].

In tumor–stromal cell interactions, CXCR4 and CXCL12 form an important signaling axis, with the interaction influencing adhesion, migration and invasion, reflecting the strong association of CXCR4 with the development of metastasis. In addition to being a prognostic biomarker, these findings are of clinical relevance given the emergence of new drugs targeting the CXCR4 receptor. In the context of a combination of molecular alterations, patients whose tumors overexpress CXCR4 and express the mutated KRAS gene have the worst prognosis [134,135].

Nevertheless, some studies describe the absence of significant correlation between CXCR4 expression and metastasis development. For example, Nagasawa et al., by multivariate regression analysis, found no significant association between CXCR4 transcript expression and a clinico-pathological factor in a cohort of 200 patients with CRC [136]. In the same way, work on a small cohort of liver metastases from CRC identified no difference in the level of CXCR4 expression between tumor tissue and adjacent healthy tissue [137]. Finally, Xu et al. observed that the level of CXCR4 expression in the center of tumors is not predictive of a poor prognosis, but instead its expression at the invasive border is [138].

5.3. CXCR4 as Stem Cell Marker

The following markers are considered markers of CRC stem cells (CSCs): CD133, CD144, CD24, CD166, CD44, CD29, ALDH1, LGR5, and emerging studies have also reported the involvement of the CXCL12/CXCR4 axis in several adult stem cells [131]. CD133 is one of the markers described to identify tumor-initiating cells (TICs) in several cancers and in colon cancer; it has been used to isolate CSCs [139,140]. However, CD133 expression is not only limited to CSCs [141,142], and in order to identify these cells more accurately, additional markers have been considered. This is the case, for example, in the study by Zhang et al. who demonstrated that CXCR4 expression could be used in addition to CD133 expression to characterize colorectal CSCs [143]. In addition, a high percentage of double-positive cells for these two markers in human CRCs positively correlates with the presence of lymph node metastases [144]. Another example has been described where Lgr5+/CXCR4+ colonic cancer cells respond to the properties of CSCs through a greater ability to form spheres in vitro, develop tumors in vivo and resist chemotherapy. Furthermore, high levels of Lgr5 and CXCR4 expression in resected human CRCs correlate with poor prognosis [145].

5.4. CXCR7 as a Prognostic Factor

Since its discovery in 2005 [62], the role of CXCR7 in the carcinogenesis of many cancers has been well documented, and it is expressed in a wide variety of cancers and tumor-associated blood vessels, including colon, liver, pancreatic, prostate and lung cancers [83,146]. There are conflicting observations regarding the role of CXCR7 in the nature of the site of metastasis development. The expression of CXCR7 and CXCL12 is higher in lung metastases than in primary CRC, whereas the expression of CXCR4 in both sites is not statistically different [147]. Previous studies have observed that CXCR4 expression is higher in liver metastases than in primary CRC tumor tissue [148,149] and suggest that the mechanism of development of liver and lung metastases is different. This agrees with the in vivo experience of Guillemot et al., who showed that CXCR7 is a key factor in the progression of CRC metastases specifically in the lungs, since systemic treatment of mice with CXCR7 antagonists reduces metastasis in the lungs but not in the liver, after intravenous injection of HT-29 or C26 cells expressing CXCR7 [92].

In the study by Yang et al., positive CXCR7 expression is associated with the presence of lymph node metastases, distant metastases and advanced TNM stage [85]. Sherif et al. significantly observed cytoplasmic expression of CXCR7 in 11% of colorectal adenomas and 72.4% of CRC [150]. In contrast to studies favoring a poor prognosis for high CXCR4 and CXCR7 expression in CRC, Kheirelseid et al. observe that patients with above-median expression have lower mortality (mean survival 46 months) than patients with below-median CXCR7 expression (mean survival 27 months). Similarly, lower expression of CXCR4/CXCR7 and CXCL12 is associated with increased tumor size, local invasion, poor differentiation, advanced lymph node stage, advanced tumor stage, and lymphovascular invasion [151].

Therefore, although the expression level of CXCL12, CXCR4 and CXCR7 has been considered a prognostic factor in several human tumor types (Table 1), none of the actors of this axis have yet been definitively validated as pro-tumoral factors. Studies suggest that the CXCL12 axis is a promoter rather than a tumor initiator.

Table 1. Clinical significance of CXCL12, CXCR4 and CXCR7 expression levels in CRC.

Authors	CXCL12		CXCR4		CXCR7		References
	Expression	Prognosis	Expression	Prognosis	Expression	Prognosis	
Romain, 2017	↓; ↓	If ↑; ↓ OS					[52]
Fan (meta-analysis), 2018	-				↑	↓ OS; ↓ DFS	[81]
Romain, 2014	-		↑		↑		[82]
Kim, 2005; 2006	-		↑	↓ OS			[84–121]
Yang, 2015	-				↑	↓ OS; ↓ DFS	[85]
Yang, 2015	-				↑	↓ OS; ↓ PFS	[85]
Xu, 2018	-		↑	↓ OS			[86]
Ingold, 2009	↑		vascular	↓ OS			[88]
Guillemot, 2012	↑		↑		↑		[92]
Greijer, 2008	↑						[93]
Frick, 2011	↓		↑				[94]
Amara, 2015	↑	↓ OS	↑	↓ OS			[95]
Yoshitake, 2008	If ↑	↓ OS	If ↑	↓ OS			[96]
Akishima-Fukasawa, 2009	If ↑	↓ OS					[97]
Mousavi, 2018	→	→	→	→			[98]
Wendt, 2006	↓						[99]
Mitchell, 2019	↑	↓ OS	↑	↓ OS	↑	↓ OS	[111]
Stanisavljević, 2016	↓; ↑	↓ DFS; ↑ DFS	↑	stage III, ↓ DFS			[112]
Li (meta-analysis), 2017	↑	↓ OS; ↓ DFS	↑	↓ OS; ↓ DFS			[113]
Lalos, 2021	↑	↑ OS					[120]
Schimanski, 2005	-		If ↑	↓ OS			[122]
Lv, 2014	-		↑	↓ OS; ↓ DFS			[123]
Li, 2015			↑	↓ OS			[122]
Jiang, 2019	-		↑	↓ OS			[125]
Ottaiano, 2020	-		↑	↓ OS			[124]
Yopp, 2012	↓; ↑	→	If ↑	↓ OS; ↓ DFS			[127]
Nagasawa, 2021	-		→	→			[128]
Jiao (CRC liver metastases), 2019	→	→	→	→			[129]
Xu, 2007	-		↑ invasive border	↓ OS			[132]
Kheirelseid, 2013	-				If ↑	↑ OS	[144]

↑: upregulated; ↓: downregulated; →: no change; -: not evaluated; DFS: disease-free survival; PFS: progression-free survival.

6. Mechanisms of Expression Regulation

6.1. Regulation of CXCL12 Expression

For both overexpression and loss of CXCL12 expression, several molecular mechanisms have been proposed. Intratumoral hypoxia has been shown to be a factor that promotes the overexpression of CXCL12 in vivo [152–154], ex vivo [155] and in vitro [153,155]. In these studies, CXCL12 expression is associated with hypoxic or HIF-1α-expressing areas and this association has been confirmed using siRNAs directed against HIF-1α [93,152–154]. In endothelial cells and under hypoxic conditions, the hypoxia-induced upregulation of CXCL12 expression was clearly attributed to the direct binding of HIF-1α to its specific binding sites on the CXCL12 promoter [153].

Moreover, several mechanisms have been proposed to explain the loss of CXCL12 expression. Hypermethylation of the CXCL12 promoter in CRCs has been proposed by Wendt et al. [99], as well as in cervical tumor lines and biopsies, observed by Yadav et al. [156]. In our study of a cohort of 444 MSS CRCs, we showed that the CpG islands of the CXCL12 promoter are methylated in only 30% of tumors [82]. In the same studies, we also reported that, in vitro, treatment of three colonic lines with histone deacetylases (HDAC) inhibitors such as butyrate and valproate restored CXCL12 expression and increased acetylation of histone H3 of the CXCL12 promoter [52]. In vivo, valproate treatment of APC mutant mice ($APC^{Min/+}$) decreases the number of intestinal tumors and slows down tumor growth in ectopic xenografts while restoring CXCL12 expression [52]. In these CRCs tissues, an analysis of the expression of 85 genes regulating epigenetic processes showed a loss of expression of a histone acetyltransferase, the protein P300/CBP-associated factor (PCAF), and forced expression of PCAF in colon cancer cell lines restored the expression of CXCL12 [52]. A further study in the blood–brain barrier, with endothelial cells lacking CXCL12 expression and pericytes expressing it, shows that the CXCL12 promoter is not methylated in both cell types; in contrast, ChIP experiments indicate reduced levels of histone acetylation of the promoter in endothelial cells compared with pericytes [157]. It is well documented that histone deacetylation of promoters generates a compact chromatin configuration that renders chromatin inaccessible to transcriptional factors and induces transcriptional repression [158]. Therefore, histone acetylation changes/defects associated with methylation of the CXCL12 promoter in some CRC subtypes would be involved in CXCL12 expression changes [52].

Functionally, cells with a decrease/loss in CXCL12 expression would be likely to be attracted to tissues expressing CXCL12, such as metastasis sites [47]. Moreover, this expression defect would contribute to the resistance to anoikis with, consequently, a migration and dissemination of tumor cells favoring the development of metastasis [159].

6.2. Regulation of CXCR4 Expression

The mechanisms leading to the high receptor expression are not clearly defined. A mode of regulation of gene expression is related to the intrinsic instability of transcripts due to the presence of adenylate-uridylate-rich element (AREs) in their 3′-UTRs, which are targeted by RNA-binding proteins for degradation, among which are those of cytokines or chemokines [160]. However, the role of these AREs may be compromised in cancer, largely due to a deficiency in proteins that promote mRNA degradation. These sequences were found in the 3′UTR of many labile mRNAs that encode proto-oncoproteins (c-myc, c-fos, c-jun) and cytokines [160]. Al-Souhibani et al. showed that in breast tumor cells, the CXCR4 gene harbors a functional ARE in its 3′-UTR portion, a potential target for the RNA degradation proteins, TTP and HuR [161]. They also demonstrate that overexpression of HuR combined with low expression of TTP results in increased stability of CXCR4 mRNA and consequently higher levels of protein that will promote detachment and migration of breast tumor cells to distant sites [161].

Tumor progression is associated with intratumor hypoxia, which leads to increased vascular density, and HIF-1α is a transcription factor that allows for adaptation of tumor cells to hypoxia [162]. In CRC, hypoxia has been shown to promote increased expression of

CXCR4 [82,163], and in human colonic cell lines, this effect is mediated by the transcription factor HIF-1α [82]. Many studies have shown that HIF-1α is expressed at elevated levels in highly aggressive CRCs [164] and plays a major role in regulating the expression of many genes involved in angiogenesis and chemotaxis via the CXCL12/CXCR4 axis [165]. Our team has demonstrated that CXCR4 expression is increased by hypoxia in human colonic cell lines [82]. Additionally, the combined expression of CXCR4, HIF-1α and VEGF is strongly correlated with the presence of lymph node metastasis and distant metastasis in human CRC [166]. Zong et al. performed a bioinformatics analysis of Gene Expression Omnibus (GEO) data of HCT-116 cells subjected to acute and chronic hypoxia to identify genes differentially expressed in normoxic and hypoxic conditions. Among these genes, they found CXCR4 whose expression is upregulated under these conditions [163]. Numerous publications have reported the direct involvement of the two hypoxia-inducible factors, HIF-1α and HIF-2α, on the increase in CXCR4 expression [167], as the promoter of the gene encoding CXCR4 contains a hypoxia response element (HRE) [168,169].

Studies have suggested that the ERK1/2 and PI3K/Akt pathways, mediators of chemokine-induced migration, are activated by hypoxia in many cell types [170]. In addition, in vitro treatment of endothelial progenitor cells with specific inhibitors of the ERK1/2 or PI3K/Akt pathway indicates that only Akt activation is required for hypoxia-induced increase in CXCR4 expression and increased chemotaxis [171]. Other studies report that activation of the PI3K/Akt pathway can increase translation of HIF-1α-coding mRNA and stabilization of the protein under hypoxic conditions [172], which would promote increased CXCR4 expression. Likewise, reduction of CXCR4 expression by siRNA in human colonic tumor cells cultured in hypoxia decreases CXCL12-induced phosphorylation and activation of Akt, while ERK activation is unchanged [82].

Epigenetic alterations have also been described to regulate CXCR4 expression in CRC. MicroRNAs (miRNAs or miRs) have emerged as critical regulators of carcinogenesis and tumor progression and are described to modulate cell proliferation, apoptosis, invasion, angiogenesis, and metastasis [173]. It is now evident that certain miRNAs may be involved in the activation of the CXCL12/CXCR4 axis and thus participate in the progression of CRC to metastasis by controlling CXCR4 expression. For example, miR-9 expression is decreased in late-stage CRC and low miR-9 levels are significantly associated with lymph node metastasis [174]. Furthermore, Kaplan–Meier analysis reveals that decreased miR-9 expression is significantly correlated with shorter median survival time, suggesting that miR-9 is an independent prognostic marker for overall survival of CRC patients and acts as a potential tumor suppressor gene [174]. In the same study, the authors show that in vitro, this miR inhibits cell migration and invasion. Moreover, a bioinformatics analysis of miR-9 target genes identified CXCR4, whose transcript has a possible miR-9 binding element in its 3′-UTR region. Using a Dual-Reporter assay, this observation was validated by demonstrating that miR-9 negatively modulates the transcriptional and protein expression of CXCR4 by binding directly to its 3′-UTR. In vivo, injections of colonic tumor cells overexpressing miR-9 into the tail vein of mice resulted in fewer lung metastases than with control cells, a similar effect obtained with cells deleted for CXCR4 expression [174].

Another study investigated the prognostic value of miR-126 expression level associated with that of CXCR4 in CRC, and an inverse correlation was observed between miR-126 and CXCR4 protein expression in CRC [175]. Furthermore, low miR-126 and high CXCR4 expression is associated with distant metastasis, TNM clinical stage, and poor survival, Multivariate analysis indicates that miR-126 is an independent prognostic factor for overall survival [176]. In another study, the same team showed that miR-126 negatively regulates CXCR4 through the AKT and ERK1/2 signaling pathways, and thus this miR functions as a tumor suppressor in CRC cells [175].

Another miR might be involved in the regulation of CXCR4, miR-622, which is underexpressed in CRC metastases and has been described as a potential tumor suppressor gene by slowing down KRAS-dependent tumor and metastasis formation in mice [177]. By overexpressing KRAS in cells, the authors restore normal tumor growth. The same authors subsequently showed that in vitro, overexpression of miR-622 in HUVEC cells inhibits capillary tube formation and that in vivo, this overexpression in HT29 cells xenografted to mice, slows tumor growth by strongly decreasing angiogenesis [177]. In parallel, analyses also showed that CXCR4 and VEGF-α expression is strongly decreased in these tumors. Similarly, as for miR-9, miR-622 has a binding site in the 3′-UTR region of the CXCR4 transcript and can therefore directly inhibit CXCR4 expression. Since VEGF is a target of CXCR4, the anti-angiogenic impact of miR-622 can be mediated by the repression of CXCR4 and consequently, reduces VEGF expression [178].

MiR-133b has also been described as a regulator of CXCR4 expression, with its expression being much lower in metastatic CRCs (stages C and D) than in early tumors (stages A and B) [179]. Using bioinformatics algorithms to identify targets of this miRNA, several targets including CXCR4 have emerged from the analysis, and a luciferase assay showed the existence of a binding site for miR-133b in the 3′-UTR of the CXCR4 transcript [179], as had also been described for miR-9 [174], miR-622 [177] or miR-139 [180].

Finally, the relative expression of the CXCR4 transcript and protein are significantly suppressed by transfecting DLD-1 and SW480 colonic cells with miR-140-3p, and this effect is reinforced by the existence of a binding site of this miR on the CXCR4 messenger [181]. In another context, the human miR-125b has been described to positively regulate Wnt/β-catenin signaling by targeting APC expression; however, in a positive feedback, the increase in miR-125b in turn leads to increased expression of CXCR4 [182].

Studies also highlight the possibility of modulation of CXCR4 expression by changes in the DNA methylation profile and/or histones of the promoter. A recent work by Stuckel et al. showed that the overexpression of CXCR4 in human CRCs is observed in both colonocytes and stromal cells. The authors found that this overexpression is not the result of hypermethylation of the CpG islands of the CXCR4 promoter but rather of an increase in 5-hydroxymethylcytosine (5hmC), a marker of active demethylation of a gene [183]; and in this case, the accumulation of 5hmC would reflect increased transcription of CXCR4 in the CRC [184]. This work complements other studies demonstrating the regulation of CXCR4 expression by epigenetic processes associated with genome methylation. Such 5hmC marks have been described for genomic and circulating DNA from different cancer types, including CRC, and were distributed in transcriptionally active regions. In addition, by using 5hmCs as biomarkers, it was possible to separate patients who developed CRC from those who did not, which also allowed the definition of marks to discriminate genomic DNA from tumor and healthy tissues [185].

In addition, studies carried out in vitro [186] and in vivo [187] show that cells lacking CXCR4 expression under stress conditions can begin to express the receptor. This is the case in Ewing's sarcoma cell lines, in which the CXCR4 promoter is highly enriched in activating but also repressive histone marks. These cells, once under stress, show a loss of the repressive mark H3K27me3 while the activating mark H3K4me3 is increased with a consequent increase in the expression of CXCR4 [187].

Demonstrating that increased CXCR4 expression facilitates the development of liver but not lung metastases, and that decreased CXCR4 also reduces liver metastasis without affecting lung metastasis, Urosevic et al. also showed that transcription factors of the ETS family mediate CXCR4 expression downstream of RAS-ERK1/2 signaling. ETV4 and ETV5 factors induce a strong expression of CXCR4 in human colorectal lines [188]. It is also known that the deregulation of genes of the HOX family of transcription factors facilitates the progression of cancers through various mechanisms [189]. In two independent CRC cohorts, high HOXB5 expression was positively correlated with the presence of lymph node metastases, distant metastases, poor tumor differentiation and advanced clinical stage [190]. Moreover, overexpression of HOXB5 in the Caco-2 colorectal cell line leads to changes in the expression of several genes involved in metastasis, including CXCR4, and the use of reporter gene systems shows that CXCR4 is a transcriptional target of HOXB5 [190].

6.3. Regulation of CXCR7 Expression

While the literature provides numerous studies regarding the mechanisms of regulation of CXCR4 expression in CRC, much less data are available for CXCR7. Evidence for an impact of hypoxia and the transcription factors HIF-1 and -2 exists in other cell types, such as in bone marrow-derived mesenchymal stem cells where the PI3K/Akt-HIF-1α-CXCR4/CXCR7 pathway is essential for cell migration, adhesion, and survival [168] or in glioblastoma cells [191]. The only study published to date in CRC is a work by our team that showed that in human colonic cells, hypoxia or HIF-1α silencing does not alter the expression level of CXCR7 [82].

Gene expression can be regulated by transcription factors such as the HIC1 (Hypermethylated in Cancer 1), which is hypermethylated in many tumors including CRC [192,193], and inactivation following hypermethylation is thought to be a tumorigenesis-triggering event [194]. The search for HIC1 consensus binding sites (HiRE) in the CXCR7 regulatory region identified 11 putative HiREs to which HIC1 could bind directly [195]. HIC1 gene knockdown, CXCR7 promoter HiRE mutations and ChiP-seq approaches demonstrate that CXCR7 is a direct target of HIC1, which acts as a direct repressor of CXCR7 expression [195]. This suggests that in tumors with loss of HIC1 expression, the subsequent increase in CXCR7 may participate in tumor progression.

Although, similar to its partner CXCR4, CXCR7 expression can be regulated by epigenetic mechanisms involving miRs in different tumor types [195–198]; to date, no data in the literature have demonstrated the involvement of a miR to regulate CXCR7 expression in CRC.

7. Implication of CXCL12/CXCR4/CXCR7 Axis in Metastatic Dissemination

For many years, the signaling mediated by this axis has been described to participate in the different aspects of tumor progression and dissemination (Figure 1). To better determine the respective involvement of each partner of this axis, different approaches have been used, such as interfering RNA, genetic editing by overexpression or loss of function, pharmacological inhibitors, neutralizing antibodies in vitro or in vivo. To understand the involvement of CXCL12 in tumor dissemination, it is necessary to separate the role of the chemokine itself from that of the CXCR4 and CXCR7 receptors, as well as the level of expression of CXCL12 in the primary tumor and the sites of metastatic implantation where it is highly expressed [52,78].

One hypothesis is that before metastasis develops, many CRC cells undergo DNA hypermethylation on the CXCL12 promoter [99], such that autocrine and paracrine CXCL12 signaling is reduced and tumor cells can migrate along a gradient that leads them to distant organs, known to highly express the chemokine [47]. This process would be initiated early in colonic carcinogenesis since CXCL12 expression is already lost at the adenoma stage [52]. This downregulation of CXCL12 expression also prevents colonic tumor cells from undergoing anoikis, a form of apoptosis when cell–cell contact is lost between epithelial cells [159].

7.1. CXCL12

The implication of CXCL12 has been demonstrated in different models. For example, in the dorsal skinfold chamber model of syngenic BALB/c mice, Kollmar et al. studied the effects of increasing concentrations of CXCL12 on tumor growth and angiogenesis induced by CT26 cell implantation [199]. In vivo, CXCL12 accelerates tumor growth through induction of angiogenesis, cell proliferation and inhibition of apoptosis [199]. In another study, the same authors used the same experimental model but performed a hepatectomy in mice [200]. It is known that liver resection is associated with liver regeneration and a local and systemic release of potent growth factors, including chemokines [201,202]. This model permits to understand the role of CXCL12 on the dissemination of CT26 cells in the tissues around the skinfold chamber. The authors report that neutralization of CXCL12 with an antibody promotes tumor extension to nearby tissues, accelerates angiogenesis and neovascularization, increases VEGF expression, microvascular permeability and increases CXCR7 expression [199]. Moreover, neovascularization and tumor growth are reduced after CXCR4 neutralizing treatment. Therefore, in the absence of CXCL12, signaling by CXCR4 is interrupted and an alternative pathway must be considered that would be carried by CXCR7. CXCR7 has been described to increase the production of VEGF [182], which would be the trigger of the pro-angiogenic effect observed after neutralization of CXCL12 [199].

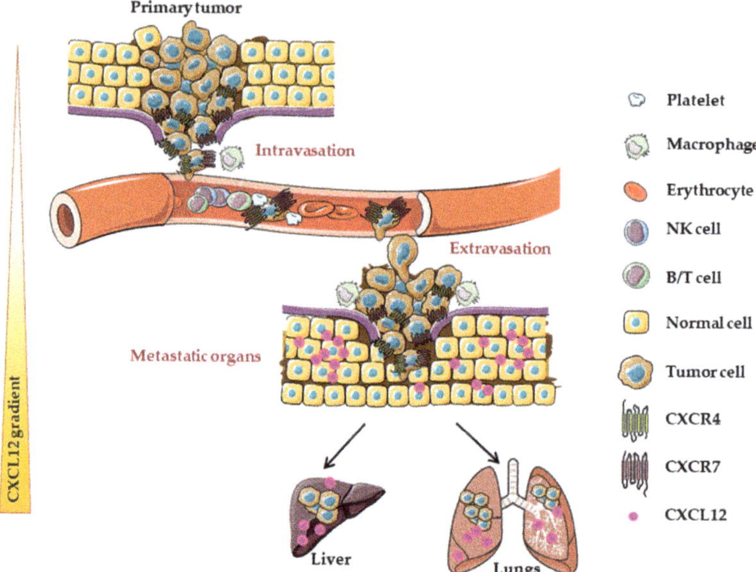

Figure 1. Process of metastatic dissemination in CRC. Tumor cells that no longer express CXCL12 will migrate to organs of metastasis via blood circulation. The expression of the CXCR4 and CXCR7 receptors allows for the intra- and extravasation of the cells through the vessels and then the implantation in the liver and the lungs where CXCL12 is strongly expressed. During intravasation of tumor cells into circulation, macrophages localized to perivascular areas within tumors help tumor cells traverse vessel barriers. In the circulation, platelets support tumor cell survival by protecting them from cytotoxic immune cell recognition.

Conversely, CXCL12 has also been described as an anti-tumor molecule in pancreatic cancer [203]. In CRC, Wendt et al. described a strong decrease in CXCL12 expression [99] and when colon cells treated with a demethylation agent to restore CXCL12, are injected into the tail vein of mice, metastatic tumor formation is greatly reduced as compared to cells lacking CXCL12. A similar situation has been observed in APC mutant mice that spontaneously develop CRCs. When these mice are treated with a histone deacetylase inhibitor, such as valproate, there is a re-expression of CXCL12 and a decrease in the number of tumors [52].

7.2. CXCL12 and CXCR4

The contribution of CXCR4 in tumor cell migration involves several cellular aspects that all converge toward the facilitation of cell migration and invasion. For example, overexpression of CXCR4 promotes the formation of pseudopodia through actin polymerization [78] and reorganization of the cytoskeleton [204]. Other processes are induced under hypoxic conditions such as epithelial–mesenchymal transition (EMT) and overexpression of $\alpha2$, $\alpha5$ and $\beta1$ integrins [205].

Migration and invasion processes also involve proteolytic activities induced by the important secretion of gelatinases such as MMP-2, MMP-9 or matrylisin-1 (MMP-7) [204,206,207].

One study showed that the ability of CXCL12-mediated cell migration and invasion is highly dependent on MMP-9 secretion and activity via Akt and ERK/MAPK signaling [204] and on β-catenin translocation in the nucleus, suggesting the interaction of the CXCL12/CXCR4 axis with the Wnt-β-catenin signaling [206].

At the signaling level, CXCR4 regulates the migratory and invasive ability of cells via the MAPK/ERK1/2 and PI3K/Akt signaling pathways activated by CXCL12-CXCR4 binding [82,204,208,209]. Similarly, CXCL12 binding to CXCR4 activates pro-metastatic signaling by decreasing E-cadherin expression but inducing ICAM-1 expression (InterCellular Adhesion Molecule) [210]; however, it has been shown that high levels of ICAM-1 in CRCs are associated with decreased tumor progression and liver metastasis [211]. Activation of the CXCR4/CXCL12 axis also involves the TGF-β pathway to promote invasion, angiogenesis, and promotion of distant metastasis by promoting differentiation of hepatic stellate cells into CAFs [212]. Furthermore, CXCR4 knockdown strongly reduces in vivo tumor growth associated with the reduction of tumor capillaries and intra-tumoral blood flow without affecting VEGF expression [213]. Moreover, in HUVEC cells, CXCR4 knockdown strongly inhibits angiogenesis after stimulation with CXCL12 ligand, by reducing EGFR, VEGF, and MMP-2, affecting MAPK/ERK, PI3K/Akt and Wnt/β-catenin pathways [207]. Another study showed that CXCL12 could stimulate the metastatic behavior of colonic cells expressing CXCR4 by increasing cell proliferation and adhesion to fibronectin [214]. These studies agree with those of Gouveia-Fernandes et al. who show that overexpression of fibronectin confers invasive and disseminative potential to HCT15 cells by promoting activation of the CXCL12/CXCR4 axis through modulation of α3 and β3 integrin expression [215].

Furthermore, Zeelenberg et al. demonstrate that CXCR4 expression is regulated positively by the tumor microenvironment, but it appears that CXCR4 is not required for tumor cell entry into metastatic sites, but rather for the establishment of micro-metastases [216]. Mice in which murine CT-26 colonic cells deficient in CXCR4 by retention of the receptor at the level of the endoplasmic reticulum, were injected into the spleen or the tail vein, indicate that CXCR4 would not play a role in invasion, but rather in the survival of the cells to form micro-metastases without impacting their proliferation [216].

In the same idea, Matsusue et al. showed that HCT116 cells stimulated by CXCL12 become resistant to apoptosis, and the use of AMD3100 reduces this CXCL12-dependent anti-apoptotic ability [217]. In vivo, these cells metastasize because CXCR4-positive cancer cells selectively survive by an anti-apoptotic effect and by the secretion of CXCL12 by stellate liver cells. Thus, these liver cells, under the action of TFG-β1 secreted by the tumor cells, will differentiate into CAFs [217]. The involvement of CXCR4 in the metastatic process could also be potentiated by another receptor, CXCR3 [218], which is activated by the chemokine CXCL10 by inducing cytoskeletal rearrangements, migration, invasion, expression of the matrix metalloproteinase MMP-2/9, and cell survival through the activation of ERK1/2, Akt and protein kinase G [219]. These ideas are supported by the findings of Tan et al. who propose that overexpression of CXCR4 by tumor cells in the hepatic metastatic microenvironment stimulates the production of CXCL12 by stellate cells, which through a paracrine action, stimulates the secretion of TGF-β1 by tumor cells, necessary for the differentiation of hepatic stellate cells into CAFs [212]. These studies suggest that modulation of the CXCL12-CXCR4 interaction can have a strong impact on tumor dissemination to target organs.

7.3. CXCL12 and CXCR7

A growing number of studies are emerging to understand the mechanisms by which CXCR7 participates in the growth and progression of colon cancer to organs of metastasis. Thus, the involvement of CXCR7 in colorectal tumorigenesis has been discussed in several models and is through the regulation of proliferation, survival, migration, invasion, angiogenesis, tumor growth and metastatic dissemination [89,218]. CXCR7 gene silencing represses cell proliferation and invasion and induces apoptosis with decreased expression of p-ERK, β-arrestin, PCNA and MMP-2 but with increased expression of caspase-3 [220].

Subcutaneous tumors induced by SW480 cells deleted for CXCR7 expression are significantly smaller than those in control groups [220]. CXCR7, but not CXCR4, expression can be increased by lipopolysaccharide treatment in cells expressing both TLR4 and the MD2 coreceptor [221]; additionally, in patients, high expression of TLR4, MD2 and CXCR7 is associated with tumor cell infiltration in lymph nodes and distant metastases [221].

The involvement of CXCR7 has also been described in transendothelial migration, and CXCR7 expression is found both in tumor cells and in tumor-associated vessels. However, using the endothelial cell line HUVEC, it was shown that this expression by vessels is not necessary for CXCL12-mediated transendothelial migration, and this process requires CXCR7 expressed by tumor cells, without involving CXCR4 [222]. CXCR7 appears to have angiogenic activity since its overexpression in colonic cells cultured with HUVEC cells promotes the formation of capillary tubes, and the stable extinction of CXCR7 in colonic cells prevents this tube formation [89]. CXCR7 exhibits low levels of expression in normal mature vascular endothelial cells but is highly expressed in endothelial cells of neovascularized tumors [223]. This effect might be a consequence of the stimulation by CXCR7 of VEGF production by endothelial cells via activation of the ERK and AKT pathways [89].

An in vivo study with transgenic mice overexpressing CXCR7 in the intestine showed that this overexpression exacerbates DSS treatment-induced inflammation by causing extensive infiltration of myeloid suppressor cells, M2-like macrophages, and Tregs in the colon, associated with elevated amounts of the proinflammatory cytokines TNF-α, IL-6, and c-Myc but decreased numbers of CD8+ T cells [205]. This CXCR7 overexpression also increases tumorigenesis in $APC^{Min/+}$ mice and these effects are amplified when mice overexpress the CXCR4/CXCR7 heterodimer [224].

Although the implication of CXCR7 in the metastatic process is well demonstrated, there are still some questions about its capacity to direct the dissemination more specifically in an organ. Guillemot et al. have shown that in mice, systemic treatment with specific CXCR7 antagonists prevents the dissemination of cells in the lungs but not in the liver [92]. Concomitantly, higher expression of CXCL12 and CXCL11 was found in tumor areas in the lung compared with the liver, indicating that distinct pathways regulate the mechanism of pulmonary and hepatic metastatic spread. In another study in human CRCs, CXCR7 expression was also found to be higher in lung metastases than in the primary tumor [138].

7.4. CXCL12, CXCR4 and CXCR7

For a possible mechanism of action of CXCL12 in promoting metastasis, numerous works have highlighted the role of matrix metalloproteinases (MMPs), proteinases responsible for degradation and remodeling of the extracellular matrix (ECM). Thus, the persistent localization of these enzymes at the interface between migrating CRC cells and the surrounding stroma has been demonstrated, supporting a role for MMPs in CRC invasion and metastasis [225]. This study shows that none of the three CRC cell lines tested express MMP-2 or MMP-9. In contrast, subcutaneous tumors induced by transplantation of these cells express limited amounts of MMP-2 and MMP-9 while caecal tumors express them in large amounts [225] showing the role of murine stromal cells in the production of these proteinases.

Similarly, in myeloma cells, CXCL12 induces the expression of matrix metalloproteinases (MMPs) such as MMP-9, membrane MMPs such as MT1-MMP, represses the expression of inhibitors such as TIMP-1, promoting cell invasion in vitro [226]. However, these observations do not support a possible role for CXCL12 in the invasiveness of colonic tumor cells that no longer express CXCL12 [53,123]. In addition, it is possible to speculate that in vivo, colonic tumor cells acquire the ability to produce MMP regulatory factors other than CXCL12, such as mutations in tumor suppressor genes or proto-oncogenes, changes in the microenvironment, extracellular matrix composition, tissue oxygenation and inflammation [227].

8. Targeting of the CXCL12/CXCR4/CXCR7 Axis in CRC

8.1. Preclinical Studies

A recent report found that CXCL12 and relative expression of the CXCL12-CXCR4 axis are independent prognostic factors for 5-year relapse-free survival [120]. Multiple preclinical studies have evaluated the efficacy of many agents; however, only a few drugs targeting this axis have been approved for clinical use. These agents include anti-CXCR4 neutralizing antibodies, interfering RNAs or antagonist molecules targeting CXCR4 or CXCR7, or small peptides specifically blocking CXCR4 (Table 2).

In the clinic, the molecules used mainly target CXCR4, and a molecule more specifically targets CXCL12 [228]. Many preclinical studies targeting the CXCL12/ CXCR4/CXCR7 axis have been published, but few have focused on CRC.

8.2. AMD3100

The best-known molecule to inhibit the biological effect of CXCR4 is the molecule commonly known as AMD3100 or plerixafor (Mozobil). The Food and Drug Administration (FDA) approved AMD3100 in 2008 for use in the mobilization of hematopoietic stem cells for transplantation in patients with non-Hodgkin's lymphoma [229,230]. AMD3100 is a specific antagonist of CXCR4 of the bicyclam family [231]. This drug acts as an antagonist by binding to one glutamine and two aspartate residues in the CXCR4 receptor, preventing the conformational change necessary to activate intracellular kinases [232]. It is the most frequently used drug in clinical trials targeting the CXCL12-CXCR4/CXCR7 axis and has been described in several studies in hematological, breast, pancreatic, lung cancer [231,233].

In an orthotopic model of liver metastasis using the murine colonic line C26, blocking CXCR4 with AMD3100 reduces the number and size of liver metastatic sites [234]. Immunohistochemical analyses revealed a significant decrease in the expression of α-SMA, a marker for hepatic stellate cells, in the liver foci of AMD3100-treated mice compared with control mice [234]. The promotion of VEGF production by stellate cells has been demonstrated in liver metastases in vivo [102], facilitating the recruitment of sinusoidal endothelial cells and the transition from avascular to vascular stage in these metastatic sites. In this context, a decrease in stellate cells induced by AMD3100 could therefore alter the angiogenic response and the blood supply of oxygen and nutrients to the tumor. However, AMD3100 has been described to also interact with CXCR7 but as an agonist [235]. AMD3100 alone can induce β-arrestin recruitment to CXCR7. Moreover, and in contrast to the antagonistic effect observed for CXCR4, AMD3100 increases ^{125}I-CXCL12 binding to HEK293 cells expressing CXCR7 and CXCL12-facilitated recruitment of β-arrestin to CXCR7, recruitment that is also possible in the absence of CXCL12, albeit at relatively high concentrations ($\geq$10 mM) [235]. To date and to the best of our knowledge, no molecular mechanism has been proposed to justify the agonistic property of AMD3100 on CXCR7.

Data about mode of action of AMD100 are limited in CRC. An in vitro study in the SW480 colon cell line demonstrated that the anti-tumor effect of AMD3100 was mediated through the reduction of VEGF and MMP-9 expression, but not MMP-2 [236]. Further evidence comes from a study on mammary stem cells that identified among the proteins showing CXCL12-induced phosphorylation, up to 22% are involved in signaling pathways related to cell adhesion and migration, actin and microtubule association in cytoskeletal remodeling. These mechanisms are known to support the involvement of CXCL12/CXCR4 in the metastatic process. By exposing cells to AMD3100, the phosphorylation of key proteins in these signaling pathways is blocked, such as the catalytic subunit of serine/threonine-protein phosphatase PP1-gamma (PPPC1) [237]. Conversely, in prostate cancer where CXCR4 strongly regulates the development of metastasis, treatment of prostate cells with dihydrotestosterone increased the expression of the androgen receptor, CXCR4, PI3K and AKT phosphorylation as well as EMT and downstream cell cycle control genes. Conversely, treatment with resveratrol and AMD3100 reversed all these changes associated with increased expression of apoptosis-related genes [238].

Taken together, these observations suggest that AMD3100 is an allosteric agonist to CXCR7. Therefore, while this antagonist has proved effective in controlling tumor progression in various cancers, these observations suggest caution in its use to understand the respective roles of CXCR4 and CXCR7 as mediators of the biological effects of CXCL12.

8.3. LY2510924

In addition to AMD3100, novel CXCR4 inhibitors have been identified, including the cyclic peptide LY2510924. From X-ray crystal structures of CXCR4 [239], LY2510924 is suggested to occupy a binding pocket and possess ligand–receptor interactions with CXCR4 residues such as Asp187, Arg188, Gln200, His113, and Tyr190 [240]. The antagonistic effect of this new molecule was confirmed in an SDF-1-induced GTP (guanine-triphosphate)-binding assay where LY2510924 completely inhibits SDF-1-mediated binding to GTPγS35 with a Kb of 0.38 nmol/L. Furthermore, LY2510924 was found to inhibit CXCL12-mediated chemotaxis by blocking SDF-1-stimulated phosphorylation of ERK and Akt in a concentration-dependent manner [240].

In vivo, its antagonistic effect has been proven by the dose-dependent decrease in tumor growth in colonic, pulmonary, renal or non-Hodgkin's lymphoma xenografts and on the formation of mammary tumor metastases after intravenous injection of mammary tumor cells [240]. In the latter model, pre-treatment of mice with LY2510924 strongly decreases lung colonization and prevents the proliferation of implanted cells.

In a separate study, the inhibitory effects of LY2510924 were evaluated in orthotopic xenografts of three human colonic lines in the rectal mucosa. While treatment with LY2510924 strongly reduces tumor size, it does not affect the size of metastases and only when combined with 5-FU reduced metastasis [241]. A possible explanation for the lack of effect of LY2510924 on metastasis is the presence of a population of TICs, which is the source of many therapeutic resistances, or else this molecule is only fully effective when combined with other conventional therapies.

8.4. PepR

Peptide R (PepR) is a new CXCR4 antagonist peptide, effective mainly in combination with conventional chemotherapies such as 5-FU and oxaliplatin. In subcutaneous xenografts of HCT116 or HT-29 cells, mice treatment with PepR potentiates the inhibitory effect of chemotherapy on the proliferation and activation of EMT [242]. As a proposed mechanism, an analysis of TCGA dataset RNA-Seq indicates that adding the PepR compound to chemotherapy reverted the increased expression of the mesenchymal markers as well as PD-L1, all markers being induced by chemotherapy alone [242]. This suggests a role for CXCR4 in controlling EMT marker expression. In addition, treatment of colon cells with chemotherapy/radiochemotherapy induced a population of CD133+CXCR4+ cells, supposed to be stem-resistant cancer cells, while adding Pep R reduced this population. In a previous study, the same authors showed that this novel antagonist enhances the efficacy of anti-PD-1 therapy in a mouse model of colon cancer induced with MC38 cells [243]. The increased efficacy of anti-PD-1 therapy by PeR results from changes in the microenvironment by recruiting Granzyme B-positive cells and decreasing Tregs cells. Thus, PeR treatment makes the microenvironment more immunosensitive to anti-PD-1 therapy [243]. Other studies have shown that Pep R reduced the expression of CXCL12 and PD-L1, probably by inhibiting the immunosuppressive effect of the microenvironment and preventing the recruitment of stromal cells (CAFs, Tumor Associated Macrophages (TAM), Myeloid-Derived Suppressor Cells (MDSCs)) responsible for the exclusion of cytotoxic T lymphocytes approximately tumor cells [244,245].

8.5. MSX-122

Unlike other inhibitors that prevent the binding of CXCL12 to its receptor, this molecule MSX-122, when binding to CXCR4 could interfere with the "lock and key" mechanism between CXCR4 and CXCL12, and modulates functional signaling such as reductions in pErbB2, pAKT, pERK and increase in cAMP production, without displacing CXCL12 from the receptor [246].

The efficacy of this CXCR4 antagonist was evaluated in $APC^{Min/+}$ mice exposed to azoxymethane (AOM) and treated with MSX-122 [247]. $APC^{Min/+}$ mice are known to develop mainly small bowel tumors, while when exposed to AOM, they develop cancers in the colon [248]. As expected, AOM induced colonic tumors in these mice, whereas treatment with MSX-122 significantly reduced the incidence of colonic tumors and tumor volume through decreased cell proliferation as assessed by Ki-67 labeling. The authors propose that MSX-122, having been well tolerated in a phase Ib clinical trial, may serve as a chemopreventive agent in individuals at increased risk of developing CRC.

8.6. CCX754 and CCX771

In contrast to CXCR4, studies describing the use of CXCR7 antagonists in CRC are uncommon despite the development of several of its inhibitors by ChemoCentryx (CCX226, CCX733, CCX754, CCX771 and CCX773). These molecules have been described as ligands that do not induce phosphorylation of AKT or ERK. CCX754 and CCX771, two of these antagonists, were tested in mouse injected with human or mouse lung carcinoma cells [249] or in models of lung metastasis by injection of murine C26 and human HT-29 colon cancer cells [92]. Systemic treatment with CCX754 or CCX771 antagonist strongly reduced tumor expansion in the lungs of mice injected with these cells but not the expansion of metastases into the liver [92]. However, CCX771 has also been described as an agonist that recruits β-arrestin-2 to CXCR7 and blocks trans-endothelial migration of human cancer cells [250]. A theory of receptor desensitization has been proposed to explain the agonist/antagonistic effect of the molecule. CCX771 would not stimulate chemotactic activity but rather induce internalization of CXCR7 from the cell surface. This has been observed for a CCR5-targeting molecule in search for anti-HIV-1 agent [251] or described for a CCL7 agonist non-glycosaminoglycans (GAGs) binding and evaluated for its anti-inflammatory effect [252].

The lack of data on the efficacy of these antagonists can be explained by the following studies showing that those molecules initially designed to inhibit CXCR7 activation also act as agonists in different pathologies [253]. Likewise, some CXCR4 receptor antagonists are agonists for the CXCR7 receptor, such as the cyclic peptide TC14012 [254].

There may be several reasons why molecules presented as antagonists/agonists, exert inverse physiological activity. One possibility is that the mode of action of the molecules is more related to CXCL12-mediated effects than to CXCR7-mediated effects. For example, CXCR7 antagonists prevent CXCL12 internalization leading to increased extracellular CXCL12 concentrations. They may therefore generate pathophysiological effects such as those of CXCR7 agonists, as described in experimental autoimmune encephalomyelitis [255].

CCX771 alone induced a concentration-dependent association of CXCR7 with β-arrestin2 CCX771 was substantially more potent than its natural protein ligand CXCL12 in triggering β-arrestin2 association

8.7. Chalcones

In 2008, a screening of 3200 molecules from a medicinal library identified a new class of molecules that bind to the chemokine CXCL12 and act as neutral inhibitor of its biological activity in a way similar to neutralizing antibodies. The most potent compound which belongs to the chalcone family and named chalcone 4, has been shown to bind the chemokine CXCL12 with high affinity thus preventing the binding of the chemokine to both CXCR4 and CXCR7, and thus blocking the downstream pathways [256]. Later on, a study from our team demonstrated that chalcone 4 was able to reduce colorectal cell migration and when combined to irinotecan, further increased the inhibition [82]. However, this compound would need further characterization, yet no data has reported its capacity to block the dissemination process in vivo. CXCL12 is efficient in solubilizing chalcone molecules with a stoichiometry 3:1 for chalcone 4: CXCL12 and that chalcone 4 binds to one high affinity site and two low affinity sites in CXCL12 [256].

8.8. NOX-A12

Noxxon Pharma has developed a molecule called NOX-A12 or olaptesed pegol [257]. This molecule is an RNA aptamer (or spiegelmer), which acts by binding to CXCL12, preventing it from linking and activating its two receptors. It binds to CXCL12 with high affinity and specificity across various species such as humans, mice, and rats. NOX-A12 has been shown to bind directly to and inhibit CXCL12 but also detach the cell-surface bound CXCL12, leading to abrogation of the CXCL12 gradient [258]. In tissues, stromal cells secrete and present CXCL12 on the surface, via GAGs, and NOX-A12 can compete with GAGs to bind CXCL12, leading to the release of CXCL12 from the cell surface and thus neutralize the chemokine [258]. A study by Zboralski et al. showed in vitro that in tumor and stromal cell spheroids that mimic a solid tumor with a CXCL12-rich microenvironment, NOX-A12 promotes spheroid infiltration by T and NK cells in a dose-dependent manner. The combination of NOX-A12 and PD-1 checkpoint inhibitor acts synergistically to facilitate T cell infiltration into spheroids. These observations were validated in vivo in a mouse model of syngeneic CRC in which treatment with NOX-A12 improved the response to anti-PD-1 therapy to reduce tumor size [259]. A Phase I/II clinical trial is underway to study the effects of the NOX-A12 and anti-PD-1 combination in patients with advanced CRC or pancreatic carcinoma (NCT03168139).

Table 2. Chemical modulators of CXCR4 and CXCR7/ACKR3 activation.

Inhibitor/Antagonist	Formula	IC50	Target	References
AMD3100	1-[[4-(1,4,8,11 tetrazacyclotetradec-1-ylmethyl)phenyl]methyl]-1,4,8,11-tetrazacyclotetradecane	37.5 nM	CXCR4	[260]
LY2510924	N(1)Phe-D-Tyr-Lys(iPr)-D-Arg-2Nal-Gly-D-Glu(1)-Lys(iPr)-NH2	0.079 nM	CXCR4	[240]
PepR	(H-Arg-Ala-[Cys-Arg-Phe-Phe-Cys]-CO2H)	nd	CXCR4	[242,243]
MSX-122	N,N-9-(1,4-phenylenebis(methylene))dipyrimidin-2-amine	10 nM	CXCR4	[260]
CCX754	nd	5 nM	CXCR7	[249]
CCX771	nd	4.1 nM	CXCR7	[260]
Chalcone 4	((E)-1-(4′-chlorophenyl)-3-(4-hydroxy-3-metoxyphenyl) prop-2-en-1-one)	150 nM	CXCL12	[256]
NOX-A12	nd	5–200 nM	CXCL12	[257]

nd: not determined; IC50: 50% inhibitory concentration.

9. Clinical Trials

While many cases of CRC are diagnosed at an early stage and are treated with curative surgery, many patients develop synchronous or metachronous metastatic disease with a five-year survival rate of roughly 13% [261]. The routine treatment of metastatic CRC is based on the combination of different treatment schedules such as Folfiri/Folfox/Folfoxiri or Capiri/Capox, which resulted in a survival of about 18 months. However, more recently, the approval of targeted therapies with EGFR or VEGF antibodies has importantly improved the overall survival, approaching 30 months in clinical trials [262], but the relative unavailability of biomarkers in metastatic CRC has slowed the progress in tumor curacy. Because of the bad prognostic value of CXCR4 overexpression across different tumors, CXCR4-inhibition-based therapies have been therapeutically evaluated in hematologic and solid malignancies, either as monotherapy or in combination with chemotherapies or immunotherapies (for review, see [263–265]). Among the drugs tested in clinical trials, CXCR4 small molecule antagonists, fully humanized anti-CXCR4 antibodies and CXCR4 or CXCL12 peptide inhibitors represent the most advanced programs of CXCR4 inhibition in solid tumors. Galsky et al. published the first in-human phase I study in patients with advanced or metastatic CRC that explored the safety and tolerability of LY25110924 among other solid tumors [266]. To date, AMD3100 is the only approved CXCR4 inhibitor drug [231], while multiple antagonists are in different stages of development. Presently, the clinical trials are mainly ongoing phase I/II trials. They mainly concern the CXCR4 peptide inhibitor LY2510924 (NCT02737072), the anti-CXCR4 antibody LY2624587 (NCT01139788), the small molecule inhibitors Plerixafor (NCT20179970, NCT03277209), MSX-122 (NCT00591682, suspended), USL311 (NCT02765165); however, for a number of these trials, the cancer type is not always indicated, as it only specified that the targeted diseases are solid tumors.

The only available data from completed phase I/II trials evaluated the application of the NOX-A12 molecule (OPERA trial, NCT03168139), first as monotherapy, and then continued with pembrolizumab in patients with advanced stage pretreated metastatic colorectal or pancreatic cancer. The NOX-A12 was well tolerated and allowed for a disease control rate of 25%, and an overall survival close to 12 months could be achieved [267]. This effect was mediated by a transformation of the tumor immune microenvironment with the expression of a specific cytokine signature consisting of IL-2, IL-16 and IFN-γ as an indicator of activation in tumor tissue. Following this success, a phase II trial is currently underway to evaluate the effect of the combination of NOX-A12 and pembrolizumab in glioblastoma and pancreatic cancer.

Conversely, treating patients with CRC for seven days with continuous infusion of the CXCR4 inhibitor AMD3100/Plerixafor induces an integrated immune response with enhanced intratumoral immune B and T cell responses as observed in paired biopsies of metastatic lesions (NCT02179970) [268], an immune response that is predictive of a clinical response to T cell checkpoint inhibition. For other trials, no results are currently available, due to the required time to exploit the data.

Although several CXCR7 antagonists (CCX771, CCX662, CCX733, CCX754, and CCX777) have been investigated in preclinical models [253,269], to date, CXCR7 modulators have not been clinically investigated.

10. Resistance to Treatment

It is well established today that the increase in cancer mortality is partly due to the resistance of tumor cells to numerous anti-cancer treatments. Thus, understanding the mechanisms at the origin of this tumor resistance would lead to the development of new approaches to maximize the effectiveness of treatments. Two types of resistance are described in cancerology: innate resistance, which is a consequence of the high molecular heterogeneity of cells within a tumor, and resistance acquired during treatment [270]. In a tumor, inhibition of apoptotic signals promoting proliferation, DNA repair, genomic amplification, a defect in drug metabolism, or epigenetic modifications can generate acquired resistance [271,272]. Given the relevance of the CXCL12/CXCR4/CXCR7 axis in the development and progression of CRC, several studies have investigated its role in resistance to anti-cancer therapies.

In tumors, it is a common knowledge that a small population of cells known as Tumor Initiating Cells with stem cell characteristics are responsible for many tumor recurrences [273]. A subpopulation of tumor cells positive for TIC marker CD133 has been isolated from patient CRCs or colonic lines, and these cells are more tumorigenic than cells not sorted on marker expression CD133 [274]. Thus, the CD133+ cell population isolated and enriched for CXCR4 expression shows significant tumorigenicity with an increased in vitro cell proliferation, tumor size and angiogenesis in vivo [274,275]. By analyzing the secretion of soluble factors by the HK stromal ganglion cells, the authors found a significant expression of CXCL12, which by a paracrine action, promotes tumor vascular development and protects the cells from the therapeutic agents 5-FU and oxaliplatin [274,275].

Another study described co-expression of CXCR4 and Lgr5, a colonic stem cell marker receptor, in patients with stage IV CR [145]. In vitro, Caco-2 and HT-29 cells isolated by flow cytometry and strongly expressing CXCR4 and Lgr5 promote sphere formation and increase cell viability when treated with cytotoxic agents. Similarly in vivo, the concomitant expression of CXCR4/Lgr5 in cells implanted subcutaneously in mice confers a more important potential to develop a tumor mass [145].

In another study, combined treatment with endostar or endostatin (an angiogenesis inhibitor) and oxaliplatin synergistically decreased the proliferation, adhesion, and invasion of Matrigel [276]. This synergy is a consequence of decreased expression of CXCR4, as well as those of the hypoxic factors HIF-1α and HIF-2α [276]. The authors showed that only the accumulation of HIF-2α is responsible for this cell resistance to oxaliplatin, and the combination of endostar with oxaliplatin overcomes this resistance by making the cells more sensitive. These data suggest that CXCR4 could be used as a marker to identify tumor stem cell populations responsible for the resistance and recurrence seen in cancers.

MiRNAs, which are also involved in cancer pathology, are either tumor suppressors or oncomiRs, largely involved in proliferation, invasion, and resistance to treatment. Some miRNAs are targets of the CXCL12/CXCR4/CXCR7 axis, and one study investigated the role of miR-125b in 5-FU resistance of cells expressing CXCR4 [182]. Expression of miR-125b, increased by the treatment of HCT116 cells with CXCL12, accelerates invasive ability and promotes EMT, which in turn increases CXCR4 expression, forming a reciprocal positive feedback loop between CXCR4 and miR-125b. Upregulation of miR-125b also activates Wnt/β-catenin signaling and the APC gene and contributes to 5-FU resistance by enhancing cellular autophagy [182].

Contrary to these studies, Heckmann et al. described that overexpression of CXCR4 in the SW480 colonic line and strong endogenous expression in HT-29 cells is associated with a higher sensitivity to treatments such as 5-FU, oxaliplatin or irinotecan. This chemosensitivity, assessed by a decrease in cell survival, cytotoxicity, and apoptosis, is further increased when one of these molecules is combined with plerixafor [277]. In this case, contrary to CXCR4, it would rather be the overexpression of CXCR7 that results in the resistance [278].

11. Conclusions

In CRC, activation of the CXCL12/CXCR4/CXCR7 axis leads to progression and development of metastases with an unfavorable disease outcome and poor patient survival. Disruption of the CXCL12-CXCR4/CXCR7 axis remains an interesting target for pharmacological treatment (Figure 2). CXCR4 and CXCR7 antagonists are being tested in several preclinical and clinical trials for the treatment of CRC, and other gastrointestinal cancers, but with limited success and the development of combined antagonists, targeting both receptors are still lacking. Therefore, tumor immunotherapy entered a phase of rapid development in cancer treatments, but there are too many patients resistant to this therapy. Furthermore, the use of inhibitors targeting the oncogenic CXCL12 axis in combination with current immunotherapies should be considered and may provide hope for improving cancer treatments.

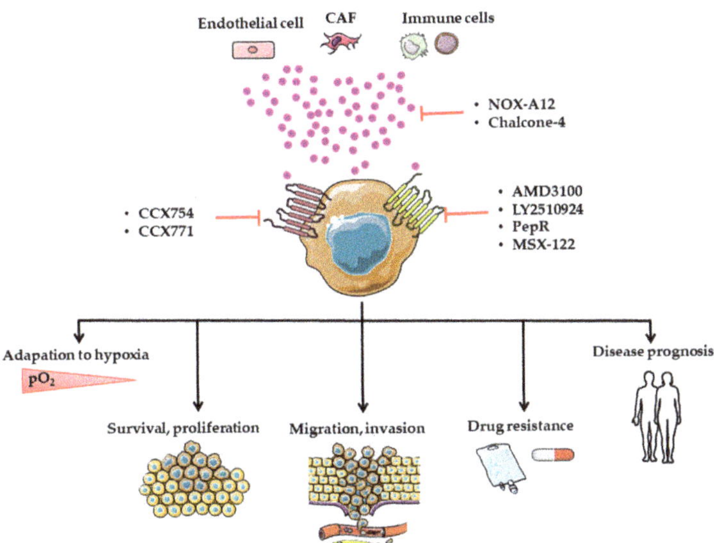

Figure 2. Involvement of the CXCL12/CXCR4/CXCR7 axis in regulating primary tumor growth and metastasis and its pharmaceutical targeting. The expression levels of either partner of this axis have a prognostic value and participate in tumor progression through the activation of multiple signaling pathways involved in cell survival, proliferation, invasion, and migration/dissemination. Each step of the process can be activated/facilitated by local hypoxia within the primary tumor. At sites of metastasis, CXCL12-producing cells (endothelial cells, CAFs, immune cells) allow for the implantation of receptor-expressing tumor cells. In colorectal cancer, several therapeutic molecules targeting receptors or chemokines are undergoing clinical trials to improve patient management and/or overcome tumor resistance. Pink dots indicate CXCL12 molecules.

Author Contributions: Both authors equally contributed to the writing of the review. All authors have read and agreed to the published version of the manuscript.

Acknowledgments: Work in the author's laboratory is supported by the Inserm (AD10 Grand Est) and the Ligue Contre le Cancer (Comité du Grand Est). Aïssata Aimée GOÏTA is a predoctoral fellow from the French Ministère de la Recherche et de l'Enseignement Supérieur. Dominique Guenot is a research director at the CNRS.

Conflicts of Interest: The authors declare no conflict of interest.

Abbreviations

ACKR3	atypical chemokine receptor 3
Akt	protein kinase B
cAMP	cyclic adenosine monophosphate
AOM	azoxymethane
APC	adenomatous polyposis coli
ARE	AU-rich element
ChIP	chromatin immunoprecipitation
CIN	chromosome instability
CIMP	CpG island methylator phenotype
CpG	cytosine-phosphate-guanine
CRC	colorectal cancer
CTLA-4	cytotoxic T lymphocyte Antigen 4
CXCL-10/12	C-X-C motif ligand 10/12

CXCR-3/4/7	C-X-C motif receptor 3/4/7
CAF	cancer-associated fibroblasts
DSS	dextran sodium sulfate
ECM	extracellular matrix
EGFR	epidermal growth factor receptor
EMT	epithelial–mesenchymal transition
ERK	extracellular signal-regulated kinases
FAP	familial adenomatous polyposis
FGF	fibroblast growth factor
5-FU	5-fluorouracil
GAG	glycosaminoglycans
GEO	Gene Expression Omnibus
GRK2	G-protein-coupled receptor kinase 2
GTP	guanine–triphosphate
HDAC	histone deacetylases
HIC1	hypermethylated in cancer 1
HIF	hypoxia-inducible factor
HiRE	HIC1-responsive elements
HIV	human immunodeficiency virus
IBD	chronic inflammatory bowel disease
ICAM	intercellular adhesion molecule
ICI	immune checkpoint inhibitor
IGF	insulin-like growth factor
IL-2/6	interleukin-2/6
ITG	integrin
Jak	Janus kinase
JNK	c-Jun N-terminal kinase
Lgr5	leucine-rich repeat-containing G-protein coupled receptor 5
MAPK	mitogen-activated protein kinases
MDSCs	myeloid-derived suppressor cells
MMP	matrix metalloproteinase
MMR	mismatch Repair
MSC	mesenchymal stromal cells
MSS	microsatellite stability
MSI	microsatellite instability
mTOR	mammalian target of rapamycin
NF-κB	nuclear factor-kappa B
PCAF	P300/CBP-associated factor
PDGF	platelet-derived growth factor
PD-1	programmed cell death protein 1
PD-L1	programmed cell death protein ligand 1
PI3K	phosphatidylinositol 3-kinase
PLC	Phospholipase C
PPPC1	serine/threonine-protein phosphatase PP1-gamma catalytic
SDF-1	stromal-cell-derived factor 1
α-SMA	alpha-smooth muscle actin
STAT	signal transducer and activator of transcription
TAM	tumor associated macrophages
TCGA	the cancer genome atlas program
TGF	tumor growth factor
TIC	tumor-initiating cell
TIMP	tissue inhibitor of metalloproteinases
TNF	tumor necrosis factor
TNM	tumor, node, metastasis
TTP	tristetraprolin
UTR	untranslated transcribed region
VEGF	vascular endothelial growth factor

References

1. Sung, H.; Ferlay, J.; Siegel, R.L.; Laversanne, M.; Soerjomataram, I.; Jemal, A.; Bray, F. Global Cancer Statistics 2020: GLOBOCAN Estimates of Incidence and Mortality Worldwide for 36 Cancers in 185 Countries. *CA Cancer J. Clin.* **2021**, *71*, 209–249. [CrossRef] [PubMed]
2. Mármol, I.; Sánchez-de-Diego, C.; Pradilla Dieste, A.; Cerrada, E.; Rodriguez Yoldi, M.J. Colorectal Carcinoma: A General Overview and Future Perspectives in Colorectal Cancer. *Int. J. Mol. Sci.* **2017**, *18*, 197. [CrossRef]
3. Ahmed, F.E. Effect of Diet, Life Style, and Other Environmental/Chemopreventive Factors on Colorectal Cancer Development, and Assessment of the Risks. *J. Environ. Sci. Health C Environ. Carcinog. Ecotoxicol. Rev.* **2004**, *22*, 91–147. [CrossRef] [PubMed]
4. Keum, N.N.; Giovannucci, E. Global Burden of Colorectal Cancer: Emerging Trends, Risk Factors and Prevention Strategies. *Nat. Rev. Gastroenterol. Hepatol.* **2019**, *16*, 713–732. [CrossRef]
5. Lukas, M. Inflammatory Bowel Disease as a Risk Factor for Colorectal Cancer. *Dig. Dis.* **2010**, *28*, 619–624. [CrossRef] [PubMed]
6. Harada, S.; Morlote, D. Molecular Pathology of Colorectal Cancer. *Adv. Anat. Pathol.* **2020**, *27*, 20–26. [CrossRef] [PubMed]
7. Migliore, L.; Migheli, F.; Spisni, R.; Coppedè, F. Genetics, Cytogenetics, and Epigenetics of Colorectal Cancer. *J. Biomed. Biotechnol.* **2011**, *2011*, 792362. [CrossRef]
8. Jass, J.R. Classification of Colorectal Cancer Based on Correlation of Clinical, Morphological and Molecular Features. *Histopathology* **2007**, *50*, 113–130. [CrossRef] [PubMed]
9. Fearon, E.R.; Vogelstein, B. A Genetic Model for Colorectal Tumorigenesis. *Cell* **1990**, *61*, 759–767. [CrossRef]
10. Nakayama, M.; Oshima, M. Mutant P53 in Colon Cancer. *J. Mol. Cell. Biol.* **2019**, *11*, 267–276. [CrossRef] [PubMed]
11. Takagi, Y.; Koumura, H.; Futamura, M.; Aoki, S.; Ymaguchi, K.; Kida, H.; Tanemura, H.; Shimokawa, K.; Saji, S. Somatic Alterations of the SMAD-2 Gene in Human Colorectal Cancers. *Br. J. Cancer* **1998**, *78*, 1152–1155. [CrossRef] [PubMed]
12. Sánchez-Pernaute, A.; Pérez-Aguirre, E.; Cerdán, F.J.; Iniesta, P.; Díez Valladares, L.; de Juan, C.; Morán, A.; García-Botella, A.; García Aranda, C.; Benito, M.; et al. Overexpression of C-Myc and Loss of Heterozygosity on 2p, 3p, 5q, 17p and 18q in Sporadic Colorectal Carcinoma. *Rev. Esp. Enferm. Dig.* **2005**, *97*, 169–178. [CrossRef] [PubMed]
13. Lynch, H.T.; Watson, P.; Smyrk, T.C.; Lanspa, S.J.; Boman, B.M.; Boland, C.R.; Lynch, J.F.; Cavalieri, R.J.; Leppert, M.; White, R. Colon Cancer Genetics. *Cancer* **1992**, *70*, 1300–1312. [CrossRef]
14. Eshleman, J.R.; Markowitz, S.D. Mismatch Repair Defects in Human Carcinogenesis. *Hum. Mol. Genet.* **1996**, *5*, 1489–1494. [CrossRef] [PubMed]
15. Lengauer, C.; Kinzler, K.W.; Vogelstein, B. Genetic Instability in Colorectal Cancers. *Nature* **1997**, *386*, 623–627. [CrossRef]
16. Yamamoto, H.; Imai, K. Microsatellite Instability: An Update. *Arch. Toxicol.* **2015**, *89*, 899–921. [CrossRef] [PubMed]
17. Lynch, H.T.; Smyrk, T. Hereditary Nonpolyposis Colorectal Cancer (Lynch Syndrome). An Updated Review. *Cancer* **1996**, *78*, 1149–1167. [CrossRef]
18. Samowitz, W.S.; Albertsen, H.; Herrick, J.; Levin, T.R.; Sweeney, C.; Murtaugh, M.A.; Wolff, R.K.; Slattery, M.L. Evaluation of a Large, Population-Based Sample Supports a CpG Island Methylator Phenotype in Colon Cancer. *Gastroenterology* **2005**, *129*, 837–845. [CrossRef]
19. Lao, V.V.; Grady, W.M. Epigenetics and Colorectal Cancer. *Nat. Rev. Gastroenterol. Hepatol.* **2011**, *8*, 686–700. [CrossRef] [PubMed]
20. Gallois, C.; Laurent-Puig, P.; Taieb, J. Methylator Phenotype in Colorectal Cancer: A Prognostic Factor or Not? *Crit. Rev. Oncol. Hematol.* **2016**, *99*, 74–80. [CrossRef] [PubMed]
21. Kazama, Y.; Watanabe, T.; Kanazawa, T.; Tanaka, J.; Tanaka, T.; Nagawa, H. Poorly Differentiated Colorectal Adenocarcinomas Show Higher Rates of Microsatellite Instability and Promoter Methylation of P16 and HMLH1: A Study Matched for T Classification and Tumor Location. *J. Surg. Oncol.* **2008**, *97*, 278–283. [CrossRef] [PubMed]
22. Popat, S.; Hubner, R.; Houlston, R.S. Systematic Review of Microsatellite Instability and Colorectal Cancer Prognosis. *J. Clin. Oncol.* **2005**, *23*, 609–618. [CrossRef]
23. Watanabe, T.; Kobunai, T.; Yamamoto, Y.; Matsuda, K.; Ishihara, S.; Nozawa, K.; Yamada, H.; Hayama, T.; Inoue, E.; Tamura, J.; et al. Chromosomal Instability (CIN) Phenotype, CIN High or CIN Low, Predicts Survival for Colorectal Cancer. *J. Clin. Oncol.* **2012**, *30*, 2256–2264. [CrossRef]
24. Guinney, J.; Dienstmann, R.; Wang, X.; de Reyniès, A.; Schlicker, A.; Soneson, C.; Marisa, L.; Roepman, P.; Nyamundanda, G.; Angelino, P.; et al. The Consensus Molecular Subtypes of Colorectal Cancer. *Nat. Med.* **2015**, *21*, 1350–1356. [CrossRef] [PubMed]
25. Thiel, A.; Ristimäki, A. Toward a Molecular Classification of Colorectal Cancer: The Role of BRAF. *Front. Oncol.* **2013**, *3*, 281. [CrossRef] [PubMed]
26. Van Cutsem, E.; Lenz, H.-J.; Köhne, C.-H.; Heinemann, V.; Tejpar, S.; Melezínek, I.; Beier, F.; Stroh, C.; Rougier, P.; van Krieken, J.H.; et al. Fluorouracil, Leucovorin, and Irinotecan plus Cetuximab Treatment and RAS Mutations in Colorectal Cancer. *J. Clin. Oncol.* **2015**, *33*, 692–700. [CrossRef] [PubMed]
27. Chalabi, M.; Fanchi, L.F.; Dijkstra, K.K.; Van den Berg, J.G.; Aalbers, A.G.; Sikorska, K.; Lopez-Yurda, M.; Grootscholten, C.; Beets, G.L.; Snaebjornsson, P.; et al. Neoadjuvant Immunotherapy Leads to Pathological Responses in MMR-Proficient and MMR-Deficient Early-Stage Colon Cancers. *Nat. Med.* **2020**, *26*, 566–576. [CrossRef] [PubMed]
28. Franke, A.J.; Skelton, W.P.; Starr, J.S.; Parekh, H.; Lee, J.J.; Overman, M.J.; Allegra, C.; George, T.J. Immunotherapy for Colorectal Cancer: A Review of Current and Novel Therapeutic Approaches. *J. Natl. Cancer. Inst.* **2019**, *111*, 1131–1141. [CrossRef] [PubMed]
29. Rollins, B.J. Chemokines. *Blood* **1997**, *90*, 909–928. [CrossRef]

30. IUIS/WHO Subcommittee on Chemokine Nomenclature Chemokine/Chemokine Receptor Nomenclature. *Cytokine* **2003**, *21*, 48–49. [CrossRef]
31. Murphy, P.M. Chemokine Receptors: Structure, Function and Role in Microbial Pathogenesis. *Cytokine Growth Factor Rev.* **1996**, *7*, 47–64. [CrossRef]
32. Zou, Y.R.; Kottmann, A.H.; Kuroda, M.; Taniuchi, I.; Littman, D.R. Function of the Chemokine Receptor CXCR4 in Haematopoiesis and in Cerebellar Development. *Nature* **1998**, *393*, 595–599. [CrossRef] [PubMed]
33. Nagasawa, T.; Hirota, S.; Tachibana, K.; Takakura, N.; Nishikawa, S.; Kitamura, Y.; Yoshida, N.; Kikutani, H.; Kishimoto, T. Defects of B-Cell Lymphopoiesis and Bone-Marrow Myelopoiesis in Mice Lacking the CXC Chemokine PBSF/SDF-1. *Nature* **1996**, *382*, 635–638. [CrossRef]
34. Yu, S.; Crawford, D.; Tsuchihashi, T.; Behrens, T.W.; Srivastava, D. The Chemokine Receptor CXCR7 Functions to Regulate Cardiac Valve Remodeling. *Dev. Dyn.* **2011**, *240*, 384–393. [CrossRef]
35. Bleul, C.C.; Fuhlbrigge, R.C.; Casasnovas, J.M.; Aiuti, A.; Springer, T.A. A Highly Efficacious Lymphocyte Chemoattractant, Stromal Cell-Derived Factor 1 (SDF-1). *J. Exp. Med.* **1996**, *184*, 1101–1109. [CrossRef] [PubMed]
36. Ekbom, A.; Helmick, C.; Zack, M.; Adami, H.O. Ulcerative Colitis and Colorectal Cancer. A Population-Based Study. *N. Engl. J. Med.* **1990**, *323*, 1228–1233. [CrossRef] [PubMed]
37. Lewellis, S.W.; Knaut, H. Attractive Guidance: How the Chemokine SDF1/CXCL12 Guides Different Cells to Different Locations. *Semin. Cell Dev. Biol.* **2012**, *23*, 333–340. [CrossRef] [PubMed]
38. Petit, I.; Jin, D.; Rafii, S. The SDF-1-CXCR4 Signaling Pathway: A Molecular Hub Modulating Neo-Angiogenesis. *Trends Immunol.* **2007**, *28*, 299–307. [CrossRef]
39. Siekmann, A.F.; Standley, C.; Fogarty, K.E.; Wolfe, S.A.; Lawson, N.D. Chemokine Signaling Guides Regional Patterning of the First Embryonic Artery. *Genes Dev.* **2009**, *23*, 2272–2277. [CrossRef] [PubMed]
40. Moser, B.; Willimann, K. Chemokines: Role in Inflammation and Immune Surveillance. *Ann. Rheum. Dis.* **2004**, *63* (Suppl. 2), ii84–ii89. [CrossRef]
41. White, G.E.; Iqbal, A.J.; Greaves, D.R. CC Chemokine Receptors and Chronic Inflammation–Therapeutic Opportunities and Pharmacological Challenges. *Pharm. Rev.* **2013**, *65*, 47–89. [CrossRef] [PubMed]
42. Balkwill, F. Cancer and the Chemokine Network. *Nat. Rev. Cancer* **2004**, *4*, 540–550. [CrossRef] [PubMed]
43. Zlotnik, A. Chemokines and Cancer. *Int. J. Cancer* **2006**, *119*, 2026–2029. [CrossRef] [PubMed]
44. Raman, D.; Baugher, P.J.; Thu, Y.M.; Richmond, A. Role of Chemokines in Tumor Growth. *Cancer Lett.* **2007**, *256*, 137–165. [CrossRef]
45. Nagasawa, T.; Kikutani, H.; Kishimoto, T. Molecular Cloning and Structure of a Pre-B-Cell Growth-Stimulating Factor. *Proc. Natl. Acad. Sci. USA* **1994**, *91*, 2305–2309. [CrossRef]
46. Aiuti, A.; Webb, I.J.; Bleul, C.; Springer, T.; Gutierrez-Ramos, J.C. The Chemokine SDF-1 Is a Chemoattractant for Human CD34+ Hematopoietic Progenitor Cells and Provides a New Mechanism to Explain the Mobilization of CD34+ Progenitors to Peripheral Blood. *J. Exp. Med.* **1997**, *185*, 111–120. [CrossRef] [PubMed]
47. Zlotnik, A. Involvement of Chemokine Receptors in Organ-Specific Metastasis. *Contrib. Microbiol.* **2006**, *13*, 191–199. [CrossRef]
48. Janssens, R.; Struyf, S.; Proost, P. Pathological Roles of the Homeostatic Chemokine CXCL12. *Cytokine Growth Factor Rev..* **2018**, *44*, 51–68. [CrossRef]
49. Roy, I.; Evans, D.B.; Dwinell, M.B. Chemokines and Chemokine Receptors: Update on Utility and Challenges for the Clinician. *Surgery* **2014**, *155*, 961–973. [CrossRef] [PubMed]
50. Shirozu, M.; Nakano, T.; Inazawa, J.; Tashiro, K.; Tada, H.; Shinohara, T.; Honjo, T. Structure and Chromosomal Localization of the Human Stromal Cell-Derived Factor 1 (SDF1) Gene. *Genomics* **1995**, *28*, 495–500. [CrossRef] [PubMed]
51. Yu, L.; Cecil, J.; Peng, S.-B.; Schrementi, J.; Kovacevic, S.; Paul, D.; Su, E.W.; Wang, J. Identification and Expression of Novel Isoforms of Human Stromal Cell-Derived Factor 1. *Gene* **2006**, *374*, 174–179. [CrossRef] [PubMed]
52. Romain, B.; Benbrika-Nehmar, R.; Marisa, L.; Legrain, M.; Lobstein, V.; Oravecz, A.; Poidevin, L.; Bour, C.; Freund, J.-N.; Duluc, I.; et al. Histone Hypoacetylation Contributes to CXCL12 Downregulation in Colon Cancer: Impact on Tumor Growth and Cell Migration. *Oncotarget* **2017**, *8*, 38351–38366. [CrossRef]
53. Smith, J.M.; Johanesen, P.A.; Wendt, M.K.; Binion, D.G.; Dwinell, M.B. CXCL12 Activation of CXCR4 Regulates Mucosal Host Defense through Stimulation of Epithelial Cell Migration and Promotion of Intestinal Barrier Integrity. *Am. J. Physiol. Gastrointest. Liver Physiol.* **2005**, *288*, G316–G326. [CrossRef] [PubMed]
54. Bleul, C.C.; Farzan, M.; Choe, H.; Parolin, C.; Clark-Lewis, I.; Sodroski, J.; Springer, T.A. The Lymphocyte Chemoattractant SDF-1 Is a Ligand for LESTR/Fusin and Blocks HIV-1 Entry. *Nature* **1996**, *382*, 829–833. [CrossRef] [PubMed]
55. Loetscher, M.; Geiser, T.; O'Reilly, T.; Zwahlen, R.; Baggiolini, M.; Moser, B. Cloning of a Human Seven-Transmembrane Domain Receptor, LESTR, That Is Highly Expressed in Leukocytes. *J. Biol. Chem.* **1994**, *269*, 232–237. [CrossRef]
56. Ma, Q.; Jones, D.; Borghesani, P.R.; Segal, R.A.; Nagasawa, T.; Kishimoto, T.; Bronson, R.T.; Springer, T.A. Impaired B-Lymphopoiesis, Myelopoiesis, and Derailed Cerebellar Neuron Migration in CXCR4- and SDF-1-Deficient Mice. *Proc. Natl. Acad. Sci. USA* **1998**, *95*, 9448–9453. [CrossRef]
57. Libert, F.; Parmentier, M.; Lefort, A.; Dumont, J.E.; Vassart, G. Complete Nucleotide Sequence of a Putative G Protein Coupled Receptor: RDC1. *Nucleic Acids Res.* **1990**, *18*, 1917. [CrossRef] [PubMed]

58. Heesen, M.; Berman, M.A.; Charest, A.; Housman, D.; Gerard, C.; Dorf, M.E. Cloning and Chromosomal Mapping of an Orphan Chemokine Receptor: Mouse RDC1. *Immunogenetics* **1998**, *47*, 364–370. [CrossRef] [PubMed]
59. Mellado, M.; Rodríguez-Frade, J.M.; Mañes, S.; Martínez, A.C. Chemokine Signaling and Functional Responses: The Role of Receptor Dimerization and TK Pathway Activation. *Annu. Rev. Immunol.* **2001**, *19*, 397–421. [CrossRef] [PubMed]
60. Walenkamp, A.M.E.; Lapa, C.; Herrmann, K.; Wester, H.-J. CXCR4 Ligands: The Next Big Hit? *J. Nucl. Med.* **2017**, *58*, 77S–82S. [CrossRef] [PubMed]
61. Teixidó, J.; Martínez-Moreno, M.; Díaz-Martínez, M.; Sevilla-Movilla, S. The Good and Bad Faces of the CXCR4 Chemokine Receptor. *Int. J. Biochem. Cell Biol.* **2018**, *95*, 121–131. [CrossRef] [PubMed]
62. Balabanian, K.; Lagane, B.; Infantino, S.; Chow, K.Y.C.; Harriague, J.; Moepps, B.; Arenzana-Seisdedos, F.; Thelen, M.; Bachelerie, F. The Chemokine SDF-1/CXCL12 Binds to and Signals through the Orphan Receptor RDC1 in T Lymphocytes. *J. Biol. Chem.* **2005**, *280*, 35760–35766. [CrossRef] [PubMed]
63. Rajagopal, S.; Kim, J.; Ahn, S.; Craig, S.; Lam, C.M.; Gerard, N.P.; Gerard, C.; Lefkowitz, R.J. Beta-Arrestin- but Not G Protein-Mediated Signaling by the "Decoy" Receptor CXCR7. *Proc. Natl. Acad. Sci. USA* **2010**, *107*, 628–632. [CrossRef] [PubMed]
64. Nguyen, H.T.; Reyes-Alcaraz, A.; Yong, H.J.; Nguyen, L.P.; Park, H.-K.; Inoue, A.; Lee, C.S.; Seong, J.Y.; Hwang, J.-I. CXCR7: A β-Arrestin-Biased Receptor That Potentiates Cell Migration and Recruits β-Arrestin2 Exclusively through Gβγ Subunits and GRK2. *Cell Biosci.* **2020**, *10*, 134. [CrossRef] [PubMed]
65. Sierro, F.; Biben, C.; Martínez-Muñoz, L.; Mellado, M.; Ransohoff, R.M.; Li, M.; Woehl, B.; Leung, H.; Groom, J.; Batten, M.; et al. Disrupted Cardiac Development but Normal Hematopoiesis in Mice Deficient in the Second CXCL12/SDF-1 Receptor, CXCR7. *Proc. Natl. Acad. Sci. USA* **2007**, *104*, 14759–14764. [CrossRef]
66. Luker, K.; Gupta, M.; Luker, G. Bioluminescent CXCL12 Fusion Protein for Cellular Studies of CXCR4 and CXCR7. *Biotechniques* **2009**, *47*, 625–632. [CrossRef] [PubMed]
67. Levoye, A.; Balabanian, K.; Baleux, F.; Bachelerie, F.; Lagane, B. CXCR7 Heterodimerizes with CXCR4 and Regulates CXCL12-Mediated G Protein Signaling. *Blood* **2009**, *113*, 6085–6093. [CrossRef] [PubMed]
68. Gerrits, H.; van Ingen Schenau, D.S.; Bakker, N.E.C.; van Disseldorp, A.J.M.; Strik, A.; Hermens, L.S.; Koenen, T.B.; Krajnc-Franken, M.A.M.; Gossen, J.A. Early Postnatal Lethality and Cardiovascular Defects in CXCR7-Deficient Mice. *Genesis* **2008**, *46*, 235–245. [CrossRef]
69. Sánchez-Alcañiz, J.A.; Haege, S.; Mueller, W.; Pla, R.; Mackay, F.; Schulz, S.; López-Bendito, G.; Stumm, R.; Marín, O. Cxcr7 Controls Neuronal Migration by Regulating Chemokine Responsiveness. *Neuron* **2011**, *69*, 77–90. [CrossRef] [PubMed]
70. Dambly-Chaudière, C.; Cubedo, N.; Ghysen, A. Control of Cell Migration in the Development of the Posterior Lateral Line: Antagonistic Interactions between the Chemokine Receptors CXCR4 and CXCR7/RDC1. *BMC Dev. Biol.* **2007**, *7*, 23. [CrossRef]
71. Valentin, G.; Haas, P.; Gilmour, D. The Chemokine SDF1a Coordinates Tissue Migration through the Spatially Restricted Activation of Cxcr7 and Cxcr4b. *Curr. Biol.* **2007**, *17*, 1026–1031. [CrossRef] [PubMed]
72. Luker, K.E.; Steele, J.M.; Mihalko, L.A.; Ray, P.; Luker, G.D. Constitutive and Chemokine-Dependent Internalization and Recycling of CXCR7 in Breast Cancer Cells to Degrade Chemokine Ligands. *Oncogene* **2010**, *29*, 4599–4610. [CrossRef]
73. Koenen, J.; Bachelerie, F.; Balabanian, K.; Schlecht-Louf, G.; Gallego, C. Atypical Chemokine Receptor 3 (ACKR3): A Comprehensive Overview of Its Expression and Potential Roles in the Immune System. *Mol. Pharm.* **2019**, *96*, 809–818. [CrossRef]
74. Lau, S.; Feitzinger, A.; Venkiteswaran, G.; Wang, J.; Lewellis, S.W.; Koplinski, C.A.; Peterson, F.C.; Volkman, B.F.; Meier-Schellersheim, M.; Knaut, H. A Negative-Feedback Loop Maintains Optimal Chemokine Concentrations for Directional Cell Migration. *Nat. Cell Biol.* **2020**, *22*, 266–273. [CrossRef] [PubMed]
75. Quinn, K.E.; Mackie, D.I.; Caron, K.M. Emerging Roles of Atypical Chemokine Receptor 3 (ACKR3) in Normal Development and Physiology. *Cytokine* **2018**, *109*, 17–23. [CrossRef]
76. Balkwill, F.R. The Chemokine System and Cancer. *J. Pathol.* **2012**, *226*, 148–157. [CrossRef] [PubMed]
77. Raman, D.; Sobolik-Delmaire, T.; Richmond, A. Chemokines in Health and Disease. *Exp. Cell Res.* **2011**, *317*, 575–589. [CrossRef] [PubMed]
78. Müller, A.; Homey, B.; Soto, H.; Ge, N.; Catron, D.; Buchanan, M.E.; McClanahan, T.; Murphy, E.; Yuan, W.; Wagner, S.N.; et al. Involvement of Chemokine Receptors in Breast Cancer Metastasis. *Nature* **2001**, *410*, 50–56. [CrossRef] [PubMed]
79. Hattermann, K.; Mentlein, R. An Infernal Trio: The Chemokine CXCL12 and Its Receptors CXCR4 and CXCR7 in Tumor Biology. *Ann. Anat.* **2013**, *195*, 103–110. [CrossRef] [PubMed]
80. Zhao, H.; Guo, L.; Zhao, H.; Zhao, J.; Weng, H.; Zhao, B. CXCR4 Over-Expression and Survival in Cancer: A System Review and Meta-Analysis. *Oncotarget* **2015**, *6*, 5022–5040. [CrossRef] [PubMed]
81. Fan, H.; Wang, W.; Yan, J.; Xiao, L.; Yang, L. Prognostic Significance of CXCR7 in Cancer Patients: A Meta-Analysis. *Cancer Cell Int.* **2018**, *18*, 212. [CrossRef] [PubMed]
82. Romain, B.; Hachet-Haas, M.; Rohr, S.; Brigand, C.; Galzi, J.-L.; Gaub, M.-P.; Pencreach, E.; Guenot, D. Hypoxia Differentially Regulated CXCR4 and CXCR7 Signaling in Colon Cancer. *Mol. Cancer* **2014**, *13*, 58. [CrossRef] [PubMed]
83. Sun, X.; Cheng, G.; Hao, M.; Zheng, J.; Zhou, X.; Zhang, J.; Taichman, R.S.; Pienta, K.J.; Wang, J. CXCL12/CXCR4/CXCR7 Chemokine Axis and Cancer Progression. *Cancer Metastasis Rev.* **2010**, *29*, 709–722. [CrossRef] [PubMed]
84. Kim, J.; Takeuchi, H.; Lam, S.T.; Turner, R.R.; Wang, H.-J.; Kuo, C.; Foshag, L.; Bilchik, A.J.; Hoon, D.S.B. Chemokine Receptor CXCR4 Expression in Colorectal Cancer Patients Increases the Risk for Recurrence and for Poor Survival. *J. Clin. Oncol.* **2005**, *23*, 2744–2753. [CrossRef] [PubMed]

85. Yang, D.; Dai, T.; Xue, L.; Liu, X.; Wu, B.; Geng, J.; Mao, X.; Wang, R.; Chen, L.; Chu, X. Expression of Chemokine Receptor CXCR7 in Colorectal Carcinoma and Its Prognostic Significance. *Int. J. Clin. Exp. Pathol.* **2015**, *8*, 13051–13058. [PubMed]
86. Xu, C.; Zheng, L.; Li, D.; Chen, G.; Gu, J.; Chen, J.; Yao, Q. CXCR4 Overexpression Is Correlated with Poor Prognosis in Colorectal Cancer. *Life Sci.* **2018**, *208*, 333–340. [CrossRef]
87. Wani, N.; Nasser, M.W.; Ahirwar, D.K.; Zhao, H.; Miao, Z.; Shilo, K.; Ganju, R.K. C-X-C Motif Chemokine 12/C-X-C Chemokine Receptor Type 7 Signaling Regulates Breast Cancer Growth and Metastasis by Modulating the Tumor Microenvironment. *Breast Cancer Res.* **2014**, *16*, R54. [CrossRef]
88. Ingold, B.; Schulz, S.; Budczies, J.; Neumann, U.; Ebert, M.P.A.; Weichert, W.; Röcken, C. The Role of Vascular CXCR4 Expression in Colorectal Carcinoma. *Histopathology* **2009**, *55*, 576–586. [CrossRef]
89. Li, X.; Wang, X.; Li, Z.; Zhang, Z.; Zhang, Y. Chemokine Receptor 7 Targets the Vascular Endothelial Growth Factor via the AKT/ERK Pathway to Regulate Angiogenesis in Colon Cancer. *Cancer Med.* **2019**, *8*, 5327–5340. [CrossRef]
90. Chen, Y.; Teng, F.; Wang, G.; Nie, Z. Overexpression of CXCR7 Induces Angiogenic Capacity of Human Hepatocellular Carcinoma Cells via the AKT Signaling Pathway. *Oncol. Rep.* **2016**, *36*, 2275–2281. [CrossRef] [PubMed]
91. Gentilini, A.; Caligiuri, A.; Raggi, C.; Rombouts, K.; Pinzani, M.; Lori, G.; Correnti, M.; Invernizzi, P.; Rovida, E.; Navari, N.; et al. CXCR7 Contributes to the Aggressive Phenotype of Cholangiocarcinoma Cells. *Biochim. Biophys. Acta Mol. Basis Dis.* **2019**, *1865*, 2246–2256. [CrossRef] [PubMed]
92. Guillemot, E.; Karimdjee-Soilihi, B.; Pradelli, E.; Benchetrit, M.; Goguet-Surmenian, E.; Millet, M.-A.; Larbret, F.; Michiels, J.-F.; Birnbaum, D.; Alemanno, P.; et al. CXCR7 Receptors Facilitate the Progression of Colon Carcinoma within Lung Not within Liver. *Br. J. Cancer* **2012**, *107*, 1944–1949. [CrossRef] [PubMed]
93. Greijer, A.E.; Delis-van Diemen, P.M.; Fijneman, R.J.A.; Giles, R.H.; Voest, E.E.; van Hinsbergh, V.W.M.; Meijer, G.A. Presence of HIF-1 and Related Genes in Normal Mucosa, Adenomas and Carcinomas of the Colorectum. *Virchows Arch.* **2008**, *452*, 535–544. [CrossRef] [PubMed]
94. Oliveira Frick, V.; Rubie, C.; Ghadjar, P.; Faust, S.K.; Wagner, M.; Gräber, S.; Schilling, M.K. Changes in CXCL12/CXCR4-Chemokine Expression during Onset of Colorectal Malignancies. *Tumour Biol.* **2011**, *32*, 189–196. [CrossRef] [PubMed]
95. Amara, S.; Chaar, I.; Khiari, M.; Ounissi, D.; Weslati, M.; Boughriba, R.; Hmida, A.B.; Bouraoui, S. Stromal Cell Derived Factor-1 and CXCR4 Expression in Colorectal Cancer Promote Liver Metastasis. *Cancer Biomark* **2015**, *15*, 869–879. [CrossRef] [PubMed]
96. Yoshitake, N.; Fukui, H.; Yamagishi, H.; Sekikawa, A.; Fujii, S.; Tomita, S.; Ichikawa, K.; Imura, J.; Hiraishi, H.; Fujimori, T. Expression of SDF-1 Alpha and Nuclear CXCR4 Predicts Lymph Node Metastasis in Colorectal Cancer. *Br. J. Cancer* **2008**, *98*, 1682–1689. [CrossRef] [PubMed]
97. Akishima-Fukasawa, Y.; Nakanishi, Y.; Ino, Y.; Moriya, Y.; Kanai, Y.; Hirohashi, S. Prognostic Significance of CXCL12 Expression in Patients with Colorectal Carcinoma. *Am. J. Clin. Pathol.* **2009**, *132*, 202–210; quiz 307. [CrossRef] [PubMed]
98. Mousavi, A.; Hashemzadeh, S.; Bahrami, T.; Estiar, M.A.; Feizi, M.A.H.; Pouladi, N.; Rostamizadeh, L.; Sakhinia, E. Expression Patterns of CXCL12 and Its Receptor in Colorectal Carcinoma. *Clin. Lab.* **2018**, *64*, 871–876. [CrossRef] [PubMed]
99. Wendt, M.K.; Johanesen, P.A.; Kang-Decker, N.; Binion, D.G.; Shah, V.; Dwinell, M.B. Silencing of Epithelial CXCL12 Expression by DNA Hypermethylation Promotes Colonic Carcinoma Metastasis. *Oncogene* **2006**, *25*, 4986–4997. [CrossRef] [PubMed]
100. Kim, H.J.; Bae, S.B.; Jeong, D.; Kim, E.S.; Kim, C.-N.; Park, D.-G.; Ahn, T.S.; Cho, S.W.; Shin, E.J.; Lee, M.S.; et al. Upregulation of Stromal Cell-Derived Factor 1α Expression Is Associated with the Resistance to Neoadjuvant Chemoradiotherapy of Locally Advanced Rectal Cancer: Angiogenic Markers of Neoadjuvant Chemoradiation. *Oncol. Rep.* **2014**, *32*, 2493–2500. [CrossRef] [PubMed]
101. Tamas, K.; Domanska, U.M.; van Dijk, T.H.; Timmer-Bosscha, H.; Havenga, K.; Karrenbeld, A.; Sluiter, W.J.; Beukema, J.C.; van Vugt, M.A.T.M.; de Vries, E.G.E.; et al. CXCR4 and CXCL12 Expression in Rectal Tumors of Stage IV Patients Before and After Local Radiotherapy and Systemic Neoadjuvant Treatment. *Curr. Pharm. Des.* **2015**, *21*, 2276–2283. [CrossRef] [PubMed]
102. Olaso, E.; Salado, C.; Egilegor, E.; Gutierrez, V.; Santisteban, A.; Sancho-Bru, P.; Friedman, S.L.; Vidal-Vanaclocha, F. Proangiogenic Role of Tumor-Activated Hepatic Stellate Cells in Experimental Melanoma Metastasis. *Hepatology* **2003**, *37*, 674–685. [CrossRef] [PubMed]
103. Guo, F.; Wang, Y.; Liu, J.; Mok, S.C.; Xue, F.; Zhang, W. CXCL12/CXCR4: A Symbiotic Bridge Linking Cancer Cells and Their Stromal Neighbors in Oncogenic Communication Networks. *Oncogene* **2016**, *35*, 816–826. [CrossRef]
104. Yu, P.F.; Huang, Y.; Xu, C.L.; Lin, L.Y.; Han, Y.Y.; Sun, W.H.; Hu, G.H.; Rabson, A.B.; Wang, Y.; Shi, Y.F. Downregulation of CXCL12 in Mesenchymal Stromal Cells by TGFβ Promotes Breast Cancer Metastasis. *Oncogene* **2017**, *36*, 840–849. [CrossRef] [PubMed]
105. Shinagawa, K.; Kitadai, Y.; Tanaka, M.; Sumida, T.; Kodama, M.; Higashi, Y.; Tanaka, S.; Yasui, W.; Chayama, K. Mesenchymal Stem Cells Enhance Growth and Metastasis of Colon Cancer. *Int. J. Cancer* **2010**, *127*, 2323–2333. [CrossRef] [PubMed]
106. Ma, J.-C.; Sun, X.-W.; Su, H.; Chen, Q.; Guo, T.-K.; Li, Y.; Chen, X.-C.; Guo, J.; Gong, Z.-Q.; Zhao, X.-D.; et al. Fibroblast-Derived CXCL12/SDF-1α Promotes CXCL6 Secretion and Co-Operatively Enhances Metastatic Potential through the PI3K/Akt/MTOR Pathway in Colon Cancer. *World J. Gastroenterol.* **2017**, *23*, 5167–5178. [CrossRef]
107. Todaro, M.; Gaggianesi, M.; Catalano, V.; Benfante, A.; Iovino, F.; Biffoni, M.; Apuzzo, T.; Sperduti, I.; Volpe, S.; Cocorullo, G.; et al. CD44v6 Is a Marker of Constitutive and Reprogrammed Cancer Stem Cells Driving Colon Cancer Metastasis. *Cell Stem Cell* **2014**, *14*, 342–356. [CrossRef]
108. Cuiffo, B.G.; Karnoub, A.E. Mesenchymal Stem Cells in Tumor Development. *Cell Adh. Migr.* **2012**, *6*, 220–230. [CrossRef] [PubMed]

109. Erreni, M.; Bianchi, P.; Laghi, L.; Mirolo, M.; Fabbri, M.; Locati, M.; Mantovani, A.; Allavena, P. Expression of Chemokines and Chemokine Receptors in Human Colon Cancer. *Meth. Enzymol.* **2009**, *460*, 105–121. [CrossRef]
110. Cheng, X.-S.; Li, Y.-F.; Tan, J.; Sun, B.; Xiao, Y.-C.; Fang, X.-B.; Zhang, X.-F.; Li, Q.; Dong, J.-H.; Li, M.; et al. CCL20 and CXCL8 Synergize to Promote Progression and Poor Survival Outcome in Patients with Colorectal Cancer by Collaborative Induction of the Epithelial-Mesenchymal Transition. *Cancer Lett.* **2014**, *348*, 77–87. [CrossRef] [PubMed]
111. Oladipo, O.; Conlon, S.; O'Grady, A.; Purcell, C.; Wilson, C.; Maxwell, P.J.; Johnston, P.G.; Stevenson, M.; Kay, E.W.; Wilson, R.H.; et al. The Expression and Prognostic Impact of CXC-Chemokines in Stage II and III Colorectal Cancer Epithelial and Stromal Tissue. *Br. J. Cancer* **2011**, *104*, 480–487. [CrossRef] [PubMed]
112. Lourenco, S.; Teixeira, V.H.; Kalber, T.; Jose, R.J.; Floto, R.A.; Janes, S.M. Macrophage Migration Inhibitory Factor-CXCR4 Is the Dominant Chemotactic Axis in Human Mesenchymal Stem Cell Recruitment to Tumors. *J. Immunol.* **2015**, *194*, 3463–3474. [CrossRef] [PubMed]
113. Cojoc, M.; Peitzsch, C.; Trautmann, F.; Polishchuk, L.; Telegeev, G.D.; Dubrovska, A. Emerging Targets in Cancer Management: Role of the CXCL12/CXCR4 Axis. *Onco Targets* **2013**, *6*, 1347–1361. [CrossRef]
114. Mishra, P.; Banerjee, D.; Ben-Baruch, A. Chemokines at the Crossroads of Tumor-Fibroblast Interactions That Promote Malignancy. *J. Leukoc. Biol.* **2011**, *89*, 31–39. [CrossRef] [PubMed]
115. Gassmann, P.; Haier, J.; Schlüter, K.; Domikowsky, B.; Wendel, C.; Wiesner, U.; Kubitza, R.; Engers, R.; Schneider, S.W.; Homey, B.; et al. CXCR4 Regulates the Early Extravasation of Metastatic Tumor Cells in Vivo. *Neoplasia* **2009**, *11*, 651–661. [CrossRef]
116. Yang, J.; Li, Y.-N.; Pan, T.; Miao, R.-R.; Zhang, Y.-Y.; Wu, S.-H.; Qu, X.; Cui, S.-X. Atypical Chemokine Receptor 3 (ACKR3) Induces the Perturbation of RRNA Biogenesis in Colonic Cells: A Novel Mechanism of Colorectal Tumorigenesis. *bioRxiv* **2021**. Posted 2 September 2021. [CrossRef]
117. Miao, Z.; Luker, K.E.; Summers, B.C.; Berahovich, R.; Bhojani, M.S.; Rehemtulla, A.; Kleer, C.G.; Essner, J.J.; Nasevicius, A.; Luker, G.D.; et al. CXCR7 (RDC1) Promotes Breast and Lung Tumor Growth in Vivo and Is Expressed on Tumor-Associated Vasculature. *Proc. Natl. Acad. Sci. USA* **2007**, *104*, 15735–15740. [CrossRef]
118. Salmaggi, A.; Maderna, E.; Calatozzolo, C.; Gaviani, P.; Canazza, A.; Milanesi, I.; Silvani, A.; DiMeco, F.; Carbone, A.; Pollo, B. CXCL12, CXCR4 and CXCR7 Expression in Brain Metastases. *Cancer Biol.* **2009**, *8*, 1608–1614. [CrossRef]
119. Mitchell, A.; Hasanali, S.L.; Morera, D.S.; Baskar, R.; Wang, X.; Khan, R.; Talukder, A.; Li, C.S.; Manoharan, M.; Jordan, A.R.; et al. A Chemokine/Chemokine Receptor Signature Potentially Predicts Clinical Outcome in Colorectal Cancer Patients. *Cancer Biomark* **2019**, *26*, 291–301. [CrossRef]
120. Stanisavljević, L.; Aßmus, J.; Storli, K.E.; Leh, S.M.; Dahl, O.; Myklebust, M.P. CXCR4, CXCL12 and the Relative CXCL12-CXCR4 Expression as Prognostic Factors in Colon Cancer. *Tumour Biol.* **2016**, *37*, 7441–7452. [CrossRef]
121. Li, Y.-P.; Pang, J.; Gao, S.; Bai, P.-Y.; Wang, W.; Kong, P.; Cui, Y. Role of CXCR4 and SDF1 as Prognostic Factors for Survival and the Association with Clinicopathology in Colorectal Cancer: A Systematic Meta-Analysis. *Tumour Biol.* **2017**, *39*, 1010428317706206. [CrossRef] [PubMed]
122. Marisa, L.; de Reyniès, A.; Duval, A.; Selves, J.; Gaub, M.P.; Vescovo, L.; Etienne-Grimaldi, M.-C.; Schiappa, R.; Guenot, D.; Ayadi, M.; et al. Gene Expression Classification of Colon Cancer into Molecular Subtypes: Characterization, Validation, and Prognostic Value. *PLoS Med.* **2013**, *10*, e1001453. [CrossRef] [PubMed]
123. Fushimi, T.; O'Connor, T.P.; Crystal, R.G. Adenoviral Gene Transfer of Stromal Cell-Derived Factor-1 to Murine Tumors Induces the Accumulation of Dendritic Cells and Suppresses Tumor Growth. *Cancer Res.* **2006**, *66*, 3513–3522. [CrossRef]
124. Galon, J.; Mlecnik, B.; Bindea, G.; Angell, H.K.; Berger, A.; Lagorce, C.; Lugli, A.; Zlobec, I.; Hartmann, A.; Bifulco, C.; et al. Towards the Introduction of the "Immunoscore" in the Classification of Malignant Tumours. *J. Pathol.* **2014**, *232*, 199–209. [CrossRef] [PubMed]
125. Pagès, F.; Kirilovsky, A.; Mlecnik, B.; Asslaber, M.; Tosolini, M.; Bindea, G.; Lagorce, C.; Wind, P.; Marliot, F.; Bruneval, P.; et al. In Situ Cytotoxic and Memory T Cells Predict Outcome in Patients with Early-Stage Colorectal Cancer. *J. Clin. Oncol.* **2009**, *27*, 5944–5951. [CrossRef]
126. Prall, F.; Dührkop, T.; Weirich, V.; Ostwald, C.; Lenz, P.; Nizze, H.; Barten, M. Prognostic Role of CD8+ Tumor-Infiltrating Lymphocytes in Stage III Colorectal Cancer with and without Microsatellite Instability. *Hum. Pathol.* **2004**, *35*, 808–816. [CrossRef]
127. Chiba, T.; Ohtani, H.; Mizoi, T.; Naito, Y.; Sato, E.; Nagura, H.; Ohuchi, A.; Ohuchi, K.; Shiiba, K.; Kurokawa, Y.; et al. Intraepithelial CD8+ T-Cell-Count Becomes a Prognostic Factor after a Longer Follow-up Period in Human Colorectal Carcinoma: Possible Association with Suppression of Micrometastasis. *Br. J. Cancer* **2004**, *91*, 1711–1717. [CrossRef]
128. Lalos, A.; Tülek, A.; Tosti, N.; Mechera, R.; Wilhelm, A.; Soysal, S.; Daester, S.; Kancherla, V.; Weixler, B.; Spagnoli, G.C.; et al. Prognostic Significance of CD8+ T-Cells Density in Stage III Colorectal Cancer Depends on SDF-1 Expression. *Sci. Rep.* **2021**, *11*, 775. [CrossRef]
129. Lv, S.; Yang, Y.; Kwon, S.; Han, M.; Zhao, F.; Kang, H.; Dai, C.; Wang, R. The Association of CXCR4 Expression with Prognosis and Clinicopathological Indicators in Colorectal Carcinoma Patients: A Meta-Analysis. *Histopathology* **2014**, *64*, 701–712. [CrossRef] [PubMed]
130. Li, L.-N.; Jiang, K.-T.; Tan, P.; Wang, A.-H.; Kong, Q.-Y.; Wang, C.-Y.; Lu, H.-R.; Wang, J. Prognosis and Clinicopathology of CXCR4 in Colorectal Cancer Patients: A Meta-Analysis. *Asian Pac. J. Cancer Prev.* **2015**, *16*, 4077–4080. [CrossRef] [PubMed]
131. Jiang, Q.; Sun, Y.; Liu, X. CXCR4 as a Prognostic Biomarker in Gastrointestinal Cancer: A Meta-Analysis. *Biomarkers* **2019**, *24*, 510–516. [CrossRef] [PubMed]

132. Ottaiano, A.; Scala, S.; Normanno, N.; Botti, G.; Tatangelo, F.; Di Mauro, A.; Capozzi, M.; Facchini, S.; Tafuto, S.; Nasti, G. Prognostic and Predictive Role of CXC Chemokine Receptor 4 in Metastatic Colorectal Cancer Patients. *Appl. Immunohistochem. Mol. Morphol.* **2020**, *28*, 755–760. [CrossRef] [PubMed]
133. Yopp, A.C.; Shia, J.; Butte, J.M.; Allen, P.J.; Fong, Y.; Jarnagin, W.R.; DeMatteo, R.P.; D'Angelica, M.I. CXCR4 Expression Predicts Patient Outcome and Recurrence Patterns after Hepatic Resection for Colorectal Liver Metastases. *Ann. Surg. Oncol.* **2012**, *19* (Suppl. S3), S339–S346. [CrossRef]
134. D'Alterio, C.; Nasti, G.; Polimeno, M.; Ottaiano, A.; Conson, M.; Circelli, L.; Botti, G.; Scognamiglio, G.; Santagata, S.; De Divitiis, C.; et al. CXCR4-CXCL12-CXCR7, TLR2-TLR4, and PD-1/PD-L1 in Colorectal Cancer Liver Metastases from Neoadjuvant-Treated Patients. *Oncoimmunology* **2016**, *5*, e1254313. [CrossRef] [PubMed]
135. Alamo, P.; Gallardo, A.; Di Nicolantonio, F.; Pavón, M.A.; Casanova, I.; Trias, M.; Mangues, M.A.; Lopez-Pousa, A.; Villaverde, A.; Vázquez, E.; et al. Higher Metastatic Efficiency of KRas G12V than KRas G13D in a Colorectal Cancer Model. *FASEB J.* **2015**, *29*, 464–476. [CrossRef] [PubMed]
136. Nagasawa, S.; Tsuchida, K.; Shiozawa, M.; Hiroshima, Y.; Kimura, Y.; Hashimoto, I.; Watanabe, H.; Kano, K.; Numata, M.; Aoyama, T.; et al. Clinical Significance of Chemokine Receptor CXCR4 and CCR7 MRNA Expression in Patients With Colorectal Cancer. *Anticancer Res.* **2021**, *41*, 4489–4495. [CrossRef] [PubMed]
137. Jiao, X.; Shu, G.; Liu, H.; Zhang, Q.; Ma, Z.; Ren, C.; Guo, H.; Shi, J.; Liu, J.; Zhang, C.; et al. The Diagnostic Value of Chemokine/Chemokine Receptor Pairs in Hepatocellular Carcinoma and Colorectal Liver Metastasis. *J. Histochem. Cytochem.* **2019**, *67*, 299–308. [CrossRef]
138. Xu, F.; Wang, F.; Di, M.; Huang, Q.; Wang, M.; Hu, H.; Jin, Y.; Dong, J.; Lai, M. Classification Based on the Combination of Molecular and Pathologic Predictors Is Superior to Molecular Classification on Prognosis in Colorectal Carcinoma. *Clin. Cancer Res.* **2007**, *13*, 5082–5088. [CrossRef]
139. O'Brien, C.A.; Pollett, A.; Gallinger, S.; Dick, J.E. A Human Colon Cancer Cell Capable of Initiating Tumour Growth in Immunodeficient Mice. *Nature* **2007**, *445*, 106–110. [CrossRef]
140. Ricci-Vitiani, L.; Lombardi, D.G.; Pilozzi, E.; Biffoni, M.; Todaro, M.; Peschle, C.; De Maria, R. Identification and Expansion of Human Colon-Cancer-Initiating Cells. *Nature* **2007**, *445*, 111–115. [CrossRef]
141. Shmelkov, S.V.; Butler, J.M.; Hooper, A.T.; Hormigo, A.; Kushner, J.; Milde, T.; St Clair, R.; Baljevic, M.; White, I.; Jin, D.K.; et al. CD133 Expression Is Not Restricted to Stem Cells, and Both CD133+ and CD133- Metastatic Colon Cancer Cells Initiate Tumors. *J. Clin. Investig.* **2008**, *118*, 2111–2120. [CrossRef] [PubMed]
142. Yu, X.; Lin, Y.; Yan, X.; Tian, Q.; Li, L.; Lin, E.H. CD133, Stem Cells, and Cancer Stem Cells: Myth or Reality? *Curr. Colorectal Cancer Rep.* **2011**, *7*, 253–259. [CrossRef] [PubMed]
143. Zhang, S.; Han, Z.; Jing, Y.; Tao, S.; Li, T.; Wang, H.; Wang, Y.; Li, R.; Yang, Y.; Zhao, X.; et al. CD133(+)CXCR4(+) Colon Cancer Cells Exhibit Metastatic Potential and Predict Poor Prognosis of Patients. *BMC Med.* **2012**, *10*, 85. [CrossRef] [PubMed]
144. Silinsky, J.; Grimes, C.; Driscoll, T.; Green, H.; Cordova, J.; Davis, N.K.; Li, L.; Margolin, D.A. CD 133+ and CXCR4+ Colon Cancer Cells as a Marker for Lymph Node Metastasis. *J. Surg. Res.* **2013**, *185*, 113–118. [CrossRef] [PubMed]
145. Wu, W.; Cao, J.; Ji, Z.; Wang, J.; Jiang, T.; Ding, H. Co-Expression of Lgr5 and CXCR4 Characterizes Cancer Stem-like Cells of Colorectal Cancer. *Oncotarget* **2016**, *7*, 81144–81155. [CrossRef] [PubMed]
146. Shi, Y.; Riese, D.J.; Shen, J. The Role of the CXCL12/CXCR4/CXCR7 Chemokine Axis in Cancer. *Front. Pharm.* **2020**, *11*, 574667. [CrossRef] [PubMed]
147. Wang, M.; Yang, X.; Wei, M.; Wang, Z. The Role of CXCL12 Axis in Lung Metastasis of Colorectal Cancer. *J. Cancer* **2018**, *9*, 3898–3903. [CrossRef]
148. Yamada, S.; Shimada, M.; Utsunomiya, T.; Morine, Y.; Imura, S.; Ikemoto, T.; Mori, H.; Arakawa, Y.; Kanamoto, M.; Iwahashi, S.; et al. CXC Receptor 4 and Stromal Cell-Derived Factor 1 in Primary Tumors and Liver Metastases of Colorectal Cancer. *J. Surg. Res.* **2014**, *187*, 107–112. [CrossRef]
149. Rubie, C.; Kollmar, O.; Frick, V.O.; Wagner, M.; Brittner, B.; Gräber, S.; Schilling, M.K. Differential CXC Receptor Expression in Colorectal Carcinomas. *Scand. J. Immunol.* **2008**, *68*, 635–644. [CrossRef] [PubMed]
150. Sherif, M.F.; Ismail, I.M.; Ata, S.M.S. Expression of CXCR7 in Colorectal Adenoma and Adenocarcinoma: Correlation with Clinicopathological Parameters. *Ann. Diagn. Pathol.* **2020**, *49*, 151621. [CrossRef] [PubMed]
151. Kheirelseid, E.A.H.; Miller, N.; Chang, K.H.; Nugent, M.; Kerin, M.J. Clinical Applications of Gene Expression in Colorectal Cancer. *J. Gastrointest. Oncol.* **2013**, *4*, 144–157. [CrossRef] [PubMed]
152. Selvaraj, U.M.; Ortega, S.B.; Hu, R.; Gilchrist, R.; Kong, X.; Partin, A.; Plautz, E.J.; Klein, R.S.; Gidday, J.M.; Stowe, A.M. Preconditioning-Induced CXCL12 Upregulation Minimizes Leukocyte Infiltration after Stroke in Ischemia-Tolerant Mice. *J. Cereb. Blood Flow Metab.* **2017**, *37*, 801–813. [CrossRef] [PubMed]
153. Ceradini, D.J.; Kulkarni, A.R.; Callaghan, M.J.; Tepper, O.M.; Bastidas, N.; Kleinman, M.E.; Capla, J.M.; Galiano, R.D.; Levine, J.P.; Gurtner, G.C. Progenitor Cell Trafficking Is Regulated by Hypoxic Gradients through HIF-1 Induction of SDF-1. *Nat. Med.* **2004**, *10*, 858–864. [CrossRef] [PubMed]
154. Karshovska, E.; Zernecke, A.; Sevilmis, G.; Millet, A.; Hristov, M.; Cohen, C.D.; Schmid, H.; Krotz, F.; Sohn, H.-Y.; Klauss, V.; et al. Expression of HIF-1alpha in Injured Arteries Controls SDF-1alpha Mediated Neointima Formation in Apolipoprotein E Deficient Mice. *Arter. Thromb Vasc. Biol.* **2007**, *27*, 2540–2547. [CrossRef] [PubMed]

155. Hitchon, C.; Wong, K.; Ma, G.; Reed, J.; Lyttle, D.; El-Gabalawy, H. Hypoxia-Induced Production of Stromal Cell-Derived Factor 1 (CXCL12) and Vascular Endothelial Growth Factor by Synovial Fibroblasts. *Arthritis Rheum.* **2002**, *46*, 2587–2597. [CrossRef] [PubMed]
156. Yadav, S.S.; Prasad, S.B.; Prasad, C.B.; Pandey, L.K.; Pradhan, S.; Singh, S.; Narayan, G. CXCL12 Is a Key Regulator in Tumor Microenvironment of Cervical Cancer: An in Vitro Study. *Clin. Exp. Metastasis* **2016**, *33*, 431–439. [CrossRef] [PubMed]
157. Seo, J.; Kim, Y.-O.; Jo, I. Differential Expression of Stromal Cell-Derived Factor 1 in Human Brain Microvascular Endothelial Cells and Pericytes Involves Histone Modifications. *Biochem. Biophys. Res. Commun.* **2009**, *382*, 519–524. [CrossRef] [PubMed]
158. Kouzarides, T. Chromatin Modifications and Their Function. *Cell* **2007**, *128*, 693–705. [CrossRef] [PubMed]
159. Wendt, M.K.; Drury, L.J.; Vongsa, R.A.; Dwinell, M.B. Constitutive CXCL12 Expression Induces Anoikis in Colorectal Carcinoma Cells. *Gastroenterology* **2008**, *135*, 508–517. [CrossRef] [PubMed]
160. Halees, A.S.; El-Badrawi, R.; Khabar, K.S.A. ARED Organism: Expansion of ARED Reveals AU-Rich Element Cluster Variations between Human and Mouse. *Nucleic Acids Res.* **2008**, *36*, D137–D140. [CrossRef]
161. Al-Souhibani, N.; Al-Ghamdi, M.; Al-Ahmadi, W.; Khabar, K.S.A. Posttranscriptional Control of the Chemokine Receptor CXCR4 Expression in Cancer Cells. *Carcinogenesis* **2014**, *35*, 1983–1992. [CrossRef]
162. Semenza, G.L. Life with Oxygen. *Science* **2007**, *318*, 62–64. [CrossRef]
163. Zong, S.; Li, W.; Li, H.; Han, S.; Liu, S.; Shi, Q.; Hou, F. Identification of Hypoxia-Regulated Angiogenic Genes in Colorectal Cancer. *Biochem. Biophys. Res. Commun.* **2017**, *493*, 461–467. [CrossRef] [PubMed]
164. Zhong, H.; De Marzo, A.M.; Laughner, E.; Lim, M.; Hilton, D.A.; Zagzag, D.; Buechler, P.; Isaacs, W.B.; Semenza, G.L.; Simons, J.W. Overexpression of Hypoxia-Inducible Factor 1alpha in Common Human Cancers and Their Metastases. *Cancer Res.* **1999**, *59*, 5830–5835. [PubMed]
165. Harris, A.L. Hypoxia–a Key Regulatory Factor in Tumour Growth. *Nat. Rev. Cancer* **2002**, *2*, 38–47. [CrossRef]
166. Wu, Y.; Jin, M.; Xu, H.; Shimin, Z.; He, S.; Wang, L.; Zhang, Y. Clinicopathologic Significance of HIF-1α, CXCR4, and VEGF Expression in Colon Cancer. *Clin. Dev. Immunol.* **2010**, *2010*, 537531. [CrossRef] [PubMed]
167. Korbecki, J.; Simińska, D.; Gąssowska-Dobrowolska, M.; Listos, J.; Gutowska, I.; Chlubek, D.; Baranowska-Bosiacka, I. Chronic and Cycling Hypoxia: Drivers of Cancer Chronic Inflammation through HIF-1 and NF-KB Activation: A Review of the Molecular Mechanisms. *Int. J. Mol. Sci.* **2021**, *22*, 10701. [CrossRef]
168. Schioppa, T.; Uranchimeg, B.; Saccani, A.; Biswas, S.K.; Doni, A.; Rapisarda, A.; Bernasconi, S.; Saccani, S.; Nebuloni, M.; Vago, L.; et al. Regulation of the Chemokine Receptor CXCR4 by Hypoxia. *J. Exp. Med.* **2003**, *198*, 1391–1402. [CrossRef] [PubMed]
169. Staller, P.; Sulitkova, J.; Lisztwan, J.; Moch, H.; Oakeley, E.J.; Krek, W. Chemokine Receptor CXCR4 Downregulated by von Hippel-Lindau Tumour Suppressor PVHL. *Nature* **2003**, *425*, 307–311. [CrossRef] [PubMed]
170. Risbud, M.V.; Fertala, J.; Vresilovic, E.J.; Albert, T.J.; Shapiro, I.M. Nucleus Pulposus Cells Upregulate PI3K/Akt and MEK/ERK Signaling Pathways under Hypoxic Conditions and Resist Apoptosis Induced by Serum Withdrawal. *Spine* **2005**, *30*, 882–889. [CrossRef] [PubMed]
171. Dai, T.; Hu, Y.; Zheng, H. Hypoxia Increases Expression of CXC Chemokine Receptor 4 via Activation of PI3K/Akt Leading to Enhanced Migration of Endothelial Progenitor Cells. *Eur. Rev. Med. Pharm. Sci.* **2017**, *21*, 1820–1827.
172. Treins, C.; Giorgetti-Peraldi, S.; Murdaca, J.; Semenza, G.L.; Van Obberghen, E. Insulin Stimulates Hypoxia-Inducible Factor 1 through a Phosphatidylinositol 3-Kinase/Target of Rapamycin-Dependent Signaling Pathway. *J. Biol. Chem.* **2002**, *277*, 27975–27981. [CrossRef]
173. Calin, G.A.; Croce, C.M. MicroRNA Signatures in Human Cancers. *Nat. Rev. Cancer* **2006**, *6*, 857–866. [CrossRef] [PubMed]
174. Xiong, W.-C.; Han, N.; Ping, G.-F.; Zheng, P.-F.; Feng, H.-L.; Qin, L.; He, P. MicroRNA-9 Functions as a Tumor Suppressor in Colorectal Cancer by Targeting CXCR4. *Int. J. Clin. Exp. Pathol.* **2018**, *11*, 526–536. [PubMed]
175. Liu, Y.; Zhou, Y.; Feng, X.; An, P.; Quan, X.; Wang, H.; Ye, S.; Yu, C.; He, Y.; Luo, H. MicroRNA-126 Functions as a Tumor Suppressor in Colorectal Cancer Cells by Targeting CXCR4 via the AKT and ERK1/2 Signaling Pathways. *Int. J. Oncol.* **2014**, *44*, 203–210. [CrossRef] [PubMed]
176. Liu, Y.; Zhou, Y.; Feng, X.; Yang, P.; Yang, J.; An, P.; Wang, H.; Ye, S.; Yu, C.; He, Y.; et al. Low Expression of MicroRNA-126 Is Associated with Poor Prognosis in Colorectal Cancer. *Genes Chromosomes Cancer* **2014**, *53*, 358–365. [CrossRef] [PubMed]
177. Fang, Y.; Sun, B.; Li, Z.; Chen, Z.; Xiang, J. MiR-622 Inhibited Colorectal Cancer Occurrence and Metastasis by Suppressing K-Ras. *Mol. Carcinog.* **2016**, *55*, 1369–1377. [CrossRef] [PubMed]
178. Fang, Y.; Sun, B.; Wang, J.; Wang, Y. MiR-622 Inhibits Angiogenesis by Suppressing the CXCR4-VEGFA Axis in Colorectal Cancer. *Gene* **2019**, *699*, 37–42. [CrossRef] [PubMed]
179. Duan, F.-T.; Qian, F.; Fang, K.; Lin, K.-Y.; Wang, W.-T.; Chen, Y.-Q. MiR-133b, a Muscle-Specific MicroRNA, Is a Novel Prognostic Marker That Participates in the Progression of Human Colorectal Cancer via Regulation of CXCR4 Expression. *Mol. Cancer* **2013**, *12*, 164. [CrossRef] [PubMed]
180. Bao, W.; Fu, H.-J.; Xie, Q.-S.; Wang, L.; Zhang, R.; Guo, Z.-Y.; Zhao, J.; Meng, Y.-L.; Ren, X.-L.; Wang, T.; et al. HER2 Interacts with CD44 to Up-Regulate CXCR4 via Epigenetic Silencing of MicroRNA-139 in Gastric Cancer Cells. *Gastroenterology* **2011**, *141*, 2076–2087.e6. [CrossRef] [PubMed]
181. Zhao, Q.; Li, J.-Y.; Zhang, J.; Long, Y.-X.; Li, Y.-J.; Guo, X.-D.; Wei, M.-N.; Liu, W.-J. Role of Visfatin in Promoting Proliferation and Invasion of Colorectal Cancer Cells by Downregulating SDF-1/CXCR4-Mediated MiR-140-3p Expression. *Eur. Rev. Med. Pharm. Sci.* **2020**, *24*, 5367–5377. [CrossRef]

182. Yu, X.; Shi, W.; Zhang, Y.; Wang, X.; Sun, S.; Song, Z.; Liu, M.; Zeng, Q.; Cui, S.; Qu, X. CXCL12/CXCR4 Axis Induced MiR-125b Promotes Invasion and Confers 5-Fluorouracil Resistance through Enhancing Autophagy in Colorectal Cancer. *Sci. Rep.* **2017**, *7*, 42226. [CrossRef] [PubMed]
183. Stuckel, A.J.; Zhang, W.; Zhang, X.; Zeng, S.; Dougherty, U.; Mustafi, R.; Zhang, Q.; Perreand, E.; Khare, T.; Joshi, T.; et al. Enhanced CXCR4 Expression Associates with Increased Gene Body 5-Hydroxymethylcytosine Modification but Not Decreased Promoter Methylation in Colorectal Cancer. *Cancers* **2020**, *12*, 539. [CrossRef]
184. Branco, M.R.; Ficz, G.; Reik, W. Uncovering the Role of 5-Hydroxymethylcytosine in the Epigenome. *Nat. Rev. Genet.* **2011**, *13*, 7–13. [CrossRef] [PubMed]
185. Li, W.; Zhang, X.; Lu, X.; You, L.; Song, Y.; Luo, Z.; Zhang, J.; Nie, J.; Zheng, W.; Xu, D.; et al. 5-Hydroxymethylcytosine Signatures in Circulating Cell-Free DNA as Diagnostic Biomarkers for Human Cancers. *Cell Res.* **2017**, *27*, 1243–1257. [CrossRef]
186. Krook, M.A.; Nicholls, L.A.; Scannell, C.A.; Chugh, R.; Thomas, D.G.; Lawlor, E.R. Stress-Induced CXCR4 Promotes Migration and Invasion of Ewing Sarcoma. *Mol. Cancer Res.* **2014**, *12*, 953–964. [CrossRef] [PubMed]
187. Krook, M.A.; Hawkins, A.G.; Patel, R.M.; Lucas, D.R.; Van Noord, R.; Chugh, R.; Lawlor, E.R. A Bivalent Promoter Contributes to Stress-Induced Plasticity of CXCR4 in Ewing Sarcoma. *Oncotarget* **2016**, *7*, 61775–61788. [CrossRef]
188. Urosevic, J.; Blasco, M.T.; Llorente, A.; Bellmunt, A.; Berenguer-Llergo, A.; Guiu, M.; Cañellas, A.; Fernandez, E.; Burkov, I.; Clapés, M.; et al. ERK1/2 Signaling Induces Upregulation of ANGPT2 and CXCR4 to Mediate Liver Metastasis in Colon Cancer. *Cancer Res.* **2020**, *80*, 4668–4680. [CrossRef] [PubMed]
189. Li, B.; Huang, Q.; Wei, G.-H. The Role of HOX Transcription Factors in Cancer Predisposition and Progression. *Cancers* **2019**, *11*, 528. [CrossRef] [PubMed]
190. Feng, W.; Huang, W.; Chen, J.; Qiao, C.; Liu, D.; Ji, X.; Xie, M.; Zhang, T.; Wang, Y.; Sun, M.; et al. CXCL12-Mediated HOXB5 Overexpression Facilitates Colorectal Cancer Metastasis through Transactivating CXCR4 and ITGB3. *Theranostics* **2021**, *11*, 2612–2633. [CrossRef] [PubMed]
191. Esencay, M.; Sarfraz, Y.; Zagzag, D. CXCR7 Is Induced by Hypoxia and Mediates Glioma Cell Migration towards SDF-1α. *BMC Cancer* **2013**, *13*, 347. [CrossRef] [PubMed]
192. Nosho, K.; Kure, S.; Irahara, N.; Shima, K.; Baba, Y.; Spiegelman, D.; Meyerhardt, J.A.; Giovannucci, E.L.; Fuchs, C.S.; Ogino, S. A Prospective Cohort Study Shows Unique Epigenetic, Genetic, and Prognostic Features of Synchronous Colorectal Cancers. *Gastroenterology* **2009**, *137*, 1609–1620.e1-3. [CrossRef]
193. Bagci, B.; Sari, M.; Karadayi, K.; Turan, M.; Ozdemir, O.; Bagci, G. KRAS, BRAF Oncogene Mutations and Tissue Specific Promoter Hypermethylation of Tumor Suppressor SFRP2, DAPK1, MGMT, HIC1 and P16 Genes in Colorectal Cancer Patients. *Cancer Biomark.* **2016**, *17*, 133–143. [CrossRef]
194. Van Rechem, C.; Rood, B.R.; Touka, M.; Pinte, S.; Jenal, M.; Guérardel, C.; Ramsey, K.; Monté, D.; Bégue, A.; Tschan, M.P.; et al. Scavenger Chemokine (CXC Motif) Receptor 7 (CXCR7) Is a Direct Target Gene of HIC1 (Hypermethylated in Cancer 1). *J. Biol. Chem.* **2009**, *284*, 20927–20935. [CrossRef] [PubMed]
195. Zhou, S.-M.; Zhang, F.; Chen, X.-B.; Jun, C.-M.; Jing, X.; Wei, D.-X.; Xia, Y.; Zhou, Y.-B.; Xiao, X.-Q.; Jia, R.-Q.; et al. MiR-100 Suppresses the Proliferation and Tumor Growth of Esophageal Squamous Cancer Cells via Targeting CXCR7. *Oncol. Rep.* **2016**, *35*, 3453–3459. [CrossRef]
196. Liu, L.; Zhao, X.; Zhu, X.; Zhong, Z.; Xu, R.; Wang, Z.; Cao, J.; Hou, Y. Decreased Expression of MiR-430 Promotes the Development of Bladder Cancer via the Upregulation of CXCR7. *Mol. Med. Rep.* **2013**, *8*, 140–146. [CrossRef]
197. Ge, Y.; Shu, J.; Shi, G.; Yan, F.; Li, Y.; Ding, H. MiR-100 Suppresses the Proliferation, Invasion, and Migration of Hepatocellular Carcinoma Cells via Targeting CXCR7. *J. Immunol. Res.* **2021**, *2021*, 9920786. [CrossRef] [PubMed]
198. Cao, Y.; Song, J.; Ge, J.; Song, Z.; Chen, J.; Wu, C. MicroRNA-100 Suppresses Human Gastric Cancer Cell Proliferation by Targeting CXCR7. *Oncol. Lett.* **2018**, *15*, 453–458. [CrossRef] [PubMed]
199. Kollmar, O.; Rupertus, K.; Scheuer, C.; Nickels, R.M.; Haberl, G.C.Y.; Tilton, B.; Menger, M.D.; Schilling, M.K. CXCR4 and CXCR7 Regulate Angiogenesis and CT26.WT Tumor Growth Independent from SDF-1. *Int. J. Cancer.* **2010**, *126*, 1302–1315. [CrossRef] [PubMed]
200. Kollmar, O.; Rupertus, K.; Scheuer, C.; Junker, B.; Tilton, B.; Schilling, M.K.; Menger, M.D. Stromal Cell-Derived Factor-1 Promotes Cell Migration and Tumor Growth of Colorectal Metastasis. *Neoplasia* **2007**, *9*, 862–870. [CrossRef] [PubMed]
201. Picardo, A.; Karpoff, H.M.; Ng, B.; Lee, J.; Brennan, M.F.; Fong, Y. Partial Hepatectomy Accelerates Local Tumor Growth: Potential Roles of Local Cytokine Activation. *Surgery* **1998**, *124*, 57–64. [CrossRef]
202. Rashidi, B.; An, Z.; Sun, F.X.; Sasson, A.; Gamagammi, R.; Moossa, A.R.; Hoffman, R.M. Minimal Liver Resection Strongly Stimulates the Growth of Human Colon Cancer in the Liver of Nude Mice. *Clin. Exp. Metastasis* **1999**, *17*, 497–500. [CrossRef] [PubMed]
203. Roy, I.; Zimmerman, N.P.; Mackinnon, A.C.; Tsai, S.; Evans, D.B.; Dwinell, M.B. CXCL12 Chemokine Expression Suppresses Human Pancreatic Cancer Growth and Metastasis. *PLoS ONE* **2014**, *9*, e90400. [CrossRef] [PubMed]
204. Brand, S.; Dambacher, J.; Beigel, F.; Olszak, T.; Diebold, J.; Otte, J.-M.; Göke, B.; Eichhorst, S.T. CXCR4 and CXCL12 Are Inversely Expressed in Colorectal Cancer Cells and Modulate Cancer Cell Migration, Invasion and MMP-9 Activation. *Exp. Cell Res.* **2005**, *310*, 117–130. [CrossRef] [PubMed]

205. Hongo, K.; Tsuno, N.H.; Kawai, K.; Sasaki, K.; Kaneko, M.; Hiyoshi, M.; Murono, K.; Tada, N.; Nirei, T.; Sunami, E.; et al. Hypoxia Enhances Colon Cancer Migration and Invasion through Promotion of Epithelial-Mesenchymal Transition. *J. Surg. Res.* **2013**, *182*, 75–84. [CrossRef] [PubMed]
206. Song, Z.-Y.; Gao, Z.-H.; Chu, J.-H.; Han, X.-Z.; Qu, X.-J. Downregulation of the CXCR4/CXCL12 Axis Blocks the Activation of the Wnt/β-Catenin Pathway in Human Colon Cancer Cells. *Biomed. Pharm.* **2015**, *71*, 46–52. [CrossRef] [PubMed]
207. Song, Z.-Y.; Wang, F.; Cui, S.-X.; Qu, X.-J. Knockdown of CXCR4 Inhibits CXCL12-Induced Angiogenesis in HUVECs through Downregulation of the MAPK/ERK and PI3K/AKT and the Wnt/β-Catenin Pathways. *Cancer Investig.* **2018**, *36*, 10–18. [CrossRef] [PubMed]
208. Zheng, F.; Zhang, Z.; Flamini, V.; Jiang, W.G.; Cui, Y. The Axis of CXCR4/SDF-1 Plays a Role in Colon Cancer Cell Adhesion Through Regulation of the AKT and IGF1R Signalling Pathways. *Anticancer Res.* **2017**, *37*, 4361–4369. [CrossRef] [PubMed]
209. Wang, L.; Li, C.-L.; Wang, L.; Yu, W.-B.; Yin, H.-P.; Zhang, G.-Y.; Zhang, L.-F.; Li, S.; Hu, S.-Y. Influence of CXCR4/SDF-1 Axis on E-Cadherin/β-Catenin Complex Expression in HT29 Colon Cancer Cells. *World J. Gastroenterol.* **2011**, *17*, 625–632. [CrossRef]
210. Tung, S.-Y.; Chang, S.-F.; Chou, M.-H.; Huang, W.-S.; Hsieh, Y.-Y.; Shen, C.-H.; Kuo, H.-C.; Chen, C.-N. CXC Chemokine Ligand 12/Stromal Cell-Derived Factor-1 Regulates Cell Adhesion in Human Colon Cancer Cells by Induction of Intercellular Adhesion Molecule-1. *J. Biomed. Sci.* **2012**, *19*, 91. [CrossRef] [PubMed]
211. Maeda, K.; Kang, S.-M.; Sawada, T.; Nishiguchi, Y.; Yashiro, M.; Ogawa, Y.; Ohira, M.; Ishikawa, T.; Hirakawa-YS Chung, K. Expression of Intercellular Adhesion Molecule-1 and Prognosis in Colorectal Cancer. *Oncol. Rep.* **2002**, *9*, 511–514. [CrossRef]
212. Tan, H.-X.; Gong, W.-Z.; Zhou, K.; Xiao, Z.-G.; Hou, F.-T.; Huang, T.; Zhang, L.; Dong, H.-Y.; Zhang, W.-L.; Liu, Y.; et al. CXCR4/TGF-B1 Mediated Hepatic Stellate Cells Differentiation into Carcinoma-Associated Fibroblasts and Promoted Liver Metastasis of Colon Cancer. *Cancer Biol.* **2020**, *21*, 258–268. [CrossRef]
213. Guleng, B.; Tateishi, K.; Ohta, M.; Kanai, F.; Jazag, A.; Ijichi, H.; Tanaka, Y.; Washida, M.; Morikane, K.; Fukushima, Y.; et al. Blockade of the Stromal Cell-Derived Factor-1/CXCR4 Axis Attenuates in Vivo Tumor Growth by Inhibiting Angiogenesis in a Vascular Endothelial Growth Factor-Independent Manner. *Cancer Res.* **2005**, *65*, 5864–5871. [CrossRef]
214. Shin, H.-N.; Moon, H.-H.; Ku, J.-L. Stromal Cell-Derived Factor-1α and Macrophage Migration-Inhibitory Factor Induce Metastatic Behavior in CXCR4-Expressing Colon Cancer Cells. *Int. J. Mol. Med.* **2012**, *30*, 1537–1543. [CrossRef]
215. Gouveia-Fernandes, S. Colorectal Cancer Aggressiveness Is Related to Fibronectin Over Expression, Driving the Activation of SDF-1:CXCR4 Axis. *Int. J. Cancer Clin. Res.* **2016**, *3*, 1–9. [CrossRef]
216. Zeelenberg, I.S.; Ruuls-Van Stalle, L.; Roos, E. The Chemokine Receptor CXCR4 Is Required for Outgrowth of Colon Carcinoma Micrometastases. *Cancer Res.* **2003**, *63*, 3833–3839. [PubMed]
217. Matsusue, R.; Kubo, H.; Hisamori, S.; Okoshi, K.; Takagi, H.; Hida, K.; Nakano, K.; Itami, A.; Kawada, K.; Nagayama, S.; et al. Hepatic Stellate Cells Promote Liver Metastasis of Colon Cancer Cells by the Action of SDF-1/CXCR4 Axis. *Ann. Surg. Oncol.* **2009**, *16*, 2645–2653. [CrossRef] [PubMed]
218. Wang, H.; Tao, L.; Qi, K.E.; Zhang, H.; Feng, D.; Wei, W.; Kong, H.; Chen, T.; Lin, Q.; Chen, D. CXCR7 Functions in Colon Cancer Cell Survival and Migration. *Exp. Med.* **2015**, *10*, 1720–1724. [CrossRef] [PubMed]
219. Murakami, T.; Kawada, K.; Iwamoto, M.; Akagami, M.; Hida, K.; Nakanishi, Y.; Kanda, K.; Kawada, M.; Seno, H.; Taketo, M.M.; et al. The Role of CXCR3 and CXCR4 in Colorectal Cancer Metastasis. *Int. J. Cancer* **2013**, *132*, 276–287. [CrossRef]
220. Li, X.-X.; Zheng, H.-T.; Huang, L.-Y.; Shi, D.-B.; Peng, J.-J.; Liang, L.; Cai, S.-J. Silencing of CXCR7 Gene Represses Growth and Invasion and Induces Apoptosis in Colorectal Cancer through ERK and β-Arrestin Pathways. *Int. J. Oncol.* **2014**, *45*, 1649–1657. [CrossRef] [PubMed]
221. Xu, H.; Wu, Q.; Dang, S.; Jin, M.; Xu, J.; Cheng, Y.; Pan, M.; Wu, Y.; Zhang, C.; Zhang, Y. Alteration of CXCR7 Expression Mediated by TLR4 Promotes Tumor Cell Proliferation and Migration in Human Colorectal Carcinoma. *PLoS ONE* **2011**, *6*, e27399. [CrossRef] [PubMed]
222. Zabel, B.A.; Lewén, S.; Berahovich, R.D.; Jaén, J.C.; Schall, T.J. The Novel Chemokine Receptor CXCR7 Regulates Trans-Endothelial Migration of Cancer Cells. *Mol. Cancer* **2011**, *10*, 73. [CrossRef] [PubMed]
223. Würth, R.; Bajetto, A.; Harrison, J.K.; Barbieri, F.; Florio, T. CXCL12 Modulation of CXCR4 and CXCR7 Activity in Human Glioblastoma Stem-like Cells and Regulation of the Tumor Microenvironment. *Front. Cell. Neurosci.* **2014**, *8*, 144. [CrossRef] [PubMed]
224. Song, Z.-Y.; Wang, F.; Cui, S.-X.; Gao, Z.-H.; Qu, X.-J. CXCR7/CXCR4 Heterodimer-Induced Histone Demethylation: A New Mechanism of Colorectal Tumorigenesis. *Oncogene* **2019**, *38*, 1560–1575. [CrossRef]
225. Mc Donnell, S.; Chaudhry, V.; Mansilla-Soto, J.; Zeng, Z.S.; Shu, W.P.; Guillem, J.G. Metastatic and Non-Metastatic Colorectal Cancer (CRC) Cells Induce Host Metalloproteinase Production in Vivo. *Clin. Exp. Metastasis* **1999**, *17*, 341–349. [CrossRef] [PubMed]
226. Parmo-Cabañas, M.; Molina-Ortiz, I.; Matías-Román, S.; García-Bernal, D.; Carvajal-Vergara, X.; Valle, I.; Pandiella, A.; Arroyo, A.G.; Teixidó, J. Role of Metalloproteinases MMP-9 and MT1-MMP in CXCL12-Promoted Myeloma Cell Invasion across Basement Membranes. *J. Pathol.* **2006**, *208*, 108–118. [CrossRef] [PubMed]
227. Turunen, S.P.; Tatti-Bugaeva, O.; Lehti, K. Membrane-Type Matrix Metalloproteases as Diverse Effectors of Cancer Progression. *Biochim. Biophys. Acta Mol. Cell Res.* **2017**, *1864*, 1974–1988. [CrossRef] [PubMed]

228. Halama, N.; Pruefer, U.; Frömming, A.; Beyer, D.; Eulberg, D.; Jungnelius, J.U.B.; Mangasarian, A. Experience with CXCL12 Inhibitor NOX-A12 plus Pembrolizumab in Patients with Microsatellite-Stable, Metastatic Colorectal or Pancreatic Cancer. *JCO* **2019**, *37*, e14143. [CrossRef]
229. Micallef, I.N.; Stiff, P.J.; Nademanee, A.P.; Maziarz, R.T.; Horwitz, M.E.; Stadtmauer, E.A.; Kaufman, J.L.; McCarty, J.M.; Vargo, R.; Cheverton, P.D.; et al. Plerixafor Plus Granulocyte Colony-Stimulating Factor for Patients with Non-Hodgkin Lymphoma and Multiple Myeloma: Long-Term Follow-Up Report. *Biol. Blood Marrow Transpl.* **2018**, *24*, 1187–1195. [CrossRef] [PubMed]
230. Micallef, I.N.; Stiff, P.J.; DiPersio, J.F.; Maziarz, R.T.; McCarty, J.M.; Bridger, G.; Calandra, G. Successful Stem Cell Remobilization Using Plerixafor (Mozobil) plus Granulocyte Colony-Stimulating Factor in Patients with Non-Hodgkin Lymphoma: Results from the Plerixafor NHL Phase 3 Study Rescue Protocol. *Biol. Blood Marrow Transpl.* **2009**, *15*, 1578–1586. [CrossRef]
231. De Clercq, E. The Bicyclam AMD3100 Story. *Nat. Rev. Drug Discov.* **2003**, *2*, 581–587. [CrossRef] [PubMed]
232. De Clercq, E. AMD3100/CXCR4 Inhibitor. *Front. Immunol.* **2015**, *6*, 276. [CrossRef] [PubMed]
233. Wang, J.; Tannous, B.A.; Poznansky, M.C.; Chen, H. CXCR4 Antagonist AMD3100 (Plerixafor): From an Impurity to a Therapeutic Agent. *Pharm. Res.* **2020**, *159*, 105010. [CrossRef]
234. Benedicto, A.; Romayor, I.; Arteta, B. CXCR4 Receptor Blockage Reduces the Contribution of Tumor and Stromal Cells to the Metastatic Growth in the Liver. *Oncol. Rep.* **2018**, *39*, 2022–2030. [CrossRef]
235. Kalatskaya, I.; Berchiche, Y.A.; Gravel, S.; Limberg, B.J.; Rosenbaum, J.S.; Heveker, N. AMD3100 Is a CXCR7 Ligand with Allosteric Agonist Properties. *Mol. Pharmacol.* **2009**, *75*, 1240–1247. [CrossRef]
236. Li, J.-K.; Yu, L.; Shen, Y.; Zhou, L.-S.; Wang, Y.-C.; Zhang, J.-H. Inhibition of CXCR4 Activity with AMD3100 Decreases Invasion of Human Colorectal Cancer Cells in Vitro. *World J. Gastroenterol.* **2008**, *14*, 2308–2313. [CrossRef] [PubMed]
237. Yi, T.; Zhai, B.; Yu, Y.; Kiyotsugu, Y.; Raschle, T.; Etzkorn, M.; Seo, H.-C.; Nagiec, M.; Luna, R.E.; Reinherz, E.L.; et al. Quantitative Phosphoproteomic Analysis Reveals System-Wide Signaling Pathways Downstream of SDF-1/CXCR4 in Breast Cancer Stem Cells. *Proc. Natl. Acad. Sci. USA* **2014**, *111*, E2182–E2190. [CrossRef] [PubMed]
238. Jang, Y.-G.; Go, R.-E.; Hwang, K.-A.; Choi, K.-C. Resveratrol Inhibits DHT-Induced Progression of Prostate Cancer Cell Line through Interfering with the AR and CXCR4 Pathway. *J. Steroid Biochem. Mol. Biol.* **2019**, *192*, 105406. [CrossRef] [PubMed]
239. Wu, B.; Chien, E.Y.T.; Mol, C.D.; Fenalti, G.; Liu, W.; Katritch, V.; Abagyan, R.; Brooun, A.; Wells, P.; Bi, F.C.; et al. Structures of the CXCR4 Chemokine GPCR with Small-Molecule and Cyclic Peptide Antagonists. *Science* **2010**, *330*, 1066–1071. [CrossRef] [PubMed]
240. Peng, S.-B.; Zhang, X.; Paul, D.; Kays, L.M.; Gough, W.; Stewart, J.; Uhlik, M.T.; Chen, Q.; Hui, Y.-H.; Zamek-Gliszczynski, M.J.; et al. Identification of LY2510924, a Novel Cyclic Peptide CXCR4 Antagonist That Exhibits Antitumor Activities in Solid Tumor and Breast Cancer Metastatic Models. *Mol. Cancer* **2015**, *14*, 480–490. [CrossRef]
241. Levine, D.; Hellmers, L.; Maresh, G.; Zhang, X.; Moret, R.; Green, H.; Margolin, D.; Li, L. Evaluating the Inhibitory Effects of LY2510924, a Cyclic Peptide CXCR4 Antagonist, in Human Colon Cancer Metastasis Using an Orthotopic Xenograft Model. *J. Am. Coll. Surg.* **2018**, *227*, S64–S65. [CrossRef]
242. D'Alterio, C.; Zannetti, A.; Trotta, A.M.; Ieranò, C.; Napolitano, M.; Rea, G.; Greco, A.; Maiolino, P.; Albanese, S.; Scognamiglio, G.; et al. New CXCR4 Antagonist Peptide R (Pep R) Improves Standard Therapy in Colorectal Cancer. *Cancers* **2020**, *12*, 1952. [CrossRef]
243. D'Alterio, C.; Buoncervello, M.; Ieranò, C.; Napolitano, M.; Portella, L.; Rea, G.; Barbieri, A.; Luciano, A.; Scognamiglio, G.; Tatangelo, F.; et al. Targeting CXCR4 Potentiates Anti-PD-1 Efficacy Modifying the Tumor Microenvironment and Inhibiting Neoplastic PD-1. *J. Exp. Clin. Cancer Res.* **2019**, *38*, 432. [CrossRef] [PubMed]
244. Joyce, J.A.; Fearon, D.T. T Cell Exclusion, Immune Privilege, and the Tumor Microenvironment. *Science* **2015**, *348*, 74–80. [CrossRef] [PubMed]
245. Lesokhin, A.M.; Hohl, T.M.; Kitano, S.; Cortez, C.; Hirschhorn-Cymerman, D.; Avogadri, F.; Rizzuto, G.A.; Lazarus, J.J.; Pamer, E.G.; Houghton, A.N.; et al. Monocytic CCR2(+) Myeloid-Derived Suppressor Cells Promote Immune Escape by Limiting Activated CD8 T-Cell Infiltration into the Tumor Microenvironment. *Cancer Res.* **2012**, *72*, 876–886. [CrossRef] [PubMed]
246. Liang, Z.; Zhan, W.; Zhu, A.; Yoon, Y.; Lin, S.; Sasaki, M.; Klapproth, J.-M.A.; Yang, H.; Grossniklaus, H.E.; Xu, J.; et al. Development of a Unique Small Molecule Modulator of CXCR4. *PLoS ONE* **2012**, *7*, e34038. [CrossRef] [PubMed]
247. Bissonnette, B.; Dougherty, U.; Mustafi, R.; Haider, H.; Joseph, L.; Souris, J.; Hart, J.; Pewkow, J.; LI, Y. CXCR4 Inhibitor, MSX-122 Suppresses AOM-induced Colon Cancer in Apc+/Min Mouse. *FASEB J.* **2018**, *32*, 677.4. [CrossRef]
248. Suzui, M.; Okuno, M.; Tanaka, T.; Nakagama, H.; Moriwaki, H. Enhanced Colon Carcinogenesis Induced by Azoxymethane in Min Mice Occurs via a Mechanism Independent of Beta-Catenin Mutation. *Cancer Lett.* **2002**, *183*, 31–41. [CrossRef]
249. Burns, J.M.; Summers, B.C.; Wang, Y.; Melikian, A.; Berahovich, R.; Miao, Z.; Penfold, M.E.T.; Sunshine, M.J.; Littman, D.R.; Kuo, C.J.; et al. A Novel Chemokine Receptor for SDF-1 and I-TAC Involved in Cell Survival, Cell Adhesion, and Tumor Development. *J. Exp. Med.* **2006**, *203*, 2201–2213. [CrossRef] [PubMed]
250. Zabel, B.A.; Wang, Y.; Lewén, S.; Berahovich, R.D.; Penfold, M.E.T.; Zhang, P.; Powers, J.; Summers, B.C.; Miao, Z.; Zhao, B.; et al. Elucidation of CXCR7-Mediated Signaling Events and Inhibition of CXCR4-Mediated Tumor Cell Transendothelial Migration by CXCR7 Ligands. *J. Immunol.* **2009**, *183*, 3204–3211. [CrossRef] [PubMed]
251. Saita, Y.; Kodama, E.; Orita, M.; Kondo, M.; Miyazaki, T.; Sudo, K.; Kajiwara, K.; Matsuoka, M.; Shimizu, Y. Structural Basis for the Interaction of CCR5 with a Small Molecule, Functionally Selective CCR5 Agonist. *J. Immunol.* **2006**, *177*, 3116–3122. [CrossRef]

252. Ali, S.; O'Boyle, G.; Mellor, P.; Kirby, J.A. An Apparent Paradox: Chemokine Receptor Agonists Can Be Used for Anti-Inflammatory Therapy. *Mol. Immunol.* **2007**, *44*, 1477–1482. [CrossRef] [PubMed]
253. Lounsbury, N. Advances in CXCR7 Modulators. *Pharmaceuticals* **2020**, *13*, 33. [CrossRef] [PubMed]
254. Gravel, S.; Malouf, C.; Boulais, P.E.; Berchiche, Y.A.; Oishi, S.; Fujii, N.; Leduc, R.; Sinnett, D.; Heveker, N. The Peptidomimetic CXCR4 Antagonist TC14012 Recruits Beta-Arrestin to CXCR7: Roles of Receptor Domains. *J. Biol. Chem.* **2010**, *285*, 37939–37943. [CrossRef] [PubMed]
255. Cruz-Orengo, L.; Holman, D.W.; Dorsey, D.; Zhou, L.; Zhang, P.; Wright, M.; McCandless, E.E.; Patel, J.R.; Luker, G.D.; Littman, D.R.; et al. CXCR7 Influences Leukocyte Entry into the CNS Parenchyma by Controlling Abluminal CXCL12 Abundance during Autoimmunity. *J. Exp. Med.* **2011**, *208*, 327–339. [CrossRef] [PubMed]
256. Hachet-Haas, M.; Balabanian, K.; Rohmer, F.; Pons, F.; Franchet, C.; Lecat, S.; Chow, K.Y.C.; Dagher, R.; Gizzi, P.; Didier, B.; et al. Small Neutralizing Molecules to Inhibit Actions of the Chemokine CXCL12. *J. Biol. Chem.* **2008**, *283*, 23189–23199. [CrossRef] [PubMed]
257. Roccaro, A.M.; Sacco, A.; Purschke, W.G.; Moschetta, M.; Buchner, K.; Maasch, C.; Zboralski, D.; Zöllner, S.; Vonhoff, S.; Mishima, Y.; et al. SDF-1 Inhibition Targets the Bone Marrow Niche for Cancer Therapy. *Cell Rep.* **2014**, *9*, 118–128. [CrossRef]
258. Hoellenriegel, J.; Zboralski, D.; Maasch, C.; Rosin, N.Y.; Wierda, W.G.; Keating, M.J.; Kruschinski, A.; Burger, J.A. The Spiegelmer NOX-A12, a Novel CXCL12 Inhibitor, Interferes with Chronic Lymphocytic Leukemia Cell Motility and Causes Chemosensitization. *Blood* **2014**, *123*, 1032–1039. [CrossRef] [PubMed]
259. Zboralski, D.; Hoehlig, K.; Eulberg, D.; Frömming, A.; Vater, A. Increasing Tumor-Infiltrating T Cells through Inhibition of CXCL12 with NOX-A12 Synergizes with PD-1 Blockade. *Cancer Immunol. Res.* **2017**, *5*, 950–956. [CrossRef] [PubMed]
260. Adlere, I.; Caspar, B.; Arimont, M.; Dekkers, S.; Visser, K.; Stuijt, J.; de Graaf, C.; Stocks, M.; Kellam, B.; Briddon, S.; et al. Modulators of CXCR4 and CXCR7/ACKR3 Function. *Mol. Pharm.* **2019**, *96*, 737–752. [CrossRef] [PubMed]
261. Gómez-España, M.A.; Gallego, J.; González-Flores, E.; Maurel, J.; Páez, D.; Sastre, J.; Aparicio, J.; Benavides, M.; Feliu, J.; Vera, R. SEOM Clinical Guidelines for Diagnosis and Treatment of Metastatic Colorectal Cancer (2018). *Clin. Transl. Oncol.* **2019**, *21*, 46–54. [CrossRef] [PubMed]
262. Formica, V.; Roselli, M. Targeted Therapy in First Line Treatment of RAS Wild Type Colorectal Cancer. *World J. Gastroenterol.* **2015**, *21*, 2871–2874. [CrossRef] [PubMed]
263. Khare, T.; Bissonnette, M.; Khare, S. CXCL12-CXCR4/CXCR7 Axis in Colorectal Cancer: Therapeutic Target in Preclinical and Clinical Studies. *Int. J. Mol. Sci.* **2021**, *22*, 7371. [CrossRef] [PubMed]
264. Martin, M.; Mayer, I.A.; Walenkamp, A.M.E.; Lapa, C.; Andreeff, M.; Bobirca, A. At the Bedside: Profiling and Treating Patients with CXCR4-Expressing Cancers. *J. Leukoc. Biol.* **2021**, *109*, 953–967. [CrossRef] [PubMed]
265. Santagata, S.; Ieranò, C.; Trotta, A.M.; Capiluongo, A.; Auletta, F.; Guardascione, G.; Scala, S. CXCR4 and CXCR7 Signaling Pathways: A Focus on the Cross-Talk Between Cancer Cells and Tumor Microenvironment. *Front. Oncol.* **2021**, *11*, 591386. [CrossRef] [PubMed]
266. Galsky, M.D.; Vogelzang, N.J.; Conkling, P.; Raddad, E.; Polzer, J.; Roberson, S.; Stille, J.R.; Saleh, M.; Thornton, D. A Phase I Trial of LY2510924, a CXCR4 Peptide Antagonist, in Patients with Advanced Cancer. *Clin. Cancer Res.* **2014**, *20*, 3581–3588. [CrossRef] [PubMed]
267. Suarez-Carmona, M.; Williams, A.; Schreiber, J.; Hohmann, N.; Pruefer, U.; Krauss, J.; Jäger, D.; Frömming, A.; Beyer, D.; Eulberg, D.; et al. Combined Inhibition of CXCL12 and PD-1 in MSS Colorectal and Pancreatic Cancer: Modulation of the Microenvironment and Clinical Effects. *J. Immunother. Cancer* **2021**, *9*, e002505. [CrossRef] [PubMed]
268. Biasci, D.; Smoragiewicz, M.; Connell, C.M.; Wang, Z.; Gao, Y.; Thaventhiran, J.E.D.; Basu, B.; Magiera, L.; Johnson, T.I.; Bax, L.; et al. CXCR4 Inhibition in Human Pancreatic and Colorectal Cancers Induces an Integrated Immune Response. *Proc. Natl. Acad. Sci. USA* **2020**, *117*, 28960–28970. [CrossRef]
269. Shimizu, S.; Brown, M.; Sengupta, R.; Penfold, M.E.; Meucci, O. CXCR7 Protein Expression in Human Adult Brain and Differentiated Neurons. *PLoS ONE* **2011**, *6*, e20680. [CrossRef] [PubMed]
270. Lubner, S.J.; Uboha, N.V.; Deming, D.A. Primary and Acquired Resistance to Biologic Therapies in Gastrointestinal Cancers. *J. Gastrointest Oncol.* **2017**, *8*, 499–512. [CrossRef]
271. Gottesman, M.M. Mechanisms of Cancer Drug Resistance. *Annu. Rev. Med.* **2002**, *53*, 615–627. [CrossRef] [PubMed]
272. Mansoori, B.; Mohammadi, A.; Davudian, S.; Shirjang, S.; Baradaran, B. The Different Mechanisms of Cancer Drug Resistance: A Brief Review. *Adv. Pharm. Bull.* **2017**, *7*, 339–348. [CrossRef] [PubMed]
273. Reya, T.; Morrison, S.J.; Clarke, M.F.; Weissman, I.L. Stem Cells, Cancer, and Cancer Stem Cells. *Nature* **2001**, *414*, 105–111. [CrossRef]
274. Margolin, D.A.; Silinsky, J.; Grimes, C.; Spencer, N.; Aycock, M.; Green, H.; Cordova, J.; Davis, N.K.; Driscoll, T.; Li, L. Lymph Node Stromal Cells Enhance Drug-Resistant Colon Cancer Cell Tumor Formation through SDF-1α/CXCR4 Paracrine Signaling. *Neoplasia* **2011**, *13*, 874–886. [CrossRef] [PubMed]
275. Margolin, D.A.; Myers, T.; Zhang, X.; Bertoni, D.M.; Reuter, B.A.; Obokhare, I.; Borgovan, T.; Grimes, C.; Green, H.; Driscoll, T.; et al. The Critical Roles of Tumor-Initiating Cells and the Lymph Node Stromal Microenvironment in Human Colorectal Cancer Extranodal Metastasis Using a Unique Humanized Orthotopic Mouse Model. *FASEB J.* **2015**, *29*, 3571–3581. [CrossRef] [PubMed]

276. Jin, F.; Ji, H.; Jia, C.; Brockmeier, U.; Hermann, D.M.; Metzen, E.; Zhu, Y.; Chi, B. Synergistic Antitumor Effects of Endostar in Combination with Oxaliplatin via Inhibition of HIF and CXCR4 in the Colorectal Cell Line SW1116. *PLoS ONE* **2012**, *7*, e47161. [CrossRef] [PubMed]
277. Heckmann, D.; Maier, P.; Laufs, S.; Wenz, F.; Zeller, W.J.; Fruehauf, S.; Allgayer, H. CXCR4 Expression and Treatment with SDF-1α or Plerixafor Modulate Proliferation and Chemosensitivity of Colon Cancer Cells. *Transl. Oncol.* **2013**, *6*, 124–132. [CrossRef]
278. Heckmann, D.; Maier, P.; Laufs, S.; Li, L.; Sleeman, J.P.; Trunk, M.J.; Leupold, J.H.; Wenz, F.; Zeller, W.J.; Fruehauf, S.; et al. The Disparate Twins: A Comparative Study of CXCR4 and CXCR7 in SDF-1α-Induced Gene Expression, Invasion and Chemosensitivity of Colon Cancer. *Clin. Cancer Res.* **2014**, *20*, 604–616. [CrossRef] [PubMed]

MDPI

Review

Implications of RAS Mutations on Oncological Outcomes of Surgical Resection and Thermal Ablation Techniques in the Treatment of Colorectal Liver Metastases

Rami Rhaiem [1,2,*], Linda Rached [2], Ahmad Tashkandi [2], Olivier Bouché [1,3] and Reza Kianmanesh [1,2]

1 Faculty of Medecine, University Reims Champagne-Ardenne, 51100 Reims, France; obouche@chu-reims.fr (O.B.); rkianmanesh@chu-reims.fr (R.K.)
2 Hepatobiliary, Pancreas, Endocrine and Digestive Surgical Oncology Department, Robert Debré Hospital, CHU de Reims, 51100 Reims, France; lrached@chu-reims.fr (L.R.); atashkandi@chu-reims.fr (A.T.)
3 Digestive Oncology and Hepatogastroenterology Department, Robert Debré Hospital, CHU de Reims, 51100 Reims, France
* Correspondence: rrhaiem@chu-reims.fr

Simple Summary: Modern management of colorectal liver metastases (CRLM) requires a thorough knowledge of tumor biology and oncogenes mutations. RAS mutations are of paramount interest for the indication of targeted therapies and is increasingly considered as a negative prognostic factor for patients undergoing surgical resection or ablation for CRLM. Several studies discussed the results of specific technical considerations according to RAS mutational status on the oncological outcomes after surgical resection/ablation for CRLM. We reviewed the available data on the real impact of RAS mutations on the prognosis with special regard to the need of a tailored surgical (ablation) approach according to tumoral biology.

Citation: Rhaiem, R.; Rached, L.; Tashkandi, A.; Bouché, O.; Kianmanesh, R. Implications of RAS Mutations on Oncological Outcomes of Surgical Resection and Thermal Ablation Techniques in the Treatment of Colorectal Liver Metastases. *Cancers* **2022**, *14*, 816. https://doi.org/10.3390/cancers14030816

Academic Editors: Yasushi Sato and Susanne Merkel

Received: 18 December 2021
Accepted: 3 February 2022
Published: 5 February 2022

Abstract: Colorectal cancer (CRC) is the third most common cancer worldwide and the second leading cause of cancer-related death. More than 50% of patients with CRC will develop liver metastases (CRLM) during their disease. In the era of precision surgery for CRLM, several advances have been made in the multimodal management of this disease. Surgical treatment, combined with a modern chemotherapy regimen and targeted therapies, is the only potential curative treatment. Unfortunately, 70% of patients treated for CRLM experience recurrence. RAS mutations are associated with worse overall and recurrence-free survival. Other mutations such as BRAF, associated RAS /TP53 and APC/PIK3CA mutations are important genetic markers to evaluate tumor biology. Somatic mutations are of paramount interest for tailoring preoperative treatment, defining a surgical resection strategy and the indication for ablation techniques. Herein, the most relevant studies dealing with RAS mutations and the management of CRLM were reviewed. Controversies about the implication of this mutation in surgical and ablative treatments were also discussed.

Keywords: RAS mutations; colorectal cancer; liver metastases; surgery; resection; ablation

1. Introduction

Management of colorectal liver metastases (CRLM) has evolved considerably during recent decades. Indeed, the use of a combination of perioperative medical therapy and surgical treatment remains the standard of care. The tremendous progresses of chemo- and targeted therapies regimens have been achieved allowing a higher rate of conversion (15–30%), from initially unresectable to resectable CRLM [1–3]. Additionally, recent advances of surgical techniques, including portal vein embolization and total venous deprivation to prepare 2-stage hepatectomy, associating liver partition and portal vein ligation for staged hepatectomy (ALPPS) and ablative treatments, allowed a wider indication of surgery [4]. This explains the high survival rates after surgery in selected patients for

CRLM up to 40–65% at 5-year [2,5], and 25% at 10-year [5], while such long survival rates are uncommon after chemotherapy alone.

Tumor biology, in particular RAS mutations, is obviously among the strongest prognostic factors of CRLM. It is of paramount interest in the choice of the appropriate chemo- and targeted therapy regimens. Several recent studies suggested a clear implication of tumor biology in defining the optimal surgical/ablation techniques and margins.

In this review, we will report and discuss data reporting the role of RAS mutations in tailoring the surgical and/or ablation approach.

2. RAS Mutations and Prognosis after CRLM Treatment

The rat sarcoma viral oncogenes (RAS) family (KRAS, NRAS and HRAS) plays a pivotal role in the promotion of tumoral cell growth, angiogenesis and the invasiveness of the tumor through the mitogen-activated protein kinase (MAPK) signaling pathway [6]. This latter is continuously activated in case of a RAS mutation, resulting in resistance to anti-EGFR therapies [7].

Interestingly, several studies have reported a prognostic impact of a RAS mutation in patients undergoing CRLM resection. The Memorial Sloan–Kettering Cancer Center (MSKCC) group [8] were the first to report that the KRAS mutation was associated with worse disease specific survival than the KRAS wild type after both primary (median 2.6 vs. 4.8 years; $p = 0.0003$) and liver metastases resection (median 2.7 vs. 6 years; $p = 0.004$). The presence of an additional high Ki-67 expression harbored even worst survival rates.

Since then, incidences of the KRAS mutation reported in the surgical series ranged from 15 to 50%, with a shorter overall survival (OS) and recurrence-free survival in many studies (Table 1). The MD Anderson group reported a significant association between the RAS mutation and both the shortness of the time interval to recurrence and the rate of recurrence above all local treatments [9].

Table 1. Summary of studies reporting survival outcomes of treatment of colorectal liver metastases according to RAS/KRAS mutations.

Study	N *	RAS/KRAS Mutation (%)	Overall Survival (OS)		Recurrence/Disease Free Survival (RFS/DFS)	
			Clinical Parameter	HR (95% CI); *p*-Value	Clinical Parameter	HR (95% CI); *p*-Value
Petrowsky et al., 2001 [10]	41	6 (15%)	Survival	1.39 (0.45–4.27); $p = 0.57$	N/A	N/A
Nash et al., 2010 [8]	188	51 (27%)	5-year survival	2.4 (1.4–4.0); $p = 0.001$	N/A	N/A
Teng et al., 2012 [11]	292	111 (38%)	Median OS	1.48 (0.86–2.56); $p = 0.156$	N/A	N/A
Stremitzer et al., 2012 [12]	76	15(20%)	5-year survival	3.51 (1.30–9.45); $p = 0.013$	3-year RFS	2.48 (1.26–4.89); $p = 0.009$
Karagkounis et al., 2013 [13]	202	58 (29%)	3-Year OS	1.99 (1.21–3.26); $p = 0.007$	3-Year RFS	1.68 (1.04–2.70); $p = 0.034$
Isella et al., 2013 [14]	64	21 (33%)	N/A	N/A	Median DFS	1.58 (0.79–3.16); $p = 0.19$
Vauthey et al., 2013 [15]	193	27 (14%)	3-year OS	2.26 (1.13–4.51); $p = 0.002$	3-year RFS	1.92 (1.21–3.03); $p = 0.005$
Kemeny et al., 2014 [16]	169	51 (30.2%)	3-year OS	2.0 (0.87–4.46); $p = 0.104$	3-year RFS	1.9 (1.16–3.31); $p = 0.01$
Shoji et al., 2014 [17]	108	39 (36.1%)	N/A	N/A	Median RFS	1.91 (1.163–3.123); $p = 0.01$

Table 1. *Cont.*

Study	N *	RAS/KRAS Mutation (%)	Overall Survival (OS)		Recurrence/Disease Free Survival (RFS/DFS)	
			Clinical Parameter	HR (95% CI); *p*-Value	Clinical Parameter	HR (95% CI); *p*-Value
Margonis et al., 2015 [18]	331	91 (27.5%)	Median OS Codon 12 mutant Codon 13 mutant	1.7 (1.13–2.55); $p = 0.01$ 1.61 (0.87–2.97); $p = 0.13$	Median/ 5-year RFS	$p = 0.57$
Sasaki et al., 2016 [19]	129	78 (48.8%)	Median/ 5-year OS	1.37 (0.98–1.91); $p = 0.06$	Median/ 5-year RFS	1.10 (0.85–1.44); $p = 0.47$
	297	68 (28.8%)				
Shindoh et al., 2016 [20]	163	74 (45%)	3-Year OS Disease specific survival	2.86 (1.36–6.04); $p = 0.006$	3-Year RFS Liver RFS	1.47 (1.00–2.15); $p < 0.048$ 3.5 (2.14–5.73); $p < 0.001$
Amikura et al., 2018 [21]	421	191 (43.8%)	5-Year OS	1.67 (1.19–2.38); $p = 0.0031$	5-year RFS	1.70 (1.206–2.422); $p = 0.0024$
O'Connor et al., 2018 [22]	662	174 (26.3%)	Death	1.11 (0.73–1.69); $p = 0.207$	Recurrence	1.42 (1.10–1.85); $p = 0.008$
Goffredo et al., 2019 [23]	2655	1116 (42%)	5-Year OS	1.21 (1.04–1.39); $p = 0.012$	N/A	N/A
Brunsell et al., 2020 [24]	106	53 (50%)	3-year CSS (cancer specific survival)	3.3 (1.6–6.5); $p = 0.001$	N/A	N/A
Kim et al., 2020 [25]	227	78 (34%)	Median OS	1.420 (0.902.25); $p = 0.042$	Median RFS	1.137 (0.83–1.55); $p < 0.001$
Hatta et al., 2021 [26]	500	152 (30.4%)	5-year OS	1.52 (1.14–2.03); $p = 0.004$	5-Year RFS	1.29 (1.00–1.67); $p = 0.049$
Sakai et al., 2021 [27]	101	38 (37.6%)	5-year OS	2.41 (1.36–4.25); $p = 0.003$	3-year RFS	N/A
Saadat et al., 2021 [28]	938	445 (47%)	Median OS	HR 1.67 (1.39–2); $p < 0.001$	Median RFS	1.74 (1.45–2.09); $p < 0.001$

N *: Number of patients included in the study.

Goffredo et al. [23] explored the prognostic factors in a large cohort of 2655 patients enrolled from the US National Cancer Database. All patients were treated, between 2010 and 2015, for synchronous CRLM with concomitant resection of the primary tumor and metastases. The KRAS mutation and right-sided primary tumor were among the major prognostic factors associated with worse OS [23]. NRAS mutations, more infrequently observed, were also correlated to unfavorable oncological outcomes [29,30]. The RAS mutation was integrated in two recent clinical risk scores predicting survival after CRLM resection: the genetic and morphological evaluation "GAME score" [31,32], and the "modified-Clinical risk score" (m-CRS). These scores achieved better discriminatory power than the "Fong's Clinical risk score" [33]. The MD Anderson group has recently published the "Contour prognostic model" that was designed following the concept of the "Metroticket score", previously developed to predict survival after liver transplantation for hepatocellular carcinoma beyond the Milan criteria [34]. This score was validated by an international multicentric cohort. It is based on the diameter and number of lesions considered as continuous variables along with the RAS mutation status. It showed a good prediction power for OS after the resection of CRLM [35]. However, more recently, Tsilimigras et al. [36] reported a poor prediction power of the "Tumor Burden Score" in KRAS mutated tumors. The Tumor Burden score reflected the morphologic characteristics of metastases based on the maximum tumor size and number of lesions [37]. The authors reviewed, in an international multi-institutional database, the results of 1361 patients who underwent hepatic resection for CRLM and analyzed the prognostic impact of the Tumor Burden Score depending on the KRAS status. This score was associated with worse overall survival for the KRAS wild type but not for KRAS mutated tumors [36].

Although there is a large consensus on the negative prognostic impact of RAS mutations after liver surgery for CRLM, several recent data suggested an overestimation of its value, in particular, the possibility of different biological patterns between RAS mutants with there, subsequently, being a difference in their effect on the risk of recurrence and survival after treatment of CLRM [38,39]. Xie et al. [40] reported, in a cohort of 323 patients treated for CRLM, that the prognostic impact of the KRAS mutational status was more significant when the primary tumor was left-sided. Sakai et al. [27] analyzed the results of 101 patients, among them 38 patients with the KRAS mutation, and concluded that the KRAS mutation was an independent prognostic factor only for synchronous CRLM. Several investigators assessed the impact of the mutation location on the prognosis. Frankel et al. [38] showed that NRAS and KRAS mutations were present in 43% of patients, the majority being KRAS mutations (number of KRAS mutations = 65, number of NRAS mutations = 6). The location of the mutation was in exon 2 (codon 12 or 13) in 81.6%, exon 3 in 10% and exon 4 in 8.5% of RAS mutations. According to the location of the mutation, patients exhibited various tumoral features. Indeed, the exon 2 mutation resulted in similar features as the RAS wild type, with a median size of nodules < 5 cm and an average of 2.4 tumors per resection. The exon 3 mutation seemed to be associated with multiple but smaller nodules that tend to occur early after the primary tumor resection, whereas patients with the RAS mutation in exon 4 had solitary CRLM but were larger in size, and had a longer time interval after the resection of the primary tumor than the exon 3 mutation. Authors from the same group recently actualized their data with 938 patients treated for CRLM with sufficient tumor genomic profiling. The KRAS mutation was present in 47% of patients with 91.5% of mutations in exon 2, 3.1% in exon 3, and 5.4% in exon 4. The NRAS mutation was found in only 4.2% of patients with mostly mutations in exons 2 and 3 (53% and 41.2%, respectively). K/NRAS mutations were associated with worse OS with a tendency towards more favorable oncological results in patients with the exon 4 mutation. In the same setting, Margonis et al. [18] reported a worse prognosis when the KRAS mutation was in codon 12 when compared to it in codon 13. Among all mutations of codon 12, only patients with G12S and G12V mutations seemed to have a worse oncological outcome than KRAS wild-type patients. Meanwhile, in another study of the John Hopkins Group, KRAS codon 13 mutations seemed to be associated to a higher risk of extrahepatic recurrence than codon 12 mutations, especially in the pulmonary location [41].

More importantly, other oncogenes are valuable aside the RAS mutational status to predict optimally the prognosis after CRLM resection. Indeed, as shown by Kawaguchi and the MD Anderson group, the association of RAS, TP53 and/or SMAD4 seems to be accurately correlated to worse OS and recurrence free survival (RFS) in 507 patients undergoing surgical resection for CRLM [42]. Furthermore, the authors found no difference in OS and RFS between RAS mutated with wild-type TP53—SMAD 4 and RAS wild-type patients [42].

These data suggested differences in the tumoral pattern and in oncological outcomes according to the location (exon and codon) of the mutation and to the associated mutations.

3. Implication of RAS Mutations in the Surgical Resection of CRLM

Modern surgical management of CRLM is based on the concept of "parenchymal-sparing" surgery, shifting the paradigm from anatomical and large resections to limited resections with a surgical margin ≥ 1 mm, resulting in comparable survival outcomes with lower postoperative morbidity and mortality rates [42–47]. Moreover, R1 vascular is an acceptable surgical option in case of direct contact between the nodule and major vascular structures [48,49]. Contrariwise, a positive resection margin (R1 parenchymal) is associated with a higher rate of local recurrence and worse prognosis. In this context, the surgical margin for RAS mutated CRLM is a matter of debate. Brudvik et al. [50] have reported an association between the RAS mutation and the depth of the resection margins in patients undergoing liver resection for CRLM (hazard ratio (HR) = 2.439; $p = 0.005$). Patients with liver-first recurrence of RAS-mutated CRLM had significantly narrower margins than patients with RAS wild type tumors (4 mm vs. 7 mm; $p = 0.031$) [50]. The same conclusions were recently reported by Zhang et al. [51] in a consecutive cohort of 251 patients treated for CRLM with more micrometastases, thicker margins and a higher rate of R1 resection in the KRAS mutated group [51]. To overcome this problem, Margonis et al. [41] suggested a significant benefit from anatomical resection in KRAS-mutated CRLM, as it seems to allow better liver-specific disease-free survival (DFS) than non-anatomical resections in a multicentric cohort of 389 patients with 140 patients (36%) presenting with KRAS-mutated CRLM (33.8 vs. 10.5 months; $p < 0.001$). Such a difference was not observed in the KRAS-wild type group. The main flaw of this study was a higher rate of ablation procedures in the non-anatomical group (32% vs. 8%) and the absence of analysis of the sub-group of patients treated only with liver resection. This point might alter the interpretation of the survival difference in favor of anatomical resections [52]. Meanwhile, recently, Joechle et al. [53] found no significant difference in OS and RFS between anatomical and non-anatomical resection in 622 patients treated for CRLM with a documented RAS mutation status before and after propensity score matching. In view of these results, the MD Anderson group recommends, when anatomically feasible, wider planned resection margins (≥15 mm) in the case of RAS mutated-CRLM [52], which is debatable. Conversely, the John Hopkins hospital group analyzed the impact of surgical margins after resection of CRLM according to the RAS mutation status [54]. Margonis et al. [54] compared the outcomes after the R0 and R1 resection, and subsequently subdivided the R0 resection group into 3 subgroups according to the width of the surgical margins: 1–4, 5–9 and ≥10 mm. In the KRAS wild type group, the R1 resection was associated with worse OS compared to the R0 resection, but wider margins did not confer an additional OS benefit. In the other hand, for the KRAS mutation group, the OS of the R0 resection, regardless of the width of margin, was not better than the R1 resection group. The same conclusions were drawn in a more recent study with 500 patients [26]. While the resection margin seemed to be associated to death-censored liver-specific recurrence-free survival, it did not impact survival outcomes for KRAS mutated patients [26]. These results stressed the importance of tumor biology and aggressiveness of RAS-mutated CRLM that outbalance the prognostic impact of the surgical margin width. Furthermore, R1 vascular resection seems to harbor a lower risk of local recurrence in KRAS mutated CRLM [55]. The Humanitas group evaluated the local recurrence after CRLM resection according to the quality of resection and to the KRAS mutation status [55]. KRAS mutation was not associated to a higher risk of local recurrence in R0 patients. R1 parenchymal resection, exposing the tumor edge during parenchymal dissection, was correlated to a higher rate of local recurrence in mutated KRAS tumors when compared to the KRAS wild type (respectively, local recurrence rate per patient: 25.4% vs. 18.3%; $p = 0.404$, in situ local recurrence rate: 19.5% vs. 9.9%; $p = 0.048$). Interestingly, results were different for R1 vascular resections (resections with the detachment of nodules from vascular structures). In this regard, the local recurrence rate was higher in the KRAS wild type subgroup (local recurrence rate per-patient 14.6% vs. 2%, $p = 0.043$, in situ local recurrence rate 13.3% vs. 1.9%, $p = 0.046$) [55]. These data are valuable and introduced the concept of a tailored surgical approach according to tumor biology in patients treated for CRLM.

4. Implication of RAS Mutations in Ablative Treatment of CRLM

Ablation is a valuable treatment of CRLM < 3 cm. This debate around the impact of the RAS mutational status on the oncological outcomes of surgical resection also brought the same questioning. Obviously, all published studies reported shorter local tumor progression-free survival in RAS mutated [56,57] and KRAS mutated patients with CRLM [58,59]. Even if all these studies included patients with tumors larger than 3 cm, which is questionable, in multivariate analysis, the 2 main risk factors of local tumor progression were mutational status and ablation margins (Tables 2 and 3). These data were also confirmed by a more recent study from the Amsterdam group [60]. The authors analyzed the impact of primary tumor sidedness, genetic mutations (RAS and BRAF) and the microsatellite instability status to determine the prognosis of patients treated for CRLM enrolled in the Amsterdam Colorectal Liver Met Registry. RAS mutation was associated to shorter local tumor progression-free survival and to lower local control rates after thermal ablation. In these studies, the optimal minimal ablation margin was >5 mm [56,58] and raised to 10 mm [57,59]. Wider margins seem to be necessary to reduce rates of local tumor progression in RAS/KRAS mutation patients [60].

Table 2. Results of studies reporting implications of RAS mutations on surgical resection of colorectal liver metastases.

Authors	Study Period	N * (%) of RAS/KRAS Mutation	Associated Ablation Procedures	Study Keypoint	Findings	Results
Brudvik et al. [50]	2005–2013	RAS 229/633 (36.2%)	N/A	Resection margin	RAS mutation associated: - to positive resection margin (<1 mm) -worst OS	HR: 2.439; $p = 0.005$ HR 1.629; $p = 0.044$
Zhang et al. [51]	2010–2017	KRAS 121/251 (48.2%)	N/A	Micrometastasis	KRAS mutation associated with higher rate	KRAS mut vs. KRAS wild 60.3% vs. 40.8%; $p = 0.002$
					higher number and	(median 2.0 (range 0–38.0) vs. median 0 (range: 0–15.0); $p = 0.001$)
					density of micrometastases	56% vs. 43%; $p = 0.013$
				Resection margin	Higher rate of R1 resection (tumoral cell on the resection margin)	21.5 vs. 9.2%; $p = 0.007$
					Narrower resection margin in KRAS mut	median 2.00 (range 0–40.00) vs. 4.30 (range 0–50.00) mm; $p = 0.002$
				LRFS	KRAS mut associated with worst LRFS	HR: 1.495 (95% CI: 1.069–2.092); $p = 0.019$
				OS	KRAS mut associated with worst OS	HR: 2.039 (95% CI: 1.217–3.417); $p = 0.007$
Margonis et al. [41]	2000–2015	KRAS 140/389 (36%)	NAR:53/165 (32%) AR:19/224 (8.5%)	Anatomical vs. non anatomical resection	AR was associated with better DFS in KRAS mut but not in KRAS wild	DFS: KRAS mut HR: 0.45 (95% CI: 0.27–0.74; $p = 0.002$ KRAS wild: NS
Joechle et al. [53]	2006–2016	RAS 274/622 (40%)	N/A	Anatomical vs. non anatomical resection	No difference in OS and Live specific RFS before and after PSM RFS was better in the AR before PSM but not after PSM	

Table 2. *Cont.*

Authors	Study Period	N * (%) of RAS/KRAS Mutation	Associated Ablation Procedures	Study Keypoint	Findings	Results
Margonis et al. [54]	2003–2015	KRAS 153/411(37.2%)	84 (20.4%)	Impact of resection margin width on OS according to KRAS status	KRAS wild type: R0 resection was associated to better OS than R1 resection (<1 mm) with no benefit from wider margin (1–4 mm; 5–9 mm; >9 mm)	KRAS wild:
					KRAS mut: No difference in OS between R0 and R1 resection, regardless of the width of surgical margin	R1 ref 1–4 mm: HR: 0.45, 95%CI: 0.24–0.85; $p = 0.014$) 5–9 mm: HR: 0.35, 95%CI: 0.17–0.70; $p = 0.003$ >9 mm: HR: 0.33, 95%CI: 0.16–0.68; $p = 0.002$ KRAS mut: 1–4 mm: HR: 0.80, 95%CI: 0.38–1.70; $p = 0.522$ 5–9 mm: HR: 0.68, 95%CI: 0.30–1.54; $p = 0.356$ >9 mm: HR: 1.08, 95%CI: 0.50–2.35; $p = 0.844$
Hatta et al. [26]	2011–2016	KRAS 152/500 (30.4%)	N/A	Impact of resection margin width on OS, RFS and LS-RFS according to KRAS status	KRAS wild type: Resection margin width was associated to a better OS, RFS (Death censored) and LS-RFS (Death censored) KRAS mut: No difference between R0 (regardless to the width of margin) and R1 in all studied survival parameters	
Procopio et al. [55]	2008–2016	KRAS	N/A	Impact of R1 parenchymal and R1 vascular resections on risk of local recurrence after resection according to KRAS status	Higher rates of recurrence in KRas mut after R1 parenchymal resection	R1 parenchymal resection (KRAS mut vs. KRAS wild) local recurrence rate per patient: 25.4% vs. 18.3%; $p = 0.404$ in situ local recurrence rate: 19.5% vs. 9.9%; $p = 0.048$ R1 vascular resection (KRAS mut vs. KRAS wild) local recurrence rate per patient 2% vs. 14.6%; $p = 0.043$, in situ local recurrence rate
		155/340 (46%)			Higher rates of recurrence in KRAS wild after R1 vascular resection	1.9%, vs. 13.3%; $p = 0.046$

N */(%): Number and percentage of RAS/KRAS mutations in the study, N/A: Not mentioned, HR Hazard ratio, *p*: *p*-value, OS: Overall survival, RFS: Recurrence-free survival, LS-RFS: Liver specific Recurrence-free survival, AR: Anatomical resection, NAR: Non anatomical resection.

Table 3. Results of studies reporting ablative treatment for colorectal liver metastases according to RAS mutations.

Study Year	N	Median Size of CRLM (cm)	N * of KRAS Mutation	Procedures of Ablation	OS HR (95%CI); p-Value % at 3 Years, p-Value	LTPFS HR (95%CI); p-Value	LC (Site Specific Recurrence) HR (95%CI); p-Value % at 3 Years, p-Value
Shady 2017 [58]	97	1.7 (0.6–5)	38/97 (exon 2)	Percutaneous RFA	2.0 (1.2–3.3); $p = 0.009$	1.7 (0.89–3.2) $p = 0.11$	2.0 (1.0–3.7) $p = 0.037$
Odisio 2017 [56]	92	1.6 (0.4–4.0)	36/92	Percutaneous RFA + MWA	N/A 40% vs. 82%; $p = 0.013$	3.01 (1.60–5.77) $p = 0.001$	N/A 56% vs. 43% $p = 0.013$
Calandri 2018 [57]	136	1.6 (0.5–5.2)	54/136	Percutaneous RFA, MWA, cryotherapy	N/A	2.85 (1.7–4.6) $p < 0.001$	N/A
Jiang 2019 [59]	76	2.3 (0.9–0.7)	38/76	Percutaneous RFA	Not significant $p = 0.228$	3.24 (1.41–7.41) $p = 0.005$	Not significant $p = 0.356$
Dijkstra 2021 [60]	79		36/79	Percutaneous RFA + MWA	N/A	Significantly lower $p = 0.037$	N/A

N * = Number of patients with KRAS mutation, CRLM: Colorectal liver metastasis, OS: Overall survival, HR: Hazard ratio, CI: confidence interval, LTPFS: Liver tumor progression free survival, RFA: Radiofrequency ablation, MWA: Microwave ablation.

5. Conclusions

RAS mutations seem to present a negative impact on the oncological outcome of patients treated for CRLM. Several studies pointed to the importance of a multidisciplinary "tailored" approach of CRLM according to the RAS mutational status to choose the optimal preoperative treatment and to optimize the surgical resection and/or ablation technique and planned margins. However, larger studies with genetic sequencing are required to assess a more thorough analysis of the real impact of the RAS mutational status according to the exon/codon location of the mutations, and, more specifically, to the association of other somatic mutations such BRAF, TP 53, PIK3CA and/or SMAD 4 that may harbor a poorer prognostic outcome. Interestingly the primary tumor side and RAS mutation might also be considered. Such a tailored approach, considering the whole genetic profiling of the tumor, will allow further advancement in the knowledge of tumor biology and may be valuable for the management and counseling of patients treated for metastatic colorectal cancer.

Author Contributions: Conceptualization, R.R., O.B. and R.K.; data curation, R.R., L.R. and A.T.; writing—original draft preparation, R.R., L.R. and A.T.; writing—review and editing, O.B. and R.K.; funding acquisition, O.B. and R.K. All authors have read and agreed to the published version of the manuscript.

Funding: This research received no external funding.

Conflicts of Interest: The authors declare no conflict of interest.

References

1. Osterlund, P.; Salminen, T.; Soveri, L.M.; Kallio, R.; Kellokumpu, I.; Lamminmäki, A.; Halonen, P.; Ristamäki, R.; Lantto, E.; Uutela, A.; et al. Repeated Centralized Multidisciplinary Team Assessment of Resectability, Clinical Behavior, and Outcomes in 1086 Finnish Metastatic Colorectal Cancer Patients (RAXO): A Nationwide Prospective Intervention Study. *Lancet Reg. Health–Eur.* **2021**, *3*, 100049. [CrossRef]
2. Imai, K.; Adam, R.; Baba, H. How to Increase the Resectability of Initially Unresectable Colorectal Liver Metastases: A Surgical Perspective. *Ann. Gastroenterol. Surg.* **2019**, *3*, 476–486. [CrossRef]
3. Lévi, F.A.; Boige, V.; Hebbar, M.; Smith, D.; Lepère, C.; Focan, C.; Karaboué, A.; Guimbaud, R.; Carvalho, C.; Tumolo, S.; et al. Conversion to Resection of Liver Metastases from Colorectal Cancer with Hepatic Artery Infusion of Combined Chemotherapy and Systemic Cetuximab in Multicenter Trial OPTILIV. *Ann. Oncol.* **2016**, *27*, 267–274. [CrossRef] [PubMed]
4. Petrowsky, H.; Fritsch, R.; Guckenberger, M.; De Oliveira, M.L.; Dutkowski, P.; Clavien, P.A. Modern Therapeutic Approaches for the Treatment of Malignant Liver Tumours. *Nat. Rev. Gastroenterol. Hepatol.* **2020**, *17*, 755–772. [CrossRef] [PubMed]

5. Jones, R.P.; Brudvik, K.W.; Franklin, J.M.; Poston, G.J. Precision Surgery for Colorectal Liver Metastases: Opportunities and Challenges of Omics-Based Decision Making. *Eur. J. Surg. Oncol.* **2017**, *43*, 875–883. [CrossRef]
6. Downward, J. Targeting RAS Signalling Pathways in Cancer Therapy. *Nat. Rev. Cancer* **2003**, *3*, 11–22. [CrossRef] [PubMed]
7. Misale, S.; Yaeger, R.; Hobor, S.; Scala, E.; Janakiraman, M.; Liska, D.; Valtorta, E.; Schiavo, R.; Buscarino, M.; Siravegna, G.; et al. Emergence of KRAS Mutations and Acquired Resistance to Anti-EGFR Therapy in Colorectal Cancer. *Nature* **2012**, *486*, 532–536. [CrossRef] [PubMed]
8. Nash, G.M.; Gimbel, M.; Shia, J.; Nathanson, D.R.; Ndubuisi, M.I.; Zeng, Z.S.; Kemeny, N.; Paty, P.B. KRAS Mutation Correlates with Accelerated Metastatic Progression in Patients with Colorectal Liver Metastases. *Ann. Surg. Oncol.* **2010**, *17*, 572–578. [CrossRef]
9. Okuno, M.; Goumard, C.; Kopetz, S.; Vega, E.A.; Joechle, K.; Mizuno, T.; Omichi, K.; Tzeng, C.W.D.; Chun, Y.S.; Vauthey, J.N.; et al. RAS Mutation Is Associated with Unsalvageable Recurrence Following Hepatectomy for Colorectal Cancer Liver Metastases. *Ann. Surg. Oncol.* **2018**, *25*, 2457–2466. [CrossRef]
10. Petrowsky, H.; Sturm, I.; Graubitz, O.; Kooby, D.A.; Staib-Sebler, E.; Gog, C.; Köhne, C.H.; Hillebrand, T.; Daniel, P.T.; Fong, Y.; et al. Relevance of Ki-67 Antigen Expression and K-Ras Mutation in Colorectal Liver Metastases. *Eur. J. Surg. Oncol.* **2001**, *27*, 80–87. [CrossRef]
11. Teng, H.W.; Huang, Y.C.; Lin, J.K.; Chen, W.S.; Lin, T.C.; Jiang, J.K.; Yen, C.C.; Li, A.F.Y.; Wang, H.W.; Chang, S.C.; et al. BRAF Mutation Is a Prognostic Biomarker for Colorectal Liver Metastasectomy. *J. Surg. Oncol.* **2012**, *106*, 123–129. [CrossRef] [PubMed]
12. Stremitzer, S.; Stift, J.; Gruenberger, B.; Tamandl, D.; Aschacher, T.; Wolf, B.; Wrba, F.; Gruenberger, T. KRAS Status and Outcome of Liver Resection after Neoadjuvant Chemotherapy Including Bevacizumab. *Br. J. Surg.* **2012**, *99*, 1575–1582. [CrossRef]
13. Karagkounis, G.; Torbenson, M.S.; Daniel, H.D.; Azad, N.S.; Diaz, L.A.; Donehower, R.C.; Hirose, K.; Ahuja, N.; Pawlik, T.M.; Choti, M.A. Incidence and Prognostic Impact of KRAS and BRAF Mutation in Patients Undergoing Liver Surgery for Colorectal Metastases. *Cancer* **2013**, *119*, 4137–4144. [CrossRef]
14. Isella, C.; Mellano, A.; Galimi, F.; Petti, C.; Capussotti, L.; De Simone, M.; Bertotti, A.; Medico, E.; Muratore, A. MACC1 MRNA Levels Predict Cancer Recurrence after Resection of Colorectal Cancer Liver Metastases. *Ann. Surg.* **2013**, *257*, 1089–1095. [CrossRef] [PubMed]
15. Vauthey, J.N.; Zimmitti, G.; Kopetz, S.E.; Shindoh, J.; Chen, S.S.; Andreou, A.; Curley, S.A.; Aloia, T.A.; Maru, D.M. RAS Mutation Status Predicts Survival and Patterns of Recurrence in Patients Undergoing Hepatectomy for Colorectal Liver Metastases. *Ann. Surg.* **2013**, *258*, 619–626; discussion 626–627. [CrossRef]
16. Kemeny, N.E.; Chou, J.F.; Capanu, M.; Gewirtz, A.N.; Cercek, A.; Kingham, T.P.; Jarnagin, W.R.; Fong, Y.C.; DeMatteo, R.P.; Allen, P.J.; et al. KRAS Mutation Influences Recurrence Patterns in Patients Undergoing Hepatic Resection of Colorectal Metastases. *Cancer* **2014**, *120*, 3965–3971. [CrossRef] [PubMed]
17. Shoji, H.; Yamada, Y.; Taniguchi, H.; Nagashima, K.; Okita, N.; Takashima, A.; Honma, Y.; Iwasa, S.; Kato, K.; Hamaguchi, T.; et al. Clinical Impact of C-MET Expression and Genetic Mutational Status in Colorectal Cancer Patients after Liver Resection. *Cancer Sci.* **2014**, *105*, 1002–1007. [CrossRef]
18. Margonis, G.A.; Kim, Y.; Spolverato, G.; Ejaz, A.; Gupta, R.; Cosgrove, D.; Anders, R.; Karagkounis, G.; Choti, M.A.; Pawlik, T.M. Association Between Specific Mutations in KRAS Codon 12 and Colorectal Liver Metastasis. *JAMA Surg.* **2015**, *150*, 722–729. [CrossRef]
19. Sasaki, K.; Andreatos, N.; Margonis, G.A.; He, J.; Weiss, M.; Johnston, F.; Wolfgang, C.; Antoniou, E.; Pikoulis, E.; Pawlik, T.M. The Prognostic Implications of Primary Colorectal Tumor Location on Recurrence and Overall Survival in Patients Undergoing Resection for Colorectal Liver Metastasis. *J. Surg. Oncol.* **2016**, *114*, 803–809. [CrossRef]
20. Shindoh, J.; Nishioka, Y.; Yoshioka, R.; Sugawara, T.; Sakamoto, Y.; Hasegawa, K.; Hashimoto, M.; Kokudo, N. KRAS Mutation Status Predicts Site-Specific Recurrence and Survival After Resection of Colorectal Liver Metastases Irrespective of Location of the Primary Lesion. *Ann. Surg. Oncol.* **2016**, *23*, 1890–1896. [CrossRef]
21. Amikura, K.; Akagi, K.; Ogura, T.; Takahashi, A.; Sakamoto, H. The RAS Mutation Status Predicts Survival in Patients Undergoing Hepatic Resection for Colorectal Liver Metastases: The Results from a Genetic Analysis of All-RAS. *J. Surg. Oncol.* **2018**, *117*, 745–755. [CrossRef] [PubMed]
22. O'Connor, J.; Huertas, E.; Loria, F.S.; Brancato, F.; Grondona, J.; Fauda, M.; Andriani, O.; Sanchez, P.; Schelotto, P.B.; Ardiles, V.; et al. Prognostic Impact of K-RAS Mutational Status and Primary Tumour Location in Patients Undergoing Resection for Colorectal Cancer Liver Metastases: A METHEPAR Analysis (Multicentre Study in Argentina). *Ann. Oncol.* **2018**, *29*, v76. [CrossRef]
23. Goffredo, P.; Utria, A.F.; Beck, A.C.; Chun, Y.S.; Howe, J.R.; Weigel, R.J.; Vauthey, J.N.; Hassan, I. The Prognostic Impact of KRAS Mutation in Patients Having Curative Resection of Synchronous Colorectal Liver Metastases. *J. Gastrointest. Surg.* **2019**, *23*, 1957–1963. [CrossRef] [PubMed]
24. Brunsell, T.H.; Sveen, A.; Bjørnbeth, B.A.; Røsok, B.I.; Danielsen, S.A.; Brudvik, K.W.; Berg, K.C.G.; Johannessen, B.; Cengija, V.; Abildgaard, A.; et al. High Concordance and Negative Prognostic Impact of RAS/BRAF/PIK3CA Mutations in Multiple Resected Colorectal Liver Metastases. *Clin Colorectal Cancer* **2020**, *19*, e26–e47. [CrossRef] [PubMed]
25. Kim, H.S.; Lee, J.M.; Kim, H.S.; Yang, S.Y.; Han, Y.D.; Cho, M.S.; Hur, H.; Min, B.S.; Lee, K.Y.; Kim, N.K. Prognosis of Synchronous Colorectal Liver Metastases After Simultaneous Curative-Intent Surgery According to Primary Tumor Location and KRAS Mutational Status. *Ann. Surg. Oncol.* **2020**, *27*, 5150–5158. [CrossRef]

26. Hatta, A.A.Z.; Pathanki, A.M.; Hodson, J.; Sutcliffe, R.P.; Marudanayagam, R.; Roberts, K.J.; Chatzizacharias, N.; Isaac, J.; Muiesan, P.; Taniere, P.; et al. The Effects of Resection Margin and KRAS Status on Outcomes after Resection of Colorectal Liver Metastases. *HPB* **2021**, *23*, 90–98. [CrossRef]
27. Sakai, N.; Furukawa, K.; Takayashiki, T.; Kuboki, S.; Takano, S.; Ohtsuka, M. Differential Effects of KRAS Mutational Status on Long-Term Survival According to the Timing of Colorectal Liver Metastases. *BMC Cancer* **2021**, *21*, 412. [CrossRef]
28. Saadat, L.V.; Boerner, T.; Goldman, D.A.; Gonen, M.; Frankel, T.L.; Vakiani, E.; Kingham, T.P.; Jarnagin, W.R.; Wei, A.C.; Soares, K.C.; et al. Association of RAS Mutation Location and Oncologic Outcomes After Resection of Colorectal Liver Metastases. *Ann. Surg. Oncol.* **2021**, *28*, 817–825. [CrossRef]
29. Summers, M.G.; Smith, C.G.; Maughan, T.S.; Kaplan, R.; Escott-Price, V.; Cheadle, J.P. BRAF and NRAS Locus-Specific Variants Have Different Outcomes on Survival to Colorectal Cancer. *Clin. Cancer Res.* **2017**, *23*, 2742–2749. [CrossRef]
30. Cercek, A.; Braghiroli, M.I.; Chou, J.F.; Hechtman, J.F.; Kemeny, N.; Saltz, L.; Capanu, M.; Yaeger, R. Clinical Features and Outcomes of Patients with Colorectal Cancers Harboring NRAS Mutations. *Clin. Cancer Res.* **2017**, *23*, 4753–4760. [CrossRef]
31. Margonis, G.A.; Sasaki, K.; Gholami, S.; Kim, Y.; Andreatos, N.; Rezaee, N.; Deshwar, A.; Buettner, S.; Allen, P.J.; Kingham, T.P.; et al. Genetic And Morphological Evaluation (GAME) Score for Patients with Colorectal Liver Metastases. *J. Br. Surg.* **2018**, *105*, 1210–1220. [CrossRef] [PubMed]
32. Brudvik, K.W.; Jones, R.P.; Giuliante, F.; Shindoh, J.; Passot, G.; Chung, M.H.; Song, J.; Li, L.; Dagenborg, V.J.; Fretland, Å.A.; et al. RAS Mutation Clinical Risk Score to Predict Survival After Resection of Colorectal Liver Metastases. *Ann. Surg.* **2019**, *269*, 120–126. [CrossRef]
33. Fong, Y.; Fortner, J.; Sun, R.L.; Brennan, M.F.; Blumgart, L.H. Clinical Score for Predicting Recurrence After Hepatic Resection for Metastatic Colorectal Cancer. *Ann. Surg.* **1999**, *230*, 309. [CrossRef] [PubMed]
34. Mazzaferro, V.; Llovet, J.M.; Miceli, R.; Bhoori, S.; Schiavo, M.; Mariani, L.; Camerini, T.; Roayaie, S.; Schwartz, M.E.; Grazi, G.L.; et al. Predicting Survival after Liver Transplantation in Patients with Hepatocellular Carcinoma beyond the Milan Criteria: A Retrospective, Exploratory Analysis. *Lancet Oncol.* **2009**, *10*, 35–43. [CrossRef]
35. Kawaguchi, Y.; Kopetz, S.; Tran Cao, H.S.; Panettieri, E.; De Bellis, M.; Nishioka, Y.; Hwang, H.; Wang, X.; Tzeng, C.-W.D.; Chun, Y.S.; et al. Contour Prognostic Model for Predicting Survival after Resection of Colorectal Liver Metastases: Development and Multicentre Validation Study Using Largest Diameter and Number of Metastases with RAS Mutation Status. *Br. J. Surg.* **2021**, *108*, 968–975. [CrossRef] [PubMed]
36. Tsilimigras, D.I.; Hyer, J.M.; Bagante, F.; Guglielmi, A.; Ruzzenente, A.; Alexandrescu, S.; Poultsides, G.; Sasaki, K.; Aucejo, F.; Pawlik, T.M. Resection of Colorectal Liver Metastasis: Prognostic Impact of Tumor Burden vs KRAS Mutational Status. *J. Am. Coll. Surg.* **2021**, *232*, 590–598. [CrossRef]
37. Sasaki, K.; Morioka, D.; Conci, S.; Margonis, G.A.; Sawada, Y.; Ruzzenente, A.; Kumamoto, T.; Iacono, C.; Andreatos, N.; Guglielmi, A.; et al. The Tumor Burden Score: A New "Metro-Ticket" Prognostic Tool for Colorectal Liver Metastases Based on Tumor Size and Number of Tumors. *Ann. Surg.* **2018**, *267*, 132–141. [CrossRef]
38. Frankel, T.L.; Vakiani, E.; Nathan, H.; DeMatteo, R.P.; Kingham, T.P.; Allen, P.J.; Jarnagin, W.R.; Kemeny, N.E.; Solit, D.B.; D'Angelica, M.I. Mutation Location on the RAS Oncogene Affects Pathologic Features and Survival After Resection of Colorectal Liver Metastases. *Cancer* **2017**, *123*, 568–575. [CrossRef]
39. Janakiraman, M.; Vakiani, E.; Zeng, Z.; Pratilas, C.A.; Taylor, B.S.; Chitale, D.; Halilovic, E.; Wilson, M.; Huberman, K.; Ricarte Filho, J.C.; et al. Genomic and Biological Characterization of Exon 4 KRAS Mutations in Human Cancer. *Cancer Res.* **2010**, *70*, 5901–5911. [CrossRef]
40. Xie, M.; Li, J.; Cai, Z.; Li, K.; Hu, B. Impact of Primary Colorectal Cancer Location on the KRAS Status and Its Prognostic Value. *BMC Gastroenterol.* **2019**, *19*, 46. [CrossRef]
41. Margonis, G.A.; Kim, Y.; Sasaki, K.; Samaha, M.; Amini, N.; Pawlik, T.M. Codon 13 KRAS Mutation Predicts Patterns of Recurrence in Patients Undergoing Hepatectomy for Colorectal Liver Metastases. *Cancer* **2016**, *122*, 2698–2707. [CrossRef]
42. Kawaguchi, Y.; Kopetz, S.; Newhook, T.E.; De Bellis, M.; Chun, Y.S.; Tzeng, C.W.D.; Aloia, T.A.; Vauthey, J.N. Mutation Status of RAS, TP53, and SMAD4 Is Superior to Mutation Status of RAS Alone for Predicting Prognosis after Resection of Colorectal Liver Metastases. *Clin. Cancer Res.* **2019**, *25*, 5843–5851. [CrossRef] [PubMed]
43. Memeo, R.; de Blasi, V.; Adam, R.; Goéré, D.; Azoulay, D.; Ayav, A.; Gregoire, E.; Kianmanesh, R.; Navarro, F.; Sa Cunha, A.; et al. Parenchymal-Sparing Hepatectomies (PSH) for Bilobar Colorectal Liver Metastases Are Associated with a Lower Morbidity and Similar Oncological Results: A Propensity Score Matching Analysis. *HPB* **2016**, *18*, 781–790. [CrossRef] [PubMed]
44. Andreou, A.; Gloor, S.; Inglin, J.; Di Pietro Martinelli, C.; Banz, V.; Lachenmayer, A.; Kim-Fuchs, C.; Candinas, D.; Beldi, G. Parenchymal-Sparing Hepatectomy for Colorectal Liver Metastases Reduces Postoperative Morbidity While Maintaining Equivalent Oncologic Outcomes Compared to Non-Parenchymal-Sparing Resection. *Surg. Oncol.* **2021**, *38*, 101631. [CrossRef] [PubMed]
45. Donadon, M.; Cescon, M.; Cucchetti, A.; Cimino, M.; Costa, G.; Pesi, B.; Ercolani, G.; Pinna, A.D.; Torzilli, G. Parenchymal-Sparing Surgery for the Surgical Treatment of Multiple Colorectal Liver Metastases Is a Safer Approach than Major Hepatectomy Not Impairing Patients' Prognosis: A Bi-Institutional Propensity Score-Matched Analysis. *Dig. Surg.* **2018**, *35*, 342–349. [CrossRef] [PubMed]
46. Evrard, S.; Torzilli, G.; Caballero, C.; Bonhomme, B. Parenchymal Sparing Surgery Brings Treatment of Colorectal Liver Metastases into the Precision Medicine Era. *Eur. J. Cancer* **2018**, *104*, 195–200. [CrossRef]

47. Moris, D.; Ronnekleiv-Kelly, S.; Rahnemai-Azar, A.A.; Felekouras, E.; Dillhoff, M.; Schmidt, C.; Pawlik, T.M. Parenchymal-Sparing Versus Anatomic Liver Resection for Colorectal Liver Metastases: A Systematic Review. *J. Gastrointest. Surg.* **2017**, *21*, 1076–1085. [CrossRef]
48. Viganò, L.; Procopio, F.; Cimino, M.M.; Donadon, M.; Gatti, A.; Costa, G.; Del Fabbro, D.; Torzilli, G. Is Tumor Detachment from Vascular Structures Equivalent to R0 Resection in Surgery for Colorectal Liver Metastases? An Observational Cohort. *Ann. Surg. Oncol.* **2016**, *23*, 1352–1360. [CrossRef]
49. Viganò, L.; Costa, G.; Cimino, M.M.; Procopio, F.; Donadon, M.; Del Fabbro, D.; Belghiti, J.; Kokudo, N.; Makuuchi, M.; Vauthey, J.-N.; et al. R1 Resection for Colorectal Liver Metastases: A Survey Questioning Surgeons about Its Incidence, Clinical Impact, and Management. *J. Gastrointest. Surg.* **2018**, *22*, 1752–1763. [CrossRef]
50. Brudvik, K.W.; Mise, Y.; Chung, H.; Chun, Y.S.; Kopetz, S.E.; Passot, G.; Conrad, C.; Maru, D.M.; Aloia, T.A.; Vauthey, J.N. RAS Mutation Predicts Positive Resection Margins and Narrower Resection Margins in Patients Undergoing Resection of Colorectal Liver Metastases. *Ann. Surg. Oncol.* **2016**, *23*, 2635–2643. [CrossRef]
51. Zhang, Q.; Peng, J.; Ye, M.; Weng, W.; Tan, C.; Ni, S.; Huang, D.; Sheng, W.; Wang, L. KRAS Mutation Predicted More Mirometastases and Closer Resection Margins in Patients with Colorectal Cancer Liver Metastases. *Ann. Surg. Oncol.* **2020**, *27*, 1164–1173. [CrossRef] [PubMed]
52. Brudvik, K.W.; Vauthey, J.N. Surgery: KRAS Mutations and Hepatic Recurrence after Treatment of Colorectal Liver Metastases. *Nat. Rev. Gastroenterol. Hepatol.* **2017**, *14*, 638–639. [CrossRef] [PubMed]
53. Joechle, K.; Vreeland, T.J.; Vega, E.A.; Okuno, M.; Newhook, T.E.; Panettieri, E.; Chun, Y.S.; Tzeng, C.W.D.; Aloia, T.A.; Lee, J.E.; et al. Anatomic Resection Is Not Required for Colorectal Liver Metastases with RAS Mutation. *J. Gastrointest. Surg.* **2020**, *24*, 1033–1039. [CrossRef] [PubMed]
54. Margonis, G.A.; Sasaki, K.; Andreatos, N.; Kim, Y.; Merath, K.; Wagner, D.; Wilson, A.; Buettner, S.; Amini, N.; Antoniou, E.; et al. KRAS Mutation Status Dictates Optimal Surgical Margin Width in Patients Undergoing Resection of Colorectal Liver Metastases. *Ann. Surg. Oncol.* **2017**, *24*, 264–271. [CrossRef]
55. Procopio, F.; Viganò, L.; Cimino, M.; Donadon, M.; Del Fabbro, D.; Torzilli, G. Does KRAS Mutation Status Impact the Risk of Local Recurrence after R1 Vascular Resection for Colorectal Liver Metastasis? An Observational Cohort Study. *Eur. J. Surg. Oncol.* **2020**, *46*, 818–824. [CrossRef]
56. Odisio, B.C.; Yamashita, S.; Huang, S.Y.; Harmoush, S.; Kopetz, S.E.; Ahrar, K.; Shin Chun, Y.; Conrad, C.; Aloia, T.A.; Gupta, S.; et al. Local Tumour Progression after Percutaneous Ablation of Colorectal Liver Metastases According to RAS Mutation Status. *Br. J. Surg.* **2017**, *104*, 760–768. [CrossRef]
57. Calandri, M.; Yamashita, S.; Gazzera, C.; Fonio, P.; Veltri, A.; Bustreo, S.; Sheth, R.A.; Yevich, S.M.; Vauthey, J.N.; Odisio, B.C. Ablation of Colorectal Liver Metastasis: Interaction of Ablation Margins and RAS Mutation Profiling on Local Tumour Progression-Free Survival. *Eur. Radiol.* **2018**, *28*, 2727–2734. [CrossRef]
58. Shady, W.; Petre, E.N.; Vakiani, E.; Ziv, E.; Gonen, M.; Brown, K.T.; Kemeny, N.E.; Solomon, S.B.; Solit, D.B.; Sofocleous, C.T. Kras Mutation Is a Marker of Worse Oncologic Outcomes after Percutaneous Radiofrequency Ablation of Colorectal Liver Metastases. *Oncotarget* **2017**, *8*, 66117–66127. [CrossRef]
59. Jiang, B.B.; Yan, K.; Zhang, Z.Y.; Yang, W.; Wu, W.; Yin, S.S.; Chen, M.H. The Value of KRAS Gene Status in Predicting Local Tumor Progression of Colorectal Liver Metastases Following Radiofrequency Ablation. *Int. J. Hyperth.* **2019**, *36*, 211–219. [CrossRef]
60. Dijkstra, M.; Nieuwenhuizen, S.; Puijk, R.S.; Timmer, F.E.F.; Geboers, B.; Schouten, E.A.C.; Opperman, J.; Scheffer, H.J.; de Vries, J.J.J.; Versteeg, K.S.; et al. Primary Tumor Sidedness, RAS and BRAF Mutations and MSI Status as Prognostic Factors in Patients with Colorectal Liver Metastases Treated with Surgery and Thermal Ablation: Results from the Amsterdam Colorectal Liver Met Registry (AmCORE). *Biomedicines* **2021**, *9*, 962. [CrossRef]

MDPI
St. Alban-Anlage 66
4052 Basel
Switzerland
Tel. +41 61 683 77 34
Fax +41 61 302 89 18
www.mdpi.com

Cancers Editorial Office
E-mail: cancers@mdpi.com
www.mdpi.com/journal/cancers

www.ingramcontent.com/pod-product-compliance
Lightning Source LLC
LaVergne TN
LVHW071008170726
843515LV00006B/1108

* 9 7 8 3 0 3 6 5 6 8 3 7 9 *